GAUGE THEORIES OF THE STRONG, WEAK AND ELECTROMAGNETIC INTERACTIONS

FRONTIERS IN PHYSICS

David Pines, Editor

Volumes of the Series published from 1961 to 1973 are not officially numbered. The parenthetical numbers shown are designed to aid librarians and bibliographers to check the completeness of their holdings.

FRONTIERS IN PHYSICS

David Pines, Editor (*continued*)

FRONTIERS IN PHYSICS

David Pines, Editor (*continued*)

Volumes published from 1974 onward are being numbered as an integral part of the bibliography:

Other numbers in preparation

GAUGE THEORIES OF THE STRONG, WEAK, AND ELECTROMAGNETIC INTERACTIONS

Chris Quigg

Fermi National Accelerator Laboratory
Batavia, Illinois

1983

The Benjamin/Cummings Publishing Company, Inc.
Advanced Book Program
Reading, Massachusetts

London · Amsterdam · Don Mills, Ontario · Sydney · Tokyo

Coden: FRPHA

6863-1388

PHYSICS

Library of Congress Cataloging in Publication Data

Quigg, Chris.
  Gauge theories of the strong, weak, and electromagnetic interactions.

  (Frontiers in physics; v. 56)
  Bibliography: p.
  Includes index.
  1. Gauge fields (Physics) 2. Nuclear reactions.
I. Title. II. Series.
QC793.3.F5Q53 1983   530.1'43   83-8768
ISBN 0-8053-6020-4

Manufactured in the United States of America

ABCDEFGHIJ-MA-89876543

EDITOR'S FOREWORD

The problem of communicating in a coherent fashion recent developments in the most exciting and active fields of physics seems particularly pressing today. The enormous growth in the number of physicists has tended to make the familiar channels of communication considerably less effective. It has become increasingly difficult for experts in a given field to keep up with the current literature; the novice can only be confused. What is needed is both a consistent account of a field and the presentation of a definite "point of view" concerning it. Formal monographs cannot meet such a need in a rapidly developing field, and, perhaps more important, the review article seems to have fallen into disfavor. Indeed, it would seem that the people most actively engaged in developing a given field are the people least likely to write at length about it.

FRONTIERS IN PHYSICS has been conceived in an effort to improve the situation in several ways. Leading physicists today frequently give a series of lectures, a graduate seminar, or a graduate course in their special fields of interest. Such lectures serve to summarize the present status of a rapidly developing field and may well constitute the only coherent account available at the time. Often, notes on lectures exist (prepared by the lecturer himself, by graduate students, or by postdoctoral fellows) and are distributed in mimeographed form on a limited basis. One of the principal purposes of the FRONTIERS IN PHYSICS Series is to make such notes available to a wider audience of physicists.

It should be emphasized that lecture notes are necessarily rough and informal, both in style and content; and those in the series will prove no exception. This is as it should be. The point of the series is to offer new, rapid, more informal, and, it is hoped, more effective ways for physicists to teach one another. The point is lost if only elegant notes qualify.

The publication of collections of reprints of recent articles in very active fields of physics will improve communication. Such collections are themselves useful to people working in the field. The value of the reprints will, however, be enhanced if the collection is accompanied by an introduction of moderate length which will serve to tie the collection together and, necessarily, constitute a brief survey of the present status of the field. Again, it is appropriate that such an introduction be informal, in keeping with the active character of the field.

The informal monograph, representing an intermediate step between lecture notes and formal monographs, offers an author the opportunity to present his views of a field which has developed to the point where a summation might prove extraordinarily fruitful but a formal monograph might not be feasible or desirable.

Contemporary classics constitute a particularly valuable approach to the teaching and learning of physics today. Here one thinks of fields that lie at the heart of much of present-day research, but whose essentials are by now well understood, such as quantum electrodynamics or magnetic resonance. In such fields some of the best pedagogical material is not readily available, either because it consists of papers long out of print or lectures that have never been published.

The above words, written in 1961, would seem to continue to be timely. FRONTIERS IN PHYSICS has, however, evolved, in the sense that it now seems appropriate to include within the series books which have evolved from lecture notes. In such books, which might be called informal texts or monographs, the author has taken an informal approach to the subject, but has worked through his material in a way that is much closer to the standard monograph or text than it is to an edited set of lecture notes.

The present volume by Chris Quigg is just such a book. Through his many fundamental contributions to our understanding of particle physics, Dr. Quigg has demonstrated his qualifications for writing an informal text on gauge theories of the fundamental interactions of elementary particles. In writing a coherent and elementary introduction to gauge theories he has made a considerable effort to make the subject accessible to the "nonexpert." I share his hopes that serious readers of this book will learn both to compute the consequences of local gauge theories and to gain perspective on forthcoming experimental initiatives.

DAVID PINES

PREFACE

My purpose in writing this book has been to present a coherent and elementary introduction to gauge theories of the fundamental interactions, with an emphasis on applications to the physics of elementary particles. By elementary I mean that little effort will be expended on field-theoretic technicalities and that correspondingly little formal sophistication will be demanded of the reader. The physical situations to be dealt with are, however, "advanced" topics of current interest in theoretical and experimental research. A serious reader will, it is hoped, come to understand the logic of local gauge symmetries and gauge theories, acquire the ability to compute the consequences of these theories, and gain a perspective on forthcoming experimental initiatives.

The idea of gauge theories is rooted in the classic investigations of Weyl of a half-century ago, and the notion that symmetries generate interactions was given full expression in the work of Yang and Mills some twenty-five years later. Yet it is only in the past decade that the principle of local gauge invariance has blossomed into a unifying theme that seems capable of embracing and even synthesizing all the elementary interactions. The emergence of gauge theories has been coupled with the recognition that a fundamental description of the subnuclear particles must be based upon the idea that the strongly interacting particles, or hadrons, are composed of quarks. Together with leptons, such as the electron and neutrino, quarks seem to be the elementary particles—structureless and indivisible—at least at the present limits of resolution.

Thus we possess today a coherent point of view and a single language appropriate for the description of all subnuclear phenomena. This development, which is the work of many hands, has not only made of particle physics a much more unified subject, it has also helped us to perceive common interests and to make common cause with other specialties, notably astrophysics and cosmology, the physics of condensed matter, atomic physics, and intermediate-energy nuclear physics. Although this new awareness has not instantly made Renaissance men and women of us all, it represents the welcome reversal of a trend toward overspecialization and compartmentalization.

Although much of the case for the new paradigm of quarks and leptons with interactions prescribed by gauge symmetries rests on circumstantial evidence, the experimental support is impressive in its consistency, diversity, and strength. The case for quarks consists in the spectroscopy of hadrons, the evidence for pointlike constituents within hadrons, the ψ/J and Υ families of heavy mesons with their quasi-atomic spectra, and more. In support of gauge theories and the unification of elementary interactions we may cite the triumphs of the unified theory of weak and electromagnetic interactions with its implication of neutral weak currents and corollary prediction of charm, as well as the more tentative and qualitative successes of the theory of strong interactions known as quantum chromodynamics. The similarity among quarks and leptons and the mathematical resemblance among the gauge theories of the fundamental interactions motivate an audacious program of "grand unification" in which the strong, weak, and electromagnetic interactions are different manifestations of a single, underlying symmetry. What makes our times the more exciting is that we shall soon have the means to subject these far-reaching ideas to decisive experimental tests.

The absence of discouragements has inspired still grander dreams. It is a truism that the requirement of general covariance in the theory of relativity resembles a gauge principle. Recently, the theoretical development of the beautiful concepts of supersymmetry and supergravity has raised hopes for a quantum theory of gravitation unified with the other interactions. Gauge fields are also deeply related to the mathematical theory of fiber bundles, a branch of contemporary mathematics rich in elegant ideas. This connection suggests the possibility of a geometrical interpretation of the forces of Nature.

Whatever the eventual fate of specific models, whatever the dénouement of contemporary speculations, it does not seem too much to hope that some essential elements of the paradigm will endure. In any event, it is apparent that gauge theories will remain the language of high-energy physics for some time to come, as well as the point of departure for new hypotheses.

This book has grown out of graduate courses given at The University of Chicago and l'Université de Paris, lecture series presented at Fermilab and Chicago, and courses taught in summer schools in St. Croix and Les Houches. Experience with diverse audiences has persuaded me both of the need for a nontechnical treatment for specialists in elementary particle physics and of the desirability that the gauge theory perspective be part of the education of every physics graduate student. In preparing this text, I have therefore had in mind three sorts of readers:

1. Graduate students in physics who have completed (at least) a graduate course in quantum mechanics, including a study of relativistic quantum mechanics and the rudiments of Feynman diagrams.
2. Experimental physicists working in the field of elementary particle physics.
3. Physicists who have not specialized in high-energy physics but wish to gain an appreciation of the essential ideas of gauge theories.

The book is intended to serve as a text for a one-quarter or one-semester "special topics" course for advanced graduate students and as a monograph for reference and self-study. It may also be used, selectively, to supplement a course on relativistic quantum mechanics.

The presentation is relatively self-contained, but any prior exposure to relativistic quantum field theory or elementary particle phenomenology will provide the reader with a perspective that is both broader and deeper. Although formal matters and technical fine points are not emphasized, they are acknowledged, and I have endeavored to at least call attention to the important issues in continuing research. History has not been entirely neglected, but neither is the treatment strictly historical. Although the benefits of an awareness of history are undeniable, it seems to me unnecessary in a first reading of the subject to experience all the puzzles and false steps of the past.

I have developed the subject as a logical whole, while bearing in mind that it should be possible to open the book at random and make sense of what is written. There is both a coherence and a progression, in that issues raised in early chapters are recalled, amplified, refined, and in some cases resolved later in the text. There is an accompanying evolution of the physical concepts and mathematical techniques.

I have stressed the essential interplay between theory and experiment, with respect to qualitative phenomena, symmetries, and specific numbers. Not least because of the mutual stimulation of observation and abstraction, I have included in the text detailed calculations of experimental observables and have posed a number of similarly explicit problems. Among the fondest memories of my physics infancy are those of the summer vacation during which I came upon Heitler's classic treatise on the quantum theory of radiation and Feynman's two early volumes in this series and began to understand how to compute and how to learn from experiment. Not every beginning reader of this book will experience a similar epiphany, but I have sought to provide the opportunity. There is another reason for undertaking explicit calculations, which is to be struck by some of the consequences of gauge invariance. This can be done only by witnessing at first hand the miraculous cooperation among Feynman diagrams that leads, among other things, to an acceptable high-energy behavior of the weak interactions.

At the end of each chapter I have provided an extensive reading list. These annotated bibliographies offer amplification of points of conceptual or technical interest, introduce further applications, or guide the reader to alternative presentations of the text material. It should not be necessary to consult the documents cited in order to master the contents of the chapter. Rather, they are intended as a selective guide to optional further study.

Finally, it bears repeating that gauge theories of the fundamental inter-actions are still in a state of development. Much of our present-day picture is tentative, and much remains to be accomplished even if, as seems likely, the general approach is sound. Therefore I have included in the text occa-sional remarks on the unsatisfactory, arbitrary, or incomplete features of current understanding, comments on open issues, and indications of experi-mental opportunities.

Although a detailed description of the topics covered in this book is to be found in the Contents, it is appropriate to present a brief overview here. The opening chapter contains a capsule review of elementary particle phenome-nology and introduces some of the issues to be elaborated in the rest of the book. In Chapters 2 through 5 the basic theoretical concepts are developed, proceeding from the elementary implications of symmetry principles through electrodynamics, non-Abelian gauge theories, and the notion of spontaneously broken symmetries. The emphasis in this part of the book is on conceptual matters, but applications are not entirely neglected. The last four chapters are devoted to the gauge theories that seem, on present evidence, to describe the strong, weak, and electromagnetic interactions. Two chapters are concerned with the Weinberg–Salam model of the weak and electro-magnetic interactions. In the second of these, the parton model—an impor-tant vehicle for later developments—makes its appearance. Quantum chromodynamics, a theory of the strong interactions, is the subject of Chapter 8. The final chapter deals with the program of unifying the funda-mental interactions, the principal ideas and consequences of which are studied in the context of the minimal $SU(5)$ model. Matters of convention and other technical issues are relegated to three appendices.

* * * * *

It is a pleasure to be able to express my appreciation to many people and institutions for their contributions to this work. In the course of preparing this volume and the courses that preceded it, I have enjoyed the warm hospitality of a number of institutions including the Laboratoire de Physique Théorique et Hautes Energies, Université de Paris-XI, Orsay, and CERN, the European Organization for Nuclear Research, in Geneva. I am especially grateful for the opportunity to spend the 1981–1982 academic year in the stimulating environment of the Laboratoire de Physique Théorique de l'Ecole Normale Supérieure, Paris. For the generous grant of a year's leave of absence from Fermilab I am indebted to Leon Lederman, Director, and to the Trustees of Universities Research Association.

Numerous colleagues have contributed to my understanding of gauge theories, and the reactions and questions of students have influenced my presentation of the material. Special thanks are owed to Carl Albright, Jon Rosner, Jonathan Schonfeld, Roy Schwitters, Hank Thacker, and Bruce Winstein for their valued comments on the text. I also appreciate the assistance of Sue Grommes, Pat Oleck, and especially Elaine Moore in the preparation of the manuscript.

Finally I thank my wife Elizabeth and our children David and Katherine for their encouragement, support, and kindness.

CHRIS QUIGG
Paris, 1982

CHAPTER 1

INTRODUCTION

High-energy physics, the science of the ultimate constituents of matter and the interactions among them, has undergone a remarkable development during the past decade. A host of new experimental results made accessible by a new generation of particle accelerators and the accompanying rapid convergence of theoretical ideas have brought to the subject a new coherence and have raised new aspirations. Many thoughtful scientists are expressing optimism that a grand synthesis of natural phenomena is at hand.[1] Indeed, it will be seen in the course of this volume that a unified description of the strong, weak, and electromagnetic interactions may not be just a distant dream. A qualitatively satisfactory, if not entirely satisfying, unified theory may already exist. The most sanguine observers accept this sort of "grand unification" as a *fait accompli* and argue that a complete Theory of the World awaits only the incorporation of gravitation.

Ours is hardly the first time in history that physicists have believed themselves close to an enduring understanding of the laws of Nature. What, then, are the grounds for believing that (whether or not "the end" is in sight) important progress is under way? Essential elements are the significance of the quark-model description of hadrons, the nonobservation of quark or lepton substructure, the notable successes of the Weinberg–Salam model of the weak and electromagnetic interactions, and the low abundance of free quarks.

The utility of the quark model as a classification tool that provides a systematic basis for hadron spectroscopy has long been appreciated. Recognition that the quark language also permits an apt description of the dynamics of hadronic interactions has come more recently. The quark-parton model underlies a quantitative phenomenology of deeply inelastic lepton–hadron

1

scattering and of electron–positron annihilation into hadrons. The interpretation of violent collisions of hadrons in terms of the hard scattering of pointlike constituents, although for the moment somewhat schematic, is extremely appealing as well.

An elementary particle, in the time-honored sense of the term, is structureless and indivisible. Although history cautions that the physicist's list of elementary particles is dependent upon experimental resolution, and thus subject to revision with the passage of time, it also has rewarded the hope that interactions among the elementary particles of the moment would be simpler and more fundamental than those among composite systems. Neither quarks nor leptons exhibit any structure on a scale of about 10^{-16} cm, the currently attained resolution. We thus have no experimental reason but tradition to suspect that they are not the ultimate elementary particles. For nearly all of this book they will be considered as the fundamental fermions.

The appeal of unified theories of weak and electromagnetic interactions is at once esthetic and practical. The effective weak-interaction Lagrangian that evolved from Fermi's description of nuclear β-decay, and provided a serviceable low-energy phenomenology, is now seen to be the limiting form of a renormalizable field theory. At the same time, neutral-current interactions predicted by the theory have been found to occur at approximately the strength of the long-studied charged-current interactions. The observed neutral currents are neutral not only with respect to electric charge, but with respect to all other additive quantum numbers as well. To accommodate this property in the theory requires the introduction of a new quark species, bearing a new additive quantum number known as charm. This, too, has subsequently been observed in experiments. The price for this neat picture includes the prediction of several hypothetical particles: the intermediate vector bosons W^+, W^-, and Z^0, which mediate the weak interactions, and a neutral particle H^0 known as the Higgs scalar. Definite predictions for the masses and properties of the intermediate bosons will soon be put to experimental test.

Analyses of collision phenomena suggest that quarks behave as free particles within hadrons, and yet isolated free quarks are so exceedingly rare that a convenient idealization holds them to be permanently confined within hadrons. This apparently paradoxical state of affairs requires that the strong interaction among quarks be of a rather unfamiliar sort. No rigorous theoretical demonstration of the confinement hypothesis has yet been given, but it is widely held that the necessary elements are contained in the gauge theory known as quantum chromodynamics (QCD). In common with other non-Abelian gauge theories, QCD exhibits an effective interaction strength that is small at short distances and large at large distances. This is suggestive of the desired characteristics. Recent results from Monte Carlo simulations of the gauge theory vacuum provide strong numerical evidence for this view.

This book is devoted to an exposition of the logic, structure, and phenomenology of gauge theories of the fundamental interactions. Although it is possible to define gauge theories in the abstract and thus to investigate their mathematical properties, the specific theories of physical interest require experimental motivation. For that reason it is appropriate to present a brief overview of the essential phenomenology and of the idealizations drawn from it.

1.1   Leptons

Among the fundamental particles, those that experience weak but not strong interactions are known as the *leptons*.[2] All the known members of this class are spin-1/2 particles, which are structureless at the current limits of resolution of[3] $\sim 1 \times 10^{-16}$ cm.

Three charged leptons—the electron, the muon, and the tau—are firmly established, the electron and muon by direct observation and the tau through its decay products. The gyromagnetic ratios of the electron and muon have been measured with remarkable precision. They differ from the value of 2 expected for Dirac particles only by tiny fractions, which have been calculated in quantum electrodynamics (QED). This strongly supports their identification as point particles. The electron neutrino and the muon neutrino are likewise well known, and there is mounting experimental evidence for the existence of a distinct tau neutrino.[4] Some properties[5] of the known leptons are summarized in Table 1.1.

The names of the neutrinos follow from the structure of the charged weak current, which is represented by the weak-isospin doublets

$$\psi_1 = \begin{pmatrix} v_e \\ e \end{pmatrix}_L, \qquad \psi_2 = \begin{pmatrix} v_\mu \\ \mu \end{pmatrix}_L, \qquad \psi_3 = \begin{pmatrix} v_\tau \\ \tau \end{pmatrix}_L, \qquad (1.1.1)$$

TABLE 1.1:   THE SPECTRUM OF LEPTONS

Lepton	Charge	Mass (MeV/c^2)	Lifetime (sec)
v_e	0	<60 eV/c^2	Stable
e	-1	0.5110034 ± 0.0000014	Stable ($>5 \times 10^{21}$ y)
v_μ	0	<0.57	Stable
μ	-1	105.65946 ± 0.00024	$(2.197120 \pm 0.000077) \times 10^{-6}$
(?) v_τ	0	<250	Stable
τ	-1	1784 ± 4	$(4.6 \pm 1.9) \times 10^{-13a}$

[a] G. J. Feldman, G. H. Trilling, *et al.*, *Phys. Rev. Lett.* **48**, 66 (1982).

where the subscript L denotes a left-handed or vector-minus-axial-vector Lorentz structure. The leptonic charged current thus has the form[6]

$$J_\lambda^{(\pm)} = \sum_i \bar{\psi}_i \tau_{\pm} \gamma_\lambda (1 - \gamma_5) \psi_i, \qquad (1.1.2)$$

where $\tau_{\pm} \equiv (1/2)(\tau_1 \pm i\tau_2)$ are the Pauli isospin matrices

$$\tau_+ = \begin{pmatrix} 0 & 1 \\ 0 & 0 \end{pmatrix}, \qquad \tau_- = \begin{pmatrix} 0 & 0 \\ 1 & 0 \end{pmatrix}. \qquad (1.1.3)$$

The implied universality of weak-interaction transition matrix elements holds to high accuracy for the electronic and muonic transitions and has been verified within 30% for the tau family. It will be seen that the identification of weak-isospin doublets leads in addition to an understanding of the structure of the neutral weak current.

The origin of the simple and orderly family pattern exhibited by the leptons is obscure. Indeed, many apparent facts and regularities are inadequately comprehended, and therefore many questions present themselves. Why are there three doublets of leptons? Will more be found? Is the difference between the number of leptons and antileptons in the universe a conserved quantity? What is the pattern of lepton masses? Are the neutrinos precisely massless? Is the separate conservation of an additive electron number, muon number, and tau number an exact or only an approximate statement? Do oscillations between neutrino species, analogous to those between K^0 and $\bar{K}^0$, occur in Nature?

In spite of the fact that the gauge theory approach is in many ways constraining, these questions lack definite answers. Theory can, as we shall see in good time, expose and relate possibilities, but for the moment the issues are basically experimental.

1.2   Why We Believe in Quarks

Quarks are the fundamental constituents of the strongly interacting particles. They experience all the known interactions: strong, weak and electromagnetic, and gravitational. Quarks have much in common with the leptons, being spin-1/2 particles that appear pointlike at the current limits of resolution[7] of $\sim 1 \times 10^{-16}$ cm. There is, however, a crucial distinction. Free quarks are not routinely observed in the laboratory, so it will be necessary to amass indirect evidence for their existence and their properties. This will be done evocatively, but at some length because the quark concept is central to the understanding of fundamental processes. Many of the observations made here telegraphically will be developed more completely in later chapters. The present discussion is intended as a series of reminders for the true believer, but it should also suffice to indicate to the reader inexperienced in high-energy physics the degree to which the foundation for what follows is experiment, and not merely myth.

Quarks were proposed[8] in 1964 by Gell-Mann and Zweig as a means for understanding the $SU(3)$ classification of the hadrons. The light mesons occur only in $SU(3)$ singlets and octets. For example, the nine pseudoscalars are shown in the familiar (strangeness versus third component of isospin) hexagonal array in Fig. 1-1. Similarly, the light baryons are restricted to singlets, octets, and decimets of $SU(3)$. The lowest-lying baryons are displayed in Figs. 1-2 and 1-3. The observation that no higher representations are indicated is far more restrictive than the mere fact that $SU(3)$ is a good classification symmetry, and requires explanation.

As Gell-Mann and Zweig pointed out, the observed patterns can be understood in terms of the hypothesis that hadrons are composite structures

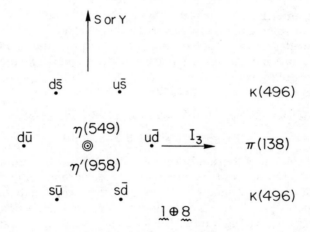

FIG. 1-1. $SU(3)$ weight diagram for the pseudoscalar meson nonet $= [1] \oplus [8]$.

FIG. 1-2. The $J^P = 1/2^+$ baryon octet.

Fig. 1-3. The $J^P = 3/2^+$ baryon decimet.

built from an elementary triplet of spin-1/2 quarks, corresponding to the fundamental representation of $SU(3)$. The three "flavors" of quarks, commonly named *up*, *down*, and *strange*, have the properties summarized in Table 1.2. A meson made up of a quark and an antiquark ($q\bar{q}$) then lies in the $SU(3)$ representations:

$$\mathbf{3} \otimes \mathbf{3^*} = \mathbf{1} \oplus \mathbf{8}, \tag{1.2.1}$$

and a baryon, composed of three quarks (qqq), must be contained in

$$\mathbf{3} \otimes \mathbf{3} \otimes \mathbf{3} = \mathbf{1} \oplus \mathbf{8} \oplus \mathbf{8} \oplus \mathbf{10}. \tag{1.2.2}$$

The quark content of the hadrons is indicated in Figs. 1-1 through 1-3. This simple model reproduces the representations seen prominently in Nature. It remains, of course, to understand why only these combinations of quarks and antiquarks are observed, or to discover the circumstances under which more complicated configurations (such as $q\bar{q}q\bar{q}$ or $qqqq\bar{q}$ or $6q$) might arise.

   Why should quarks exist in several flavors? Within the strong interactions as currently understood, flavors appear not to have any essential role but only to contribute to a richness. In contrast, the significance of flavors for the weak interactions is more readily apparent. Because of the family

TABLE 1.2: PROPERTIES OF THE LIGHT QUARKS

Quark	I	I_3	S	B	$Y = B + S$	Q	Effective mass (MeV/c^2)
u	1/2	1/2	0	1/3	1/3	2/3	~ 300
d	1/2	$-1/2$	0	1/3	1/3	$-1/3$	~ 300
s	0	0	-1	1/3	$-2/3$	$-1/3$	~ 500

patterns implied by the charged-current weak-isospin doublets such as

$$\begin{pmatrix} u \\ d_\theta \end{pmatrix}_L, \qquad (1.2.3)$$

where

$$d_\theta \equiv d \cos \theta_C + s \sin \theta_C \qquad (1.2.4)$$

and θ_C is the Cabibbo angle, flavors do appear to have specific parts to play. As we shall see shortly, a fourth flavor known as charm was needed before its discovery to fulfill a specific function in the weak interactions. From the perspective of the strong interactions, charm and other new flavors seem merely to be tolerated.

The existence of several flavors of quarks has thus been inferred from the quantum numbers of the hadrons. Let us now review the evidence for the properties imputed to the quarks.

Quarks have baryon number $1/3$; antiquarks have baryon number $-1/3$. This follows from the assertion that three quarks make up a baryon.

If baryons, which are fermions, are to be made of three identical constituents, the constituents must themselves be fermions. The observed hadron spectrum corresponds to the objects that can be formed as $(q\bar{q})$ or (qqq) composites of spin-$1/2$ quarks. It is straightforward to work out the level structure in the meson sector:

$$(q\bar{q}) \to J^{PC} = \underbrace{0^{-+}, 1^{--}}_{L=0}; \qquad \underbrace{0^{++}, 1^{++}, 1^{+-}, 2^{++}}_{L=1}; \quad \dots \qquad (1.2.5)$$

This conforms closely to the observed ordering of levels. Nowhere is this more clearly shown than in the spectrum of bound states of a charmed quark and antiquark depicted in Fig. 1-4. Combinations of spin, parity, and charge conjugation such as $J^{PC} = 0^{--}, 0^{+-}, 1^{-+}$, which cannot be formed from pairs of spin-$1/2$ quarks and antiquarks, have not been observed. The analysis of baryon multiplets is similar but more tedious. Again the observed spectrum is in agreement with the quark-model pattern.

In addition to this successful classification scheme, there are dynamical tests of the quark spin. Consider the cross sections for absorption of longitudinal or transverse virtual photons on point particles. In the Breit frame of the struck particle, illustrated in Fig. 1-5, it is easy to see that a spinless "quark" can only absorb a longitudinal (helicity $= 0$) photon, because angular momentum conservation forbids the absorption of a transverse (helicity $= \pm 1$) photon. Similarly, a spin-$1/2$ quark can absorb a transverse photon, but not a longitudinal photon. Within the parton model, deeply inelastic scattering of electrons from nucleon targets is analyzed as the scattering of electrons from noninteracting and structureless charged constituents. If the charged constituents are spin-$1/2$ quarks, the above

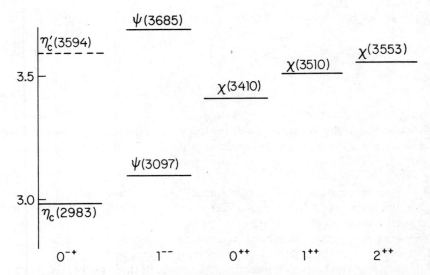

FIG. 1-4. Spectrum of charm–anticharm bound states below the threshold for dissociation into charmed particles.

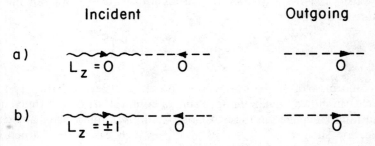

FIG. 1-5. (a) Absorption of a longitudinal photon by a spinless particle is allowed by angular momentum conservation. (b) Absorption of a transverse photon is forbidden.

analysis implies

$$\sigma_{longitudinal}/\sigma_{transverse} = 0, \tag{1.2.6}$$

in schematic agreement with experiment.[9]

A related test is provided by the angular distribution of hadron jets in electron–positron annihilations, which is observed[10] to be identical to the production angular distribution of muons in the reaction

$$e^+e^- \to \mu^+\mu^-. \tag{1.2.7}$$

That hadrons are emitted in well-collimated jets, as shown in the exemplary event in Fig. 1-6, supports the interpretation of particle production by means

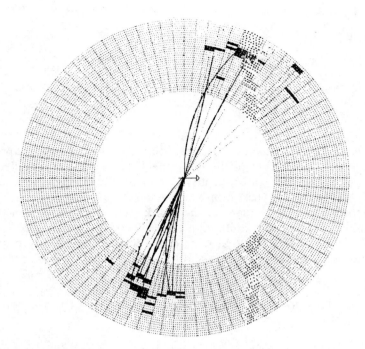

FIG. 1-6. Radius-azimuth perspective of a prototype two-jet event in the reaction $e^+e^- \rightarrow$ hadrons, as observed in the JADE detector at PETRA, at $E_{c.m.} \simeq 30$ GeV.

of the elementary process

$$e^+e^- \rightarrow q\bar{q}, \tag{1.2.8}$$

followed by the hadronization of the noninteracting quarks. The jet angular distribution then reflects the angular distribution of the spin-1/2 quarks.

The quarks also carry fractional electric charge. The Gell-Mann–Nishijima formula[11] for displaced charge multiplets,

$$Q = I_3 + \tfrac{1}{2}Y = I_3 + \tfrac{1}{2}(B + S), \tag{1.2.9}$$

implies the charges shown in Table 1.2. The same assignments follow directly from examination of members of the baryon decimet:

$$\Delta^{++} = uuu$$
$$\Delta^+ = uud$$
$$\Delta^0 = udd \qquad \Omega^- = sss.$$
$$\Delta^- = ddd$$

A number of other experimental tests of these assignments have been carried out. It is worthwhile to recall three of them.[12]

FIG. 1-7. Quark-model description of the decay of a neutral vector meson into a pair of charged leptons.

In the quark model, leptonic decays of vector mesons, $V^0 \to l^+l^-$, proceed through the annihilation of the constituent quark and antiquark into a virtual photon, which subsequently disintegrates into the lepton pair, as illustrated in Fig. 1-7. Apart from kinematic factors, the leptonic decay rate is proportional to the square of the quark charge and to the encounter probability for the quark and antiquark. The latter is given, in a nonrelativistic picture, by $|\Psi(0)|^2$, the square of the wave function at zero quark–antiquark separation. It is convenient to define a reduced decay rate

$$\tilde{\Gamma}(V^0 \to l^+l^-) \equiv M_V^2 \Gamma(V^0 \to l^+l^-) = \text{constants} \times e_q^2 \times |\Psi(0)|^2. \quad (1.2.10)$$

The mean-squared quark charge follows simply from the flavor wave functions of the mesons:

$$|\rho^0\rangle = \frac{\bar{u}u - \bar{d}d}{\sqrt{2}} \to e_q^2 = \left[\frac{1}{\sqrt{2}}\left(\frac{2}{3} + \frac{1}{3}\right)\right]^2 = \frac{1}{2}, \quad (1.2.11)$$

$$|\omega^0\rangle = \frac{\bar{u}u + \bar{d}d}{\sqrt{2}} \to e_q^2 = \left[\frac{1}{\sqrt{2}}\left(\frac{2}{3} - \frac{1}{3}\right)\right]^2 = \frac{1}{18}, \quad (1.2.12)$$

$$|\phi\rangle = \bar{s}s \to e_q^2 = \tfrac{1}{9}. \quad (1.2.13)$$

If to first approximation the spatial wave functions of $\rho(776)$, $\omega(784)$, and $\phi(1019)$ may be regarded as equal, the reduced leptonic decay rates are expected to satisfy

$$\tilde{\Gamma}(\rho^0)/9 = \tilde{\Gamma}(\omega) = \tilde{\Gamma}(\phi)/2. \quad (1.2.14)$$

This is in reasonable agreement with the measured values of 454 ± 67, 470 ± 119, and 661 ± 54 MeV^3, respectively.

A second test pertains to the production of lepton pairs in pion–nucleon collisions. This reaction may be idealized as the annihilation of an antiquark from the pion with a quark from the target, as sketched in Fig. 1-8. An isoscalar target such as carbon contains equal numbers of protons and neutrons, and thus equal numbers of up quarks and down quarks. For the reaction initiated by $\pi^-(\bar{u}d)$, the elementary process is $u\bar{u} \to l^+l^-$, for which the cross section is proportional to e_u^2. For that initiated by $\pi^+(u\bar{d})$, the elementary process is $d\bar{d} \to l^+l^-$, for which the cross section is proportional

CONTENTS

QC 793
.3
F5 Q53
1983
PHYS

vii

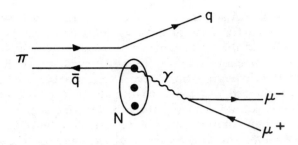

FIG. 1-8. The Drell–Yan process in the reaction $\pi N \to l^+ l^- +$ anything.

to e_d^2. The resulting expectation that

$$\frac{\sigma(\pi^+ C \to \mu^+ \mu^- + \text{anything})}{\sigma(\pi^- C \to \mu^+ \mu^- + \text{anything})} = \frac{1}{4} \qquad (1.2.15)$$

is confirmed by experiment[13] in the kinematic region to which the idealization applies.

Still another test is provided by the measurement of cross sections for deeply inelastic lepton–nucleon scattering, as will be seen in Section 7.3. It is conventional to define a structure function characteristic of the probe and target by stripping the cross section of overall coupling strengths and kinematic factors. The inclusive reaction $\nu T \to \mu^- +$ anything corresponds to the elementary process $\nu d \to \mu^- u$, so that the charged-current structure function $\mathscr{F}_2(\nu T)$ may be said to count the number of down quarks in the target. In similar fashion, the reaction $\bar{\nu} T \to \mu^+ +$ anything proceeds via the elementary interaction $\bar{\nu} u \to \mu^+ d$, so that $\mathscr{F}_2(\bar{\nu} T)$ counts the number of up quarks in the target. Deeply inelastic electron scattering results from the electromagnetic process $eq \to eq$. The structure function $F_2(eT)$ therefore counts the number of up quarks in the target weighted by $e_u^2 = 4/9$, plus the number of down quarks in the target weighted by $e_d^2 = 1/9$. Consequently, for an isoscalar target such as deuterium we expect

$$\frac{F_2(eD)}{\mathscr{F}_2(\nu D) + \mathscr{F}_2(\bar{\nu} D)} = \frac{3e_u^2 + 3e_d^2}{3 + 3} = \frac{5}{18}, \qquad (1.2.16)$$

which is in accord with the experimental findings.[14]

The characteristics of quarks that we have discussed until now—spin, flavor, baryon number, electric charge—are directly indicated by the experimental observations that originally motivated the quark model. Quarks have still another property, less obvious but of central importance for the strong interactions. This additional property is known as *color*. At first sight, the Pauli principle seems not to be respected in the wave function for the Δ^{++}. This nucleon resonance is a (*uuu*) state with spin = 3/2 and isospin $= 3/2$, in which all the quark pairs are in relative s-waves. Thus it is apparently a symmetric state of three identical fermions. Unless we are prepared to suspend the Pauli principle or to forgo the quark model, it is necessary[15] to invoke a new, hidden degree of freedom, which permits the Δ^{++} wave function to be antisymmetrized. In order that a (uuu) wave function can be antisymmetrized, each quark flavor must exist in no fewer than three distinguishable types called colors. More than three colors would raise the unpleasant possibility of distinguishable (colored) species of protons, which is contrary to common experience. There is little doubt that the introduction of color, thus motivated, appears arbitrary and artificial. However, a number of observables are sensitive to the number of distinct species of quarks, and subsequent measurements of these quantities have given strong support to the color hypothesis.

The inclusive cross section for electron–positron annihilation into hadrons is described, as earlier noted, by the elementary process $e^+e^- \to q\bar{q}$, where the quark and antiquark materialize with unit probability into the observed hadron jets. At a particular collision energy, the ratio

$$R \equiv \frac{\sigma(e^+e^- \to \text{hadrons})}{\sigma(e^+e^- \to \mu^+\mu^-)} \qquad (1.2.17)$$

is then simply given as

$$R = \sum_{\substack{\text{quark} \\ \text{species}}} e_q^2. \qquad (1.2.18)$$

At barycentric energies between approximately 1.5 and 3.6 GeV, pairs of up, down, and strange quarks are kinematically accessible. In the absence of hadronic color we would therefore expect

$$R_1 = e_u^2 + e_d^2 + e_s^2 = \tfrac{2}{3}, \qquad (1.2.19)$$

but if each quark flavor exists in three colors, we should have

$$R_3 = 3(e_u^2 + e_d^2 + e_s^2) = 2. \qquad (1.2.20)$$

Experiment decisively favors the color triplet hypothesis, as shown in Fig. 1-9. At still higher energies the heavier c- and b-quarks may be produced in the semifinal state. Their properties are summarized in Table 1.3. For the highest energies yet explored, the color triplet model thus predicts

$$R = 3(e_u^2 + e_d^2 + e_s^2 + e_c^2 + e_b^2) = \tfrac{11}{3}, \qquad (1.2.21)$$

in excellent agreement with the data.

A similar count of the number of distinguishable quarks of each flavor is provided by the decay branching ratios of the tau lepton. Within the quark

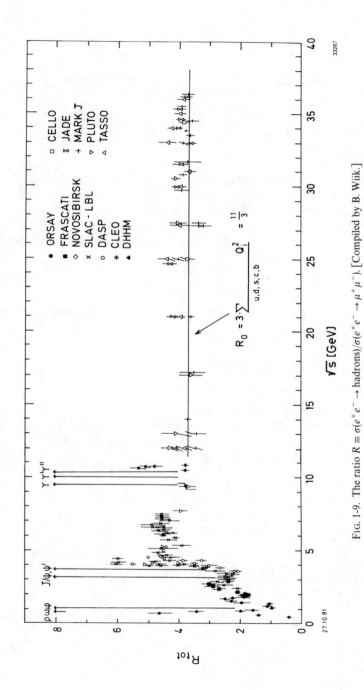

Fig. 1-9. The ratio $R \equiv \sigma(e^+e^- \rightarrow \text{hadrons})/\sigma(e^+e^- \rightarrow \mu^+\mu^-)$. [Compiled by B. Wiik.]

TABLE 1.3:   SOME PROPERTIES OF THE HEAVY QUARKS

Quark	I	Q	Charm	Beauty	Truth	Effective mass (GeV/c²)
c	0	2/3	1	0	0	~1.5
b	0	−1/3	0	1	0	~4.8
(?) t	0	2/3	0	0	1	>18

model, τ-decays may be described as shown in Fig. 1-10, namely by the decay of τ into ν_τ plus a virtual intermediate boson W^-. The intermediate boson may then disintegrate into all kinematically accessible fermion–antifermion pairs: $(e^-\bar{\nu}_e)$, $(\mu^-\bar{\nu}_\mu)$, $(\bar{u}d_\theta)$. The universality of charged-current weak interactions implies equal rates for each of these decays. Therefore, in the absence of color, we expect

$$B_1 = \frac{\Gamma(\tau \to e^-\bar{\nu}_e\nu_\tau)}{\Gamma(\tau \to \text{all})} = \frac{1}{3}. \tag{1.2.22}$$

If the quarks are color triplets, $(\bar{u}d_\theta)$ is increased to $3(\bar{u}d_\theta)$, and the leptonic branching ratio becomes[16]

$$B_3 = \tfrac{1}{5}. \tag{1.2.23}$$

The experimentally measured branching ratio,

$$B_{\text{exp}} = (17.44 \pm 0.85)\% \tag{1.2.24}$$

is in accord with the color hypothesis.

The other important measure of the number of quark colors is the $\pi^0 \to \gamma\gamma$ decay rate.[17] A calculation for π^0-decay via a quark–antiquark loop is straightforward to carry out, although the required justification is somewhat

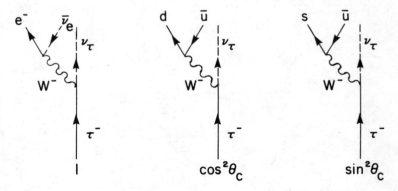

FIG. 1-10. Semileptonic decays (weight 1), and Cabibbo-favored (weight $\cos^2 \theta_C$) and Cabibbo-suppressed (weight $\sin^2 \theta_C$) nonleptonic decays of the tau lepton.

subtle. The result is that

$$\Gamma(\pi^0 \to \gamma\gamma) = \left(\frac{\alpha}{2\pi}\right)^2 [N_c(e_u^2 - e_d^2)]^2 \frac{M_\pi^3}{8\pi f_\pi}, \qquad (1.2.25)$$

where N_c is the number of colors and $f_\pi \simeq 130$ MeV is the pion decay constant, determined from the charged pion lifetime. The predicted rate is then

$$\Gamma(\pi^0 \to \gamma\gamma) = \begin{cases} 0.86 \text{ eV}, N_c = 1, \\ 7.75 \text{ eV}, N_c = 3, \end{cases} \qquad (1.2.26)$$

to be compared with the measured decay rate of (7.86 ± 0.54) eV.

From all these experimental indications, and from a further theoretical argument to be given in Section 6.6, we may conclude that the hidden color degree of freedom is indeed present. It is then tempting to suppose that color is the property that distinguishes quarks from leptons and might play the role of a strong-interaction charge. This speculation will lead eventually to the gauge theory of strong interactions, quantum chromodynamics.

The observation of hadrons with various internal quantum numbers such as isospin, strangeness, and charm has led to the application of the flavor symmetries $SU(2)$, $SU(3)$, etc., to the strong interactions. These internal symmetry groups serve both for the classification of hadrons (whence the inspiration for the quark model) and for dynamical relations among strong-interaction amplitudes. Both isospin and $SU(3)$ are excellent, but not exact, strong-interaction symmetries. Isospin invariance holds within a few percent, and $SU(3)$ is respected at the 10–20% level. All the same, the outcome of the preceding discussion has been to minimize the direct importance of flavors in the strong interactions and to emphasize the significance of color. Reasons for dismissing a strong-interaction theory based on flavor symmetry will be developed in Chapter 4, but the problem of accounting for the existence and the breaking of flavor symmetries will remain. According to an evolving view, the breaking of $SU(N)$ flavor symmetry is a consequence of the quark mass differences

$$m_u < m_d < m_s < m_c < m_b < \cdots \qquad (1.2.27)$$

which themselves follow, in a manner not fully understood, from the spontaneous symmetry breaking of the weak and electromagnetic interactions. The goodness of isospin and $SU(3)$ symmetry is then owed to the smallness and near degeneracy of the up, down, and strange quark masses.

This line of thought obviously has not yet led to a complete understanding of the strong-interaction symmetries. It is unsatisfying to conclude that isospin invariance, the oldest and most exact of strong-interaction symmetries, is merely coincidental. Quite evidently, we do not know why the pattern of quark masses and mixing angles should be as it is, nor even why so many "fundamental" fermions should exist. The similarity of quarks and leptons suggests the possibility of identifying extended families containing both types of elementary particles. It may be hoped that the implied further unification of elementary forces will provide at least partial answers to these and other questions.

1.3   The Fundamental Interactions

We now believe that the elementary interactions of the quarks and leptons can be understood as consequences of gauge symmetries. The notions of local gauge invariance and gauge theories and the construction and application of specific theories will be developed in logical sequence in the succeeding chapters. As prologue, let us merely recall here some superficial aspects of the familiar gauge theories and their practical consequences.

Some essential features of unified theories of the weak and electromagnetic interactions are exhibited by the Weinberg–Salam theory of leptons[18] in a world populated by the electron and muon doublets. In such a world, the electromagnetic current is given by

$$J_\lambda^{(\text{em})} = -\bar{e}\gamma_\lambda e - \bar{\mu}\gamma_\lambda\mu, \tag{1.3.1}$$

which evidently leaves all additive quantum numbers unchanged. The charged weak current is that indicated by the weak-isospin doublets (1.1.2), namely

$$J_\lambda^{(+)} = \bar{v}_e\gamma_\lambda(1 - \gamma_5)e + \bar{v}_\mu\gamma_\lambda(1 - \gamma_5)\mu. \tag{1.3.2}$$

If the idea of weak-isospin symmetry is to be taken seriously, these long-studied leptonic currents must be supplemented by another current, which completes the weak isovector:

$$\begin{aligned}
J_\lambda^{(3)} &= \tfrac{1}{2}\sum_i \bar{\psi}_i\tau_3\gamma_\lambda(1 - \gamma_5)\psi_i \\
&= \tfrac{1}{2}[\bar{v}_e\gamma_\lambda(1 - \gamma_5)v_e - \bar{e}\gamma_\lambda(1 - \gamma_5)e \\
&\quad + \bar{v}_\mu\gamma_\lambda(1 - \gamma_5)v_\mu - \bar{\mu}\gamma_\lambda(1 - \gamma_5)\mu].
\end{aligned} \tag{1.3.3}$$

Like the electromagnetic current, this third component of weak isospin leaves additive quantum numbers unchanged. In the standard model, the weak neutral current that results as a new prediction may be expressed as a linear combination of $J_\lambda^{(3)}$ and $J_\lambda^{(\text{em})}$,

$$J_\lambda^{(0)} = J_\lambda^{(3)} - 2\sin^2\theta_\text{W}J_\lambda^{(\text{em})}, \tag{1.3.4}$$

with the strength of the electromagnetic term governed by the weak mixing angle θ_W. The explicit form is

$$J_\lambda^{(0)} = \tfrac{1}{2}\bar{v}_e\gamma_\lambda(1 - \gamma_5)v_e + L\bar{e}\gamma_\lambda(1 - \gamma_5)e + R\bar{e}\gamma_\lambda(1 + \gamma_5)e + (e \to \mu), \tag{1.3.5}$$

where the coupling strengths L and R depend on θ_W. Notice that the electron and muon sectors remain disconnected, as required by the separate conservation of electron number and muon number. The leptonic neutral current is said to be flavor-conserving, or diagonal in flavors.

What of the hadronic sector? Consider the three quarks u, d, and s that constitute the light hadrons. According to Cabibbo's hypothesis[19] of weak-interaction universality, the hadronic charged current is represented by the weak-isospin doublet (1.2.3). It thus has the explicit form

$$J_\lambda^{(+)} = \bar{u}\gamma_\lambda(1 - \gamma_5)d \cdot \cos\theta_C + \bar{u}\gamma_\lambda(1 - \gamma_5)s \cdot \sin\theta_C. \qquad (1.3.6)$$

Even before entering the domain of unified theories, we may ask why the hadron sector should have a superfluous quark or, in other words, why the orthogonal combination of s- and d-quarks

$$s_\theta = s\cos\theta_C - d\sin\theta_C \qquad (1.3.7)$$

does not appear in the charged weak current. To pose the same question differently, why do quarks and leptons not enter more symmetrically?

These apparently idle questions become urgent in the framework of unified theories, where they also find their answers. In the three-quark theory, the weak neutral current is

$$
\begin{aligned}
J_\lambda^{(0)} &= J_\lambda^{(3)} - 2\sin^2\theta_W J_\lambda^{(em)} \\
&= \tfrac{1}{2}\big[\bar{u}\gamma_\lambda(1 - \gamma_5)u - \bar{d}\gamma_\lambda(1 - \gamma_5)d \cdot \cos^2\theta_C \\
&\quad - \bar{s}\gamma_\lambda(1 - \gamma_5)s \cdot \sin^2\theta_C - \bar{s}\gamma_\lambda(1 - \gamma_5)d \cdot \sin\theta_C\cos\theta_C \\
&\quad - \bar{d}\gamma_\lambda(1 - \gamma_5)s \cdot \sin\theta_C\cos\theta_C\big] \\
&\quad - 2\sin^2\theta_W(\tfrac{2}{3}\bar{u}\gamma_\lambda u - \tfrac{1}{3}\bar{d}\gamma_\lambda d - \tfrac{1}{3}\bar{s}\gamma_\lambda s).
\end{aligned} \qquad (1.3.8)
$$

Unlike the leptonic neutral current, the hadronic neutral current contains flavor-changing $(d \to s)$ terms. Esthetics aside, this is experimentally unacceptable because of the stringent upper limit on the decay rate for $K^+ \to \pi^+ \nu\bar{\nu}$ and the small rate observed for $K_L \to \mu^+\mu^-$. It was shown by Glashow, Iliopoulos, and Maiani[20] that lepton–quark symmetry could be arranged and the flavor-changing neutral currents could be eliminated by the addition of a second weak-isospin doublet

$$\begin{pmatrix} c \\ s_\theta \end{pmatrix}_L \qquad (1.3.9)$$

involving the charmed quark.[21] The hadronic neutral current would then be

$$
\begin{aligned}
J_\lambda^{(0)} &= \tfrac{1}{2}\big[\bar{u}\gamma_\lambda(1 - \gamma_5)u + \bar{c}\gamma_\lambda(1 - \gamma_5)c - \bar{d}\gamma_\lambda(1 - \gamma_5)d \\
&\quad - \bar{s}\gamma_\lambda(1 - \gamma_5)s\big] - 2\sin^2\theta_W J_\lambda^{(em)},
\end{aligned} \qquad (1.3.10)
$$

which is manifestly flavor-diagonal. The discovery[22] of the family of $(c\bar{c})$ bound states known as psions and the subsequent observation[23] of charmed particles that decay according to the $(c, s_\theta)_L$ pattern constitute a striking confirmation of the GIM hypothesis, and an important psychological success for the idea of unification.

The gauge theory known as quantum chromodynamics, in which colored quarks interact by means of massless colored gauge bosons named gluons, shows considerable promise as a theory of the strong interactions among quarks. The three quark colors, for which evidence has been reviewed in Section 1.2, are regarded as the basis of the color-symmetry group $SU(3)_c$, distinct from the flavor $SU(3)$ group of up, down, and strange quarks. An $SU(3)_c$ octet of vector gluons mediate the interactions among all colored objects, including the quarks and the gluons themselves.

It will be found appealing to argue that only color-singlet objects may exist in isolation. Color confinement, as it is called, would then imply quark confinement, which would in turn explain the low abundance of free quarks: free quarks and gluons would simply not exist. It will be shown in Chapter 8 that the coupling "constant" of the color interaction, far from being a constant, decreases at short distances. This property of QCD, which is known as asymptotic freedom, may help to explain why permanently confined quarks behave within hadrons as if they are free particles.

Before leaving this introductory chapter, let us review the experimental evidence for the existence of the color gauge bosons, the gluons. It is basically of two kinds. First, energy-momentum sum rules in lepton–nucleon scattering indicate that the partons that interact electromagnetically or weakly, namely the quarks, carry only about half the momentum of a nucleon. Something else, electrically neutral and inert with respect to the weak interactions, must carry the remainder. This is a role for which gluons are ideally suited. Second, at barycentric energies exceeding about 17 GeV, a fraction of hadronic events produced in electron–positron annihilations display a three-jet structure instead of the familiar two-jet $(e^+e^- \to q\bar{q})$ structure.[24] A typical event of this type is shown in Fig. 1-11. This is interpreted as evidence for the process

$$e^+e^- \to q\bar{q} + \text{gluon}, \tag{1.3.11}$$

in which the gluon is radiated from the outgoing quark in a hadronic analog of electromagnetic bremsstrahlung. This interpretation is supported by further scrutiny.

This brief survey has presented some evidence that quarks and leptons may properly be regarded as elementary particles and has introduced some of the symmetries and relationships among them. In addition, we have recalled some of the basic ideas and properties of gauge theories and have indicated their connection with the current understanding of the fundamental

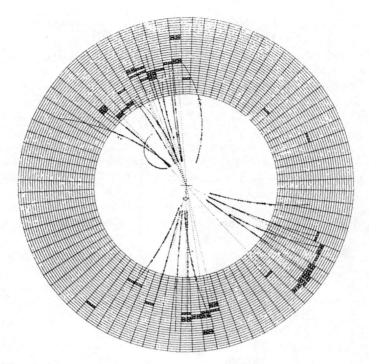

FIG. 1-11. Three-jet event observed in the reaction $e^+e^- \to$ hadrons at 31 GeV, in the JADE detector at PETRA. Such events, which are commonplace at high energies, are interpreted as evidence for the mechanism $e^+e^- \to q\bar{q}g$.

interactions. With these concepts and aspirations as background, we now turn to the foundations that underlie gauge theories. The first of these is the notion of symmetry in Lagrangian field theory.

Problems

1-1. Consider bound states composed of fundamental scalar particles (denoted σ). The quantum numbers of σ are $J^{PC} = 0^{++}$. For $(\sigma\sigma)$ composites,
(a) Show that a bound state with angular momentum L (i.e., an orbital excitation) must have quantum numbers

$$C = (-1)^L; \qquad P = (-1)^L.$$

(b) Allowing for both orbital and radial excitations, construct a schematic mass spectrum of $(\sigma\sigma)$ bound states. Label each state with its quantum numbers J^{PC}.

(c) Now suppose that the fundamental scalars have isospin I. Compute C, P, and G for $(\sigma\sigma)$ bound states, and redo part (b).

1-2.   Consider bound states composed of fundamental spin-1/2 particles (denoted f), with isospin $= 1/2$. For $(f\bar{f})$ composites,

(a) Show that a bound state with angular momentum L must have quantum numbers

$$C = (-1)^{L+s}; \qquad P = (-1)^{L+1}; \qquad G = (-1)^{L+s+I},$$

where s is the spin of the composite system, and I is its isospin.

(b) Allowing for both orbital and radial excitations, construct a schematic mass spectrum of $(f\bar{f})$ bound states. Label each state with its quantum numbers J^{PC}.

1-3.   Analyze the absorption of a virtual photon by a spin-1/2 quark in the Breit frame (brick-wall frame) of the quark. Kinematics:

incident:

$$(E;p_z) = (0;Q) \qquad (\tfrac{Q}{2}; -\tfrac{Q}{2})$$

outgoing:

$$(\tfrac{Q}{2}; \tfrac{Q}{2})$$

(a) Show that the squared matrix element for the absorption of a longitudinal photon vanishes.

(b) Compute the square of the matrix element for absorption of a photon with helicity $= +1$—i.e., a transverse photon.

(c) How would your result for a longitudinal photon differ if the incident quark and photon were not precisely (anti)collinear?

1-4.   Using the Feynman rules given in Appendix B.5, compute the differential cross section $d\sigma/d\Omega$ and the total (integrated) cross section $\sigma \equiv \int d\Omega (d\sigma/d\Omega)$ for the reaction $e^+e^- \to \sigma^+\sigma^-$, where $\sigma^\pm$ is a charged scalar particle. Work in the center of momentum frame, and in the high-energy limit (where all the masses may be neglected). Assume that the colliding beams are unpolarized.

1-5.   (a) Again referring to Appendix B.5 for the Feynman rules, compute the differential and total cross sections for the reaction $e^+e^- \to \mu^+\mu^-$.

Calculate in the c.m. frame and in the high-energy limit, neglecting the lepton masses. Assume that the colliding beams are unpolarized, and sum over the spins of the produced muons.

(b) Look up the original evidence for quark–antiquark jets in the inclusive reaction $e^+e^- \to$ hadrons [G. J. Hanson *et al.*, *Phys. Rev. Lett.* **35**, 1609 (1975)]. Now recompute the differential cross section for the reaction $e^+e^- \to \mu^+\mu^-$, assuming the initial beams to be transversely polarized. See also R. F. Schwitters *et al.*, *ibid.* **35**, 1320 (1975).

1-6. Define the requirements for an experiment to measure the gyro-magnetic ratio of the tau lepton, taking into account the τ-lifetime and the anticipated result $g_\tau \simeq 2$. For background, become acquainted with the methods used to measure the magnetic anomalies of the electron [A. Rich and J. Wesley, *Rev. Mod. Phys.* **44**, 250 (1972); R. S. Van Dyck, Jr., P. B. Schwinberg, and H. G. Dehmelt, *Phys. Rev. Lett.* **38**, 310 (1977)] and muon [F. Combley, F. J. M. Farley, and E. Picasso, *Phys. Rep.* **68**, 93 (1981)], and the magnetic dipole moments of the nucleons [N. F. Ramsey, *Molecular Beams*, Oxford University Press, Oxford, 1956] and unstable hyperons [L. Schachinger *et al.*, *Phys. Rev. Lett.* **41**, 1348 (1978)].

1-7. Assume that the charged weak current has the left-handed form of equation (1.1.2) and that the interaction Hamiltonian is of the "current–current" form $\mathscr{H}_W \sim JJ^\dagger + J^\dagger J$.

(a) Enumerate the kinds of interactions (i.e., terms) in the Hamiltonian that may occur in a world composed of the electron and muon doublets.
(b) List the leptonic processes such as $\nu_\mu e \to \nu_\mu e$ that are consistent with the known selection rules, but do not appear in the charged-current Hamiltonian.

1-8. Derive the connection between $|\Psi(0)|^2$ and the leptonic decay rate of a $(q\bar{q})$ vector meson. It is convenient to proceed by the following steps:

(a) Compute the spin-averaged cross section for the reaction $q\bar{q} \to e^+e^-$. Show that it is

$$\sigma = \frac{\pi\alpha^2 e_q^2}{12E^2} \cdot \frac{\beta_l}{\beta_q} (3 - \beta_l^2)(3 - \beta_q^2),$$

where E is the c.m. energy of a quark and β is the speed of a particle.
(b) The annihilation rate in a $^3S_1$ vector meson is the density $\times$ relative velocity $\times$ 4/3 (to undo the spin average) $\times$ σ, or

$$\Gamma = |\Psi(0)|^2 \times 2\beta_q \times \tfrac{4}{3} \times \sigma.$$

(c) How is the result modified if the vector meson wave function is

$$|V^0\rangle = \sum_i c_i |q_i \bar{q}_i\rangle?$$

(d) Now neglect the lepton mass and the quark binding energy and assume that the quarks move nonrelativistically. Show that

$$\Gamma(V^0 \rightarrow e^+ e^-) = \frac{16\pi\alpha^2}{3M_V^2} |\Psi(0)|^2 \left(\sum_i c_i e_i\right)^2.$$

(e) How is the result modified if quarks come in N_c colors and hadrons are color singlets? [This result is due to R. Van Royen and V. F. Weisskopf, *Nuovo Cim.* **50**, 617 (1967); **51**, 583 (1967), and to H. Pietschmann and W. Thirring, *Phys. Lett.* **21**, 713 (1966).]

For Further Reading

GENERALITIES. Among the standard textbooks on particle physics, the following contain excellent summaries of experimental systematics:

H. Frauenfelder and E. M. Henley, *Subatomic Physics*, Prentice-Hall, Englewood Cliffs, New Jersey, 1974.

S. Gasiorowicz, *Elementary Particle Physics*, Wiley, New York, 1966.

D. H. Perkins, *Introduction to High-Energy Physics*, second edition, Addison-Wesley, Reading, Massachusetts, 1982.

M. L. Perl, *High Energy Hadron Physics*, Wiley, New York, 1974.

$SU(3)$ FLAVOR SYMMETRY. The essentials are explained in many places, including

P. Carruthers, *Introduction to Unitary Symmetry*, Interscience, New York, 1966.

M. Gell-Mann and Y. Ne'eman, *The Eightfold Way*, Benjamin, New York, 1964.

M. Gourdin, *Unitary Symmetries and Their Applications to High Energy Physics*, North-Holland, Amsterdam, 1967.

D. B. Lichtenberg, *Unitary Symmetry and Elementary Particles*, Academic, New York, 1978.

H. J. Lipkin, *Lie Groups for Pedestrians*, second edition, North-Holland, Amsterdam, 1966.

QUARK MODEL. The standard reference for early applications is the reprint volume by

J. J. J. Kokkedee, *The Quark Model*, Benjamin, Reading, Massachusetts, 1969.

Subsequent developments are treated in detail in the book by

F. E. Close, *An Introduction to Quarks and Partons*, Academic, New York, 1979.

in review articles by

O. W. Greenberg, *Ann. Rev. Nucl. Part. Sci.* **28**, 327 (1978).

A. W. Hendry and D. B. Lichtenberg, *Rep. Prog. Phys.* **41**, 1707 (1978).

H. J. Lipkin, *Phys. Rep.* **8C**, 173 (1973).

J. L. Rosner, *Phys. Rep.* **11C**, 189 (1974).

and in summer school lectures by

R. H. Dalitz, *Fundamentals of Quark Models*, Scottish Universities Summer School in Physics, 1976, edited by I. M. Barbour and A. T. Davies, SUSSP, Edinburgh, 1977, p. 151.

C. Quigg, *Gauge Theories in High Energy Physics*, 1981 Les Houches Summer School, edited by M. K. Gaillard and R. Stora, North-Holland, Amsterdam, 1983.

J. L. Rosner, *Techniques and Concepts of High-Energy Physics*, St. Croix Advanced Study Institute, 1980, edited by T. Ferbel, Plenum, New York, 1981, p. 1.

In addition, condensed summaries and extensive lists of references are to be found in

S. Gasiorowicz and J. L. Rosner, *Am. J. Phys.* **49**, 954 (1981).

O. W. Greenberg, *Am. J. Phys.* **50**, 1074 (1982).

J. L. Rosner, *Am. J. Phys.* **48**, 90 (1980).

Much of the phenomenology is treated in

E. Leader and E. Predazzi, *An Introduction to Gauge Theories and the New Physics*, Cambridge University Press, Cambridge, 1982.

QUARK ABUNDANCE. Searches for free quarks are reviewed by

L. W. Jones, *Rev. Mod. Phys.* **49**, 417 (1977).

L. Lyons, *Progress in Particle and Nuclear Physics*, Vol. 7, edited by D. Wilkinson, Pergamon, Oxford, 1981, p. 169.

M. Marinelli and G. Morpurgo, *Phys. Rep.* **85**, 161 (1982).

Provocative evidence for fractionally charged matter has been reported by

G. S. La Rue, J. D. Phillips, and W. M. Fairbank, *Phys. Rev. Lett.* **46**, 967 (1981).

and earlier publications cited therein.

PARTON MODEL. Full treatments appear in the book by Close, *op. cit.*, and in that by

R. P. Feynman, *Photon–Hadron Interactions*, Benjamin, Reading, Massachusetts, 1972.

COLOR. A thorough historical review appears in

O. W. Greenberg and C. A. Nelson, *Phys. Rep.* **32C**, 69 (1977).

GAUGE THEORIES. An extremely accessible introduction has been given by

I. J. R. Aitchison and A. J. G. Hey, *Gauge Theories in Particle Physics*, Adam Hilger, Bristol, 1982.

The perspective summarized in this introductory chapter is further developed in

L. B. Okun, *Leptons and Quarks*, North-Holland, Amsterdam, 1981.

Past, present, and future of gauge theories are evoked in the 1979 Nobel lectures by

S. L. Glashow, *Rev. Mod. Phys.* **52**, 539 (1980).

A. Salam, *Rev. Mod. Phys.* **52**, 525 (1980).

S. Weinberg, *Rev. Mod. Phys.* **52**, 515 (1980).

The following popularizations may also be read with profit at this point:

S. Weinberg, "Unified Theories of Elementary Particle Interactions," *Sci. Am.* **231**, 50 (July, 1974).

S. L. Glashow, "Quarks with Color and Flavor," *Sci. Am.* **233**, 38 (October, 1975).

Y. Nambu, "The Confinement of Quarks," *Sci, Am.* **235**, 48 (November, 1976).

G. 't Hooft, "Gauge Theories of the Forces between Elementary Particles," *Sci. Am.* **242**, 104 (June, 1980).

H. Georgi, "A Unified Theory of Elementary Particles and Forces," *Sci. Am.* **244**, 40 (April, 1981).

R. Jaffe, "Quark Confinement," *Nature (London)* **268**, 201 (1977).

W. J. Marciano and H. Pagels, "QCD," *Nature (London)* **279**, 479 (1979).

S. D. Drell, "When is a particle?" *Am. J. Phys.* **46**, 597 (1978).

S. D. Drell, "Elementary Particle Physics," *Daedalus, Proc. Am. Acad. Arts Sci.* **106**, No. 3, (Summer, 1977).

References

[1] For an interesting development of this theme, see the essay by Stephen Hawking, *Is the End in Sight for Theoretical Physics?*, Cambridge University Press, Cambridge, 1980; reprinted in *CERN Courier* **21**, 3,71 (1981).

[2] A recent survey of heavy lepton searches has been given by M. L. Perl, *Physics in Collision*, Vol. 1, edited by W. Peter Trower and Gianpaolo Bellini, Plenum, New York, 1982, p. 1.

[3] See the compilation by J. G. Branson, in *Proceedings 1981 International Symposium on Lepton and Photon Interactions at High Energies*, edited by W. Pfeil, Physikalisches Institut Universität Bonn, Bonn, 1981, p. 279. A representative measurement is reported by H. J. Behrend *et al.* (CELLO Collaboration), *Z. Phys.* **C14**, 283 (1982).

[4] For a summary of the evidence, see G. J. Feldman, in *Particles and Fields—1981*, edited by C. A. Heusch and W. T. Kirk, American Institute of Physics, New York, 1982, p. 280.

[5] Throughout this volume, otherwise unattributed experimental results are to be found in Particle Data Group, *Phys. Lett.* **111**, 1 (1982).

[6] Perturbation theory conventions are given in Appendix A.

[7] See the compilation by Branson, ref. 3. A representative experiment is reported in R. Brandelik *et al.* (TASSO Collaboration), *Phys. Lett.* **113B**, 499 (1982).

[8] M. Gell-Mann, *Phys. Lett.* **8**, 214 (1964); G. Zweig, CERN Report 8182/TH.401 (1964, unpublished); CERN Report 8419/TH.412 (1964, unpublished), reprinted in *Developments in the Quark Theory of Hadrons, Vol. 1: 1964–1978*, edited by D. B. Lichtenberg and S. P. Rosen, Hadronic Press, Nonantum, Massachusetts, 1980, p. 22. Some historical perspective is provided by G. Zweig, *Baryon 1980*, IVth International Conference on Baryon Resonances, edited by N. Isgur, University of Toronto, Toronto, 1980, p. 439.

[9] The measurements of the SLAC–MIT Collaboration are presented in the doctoral thesis of M. D. Mestayer, *et al.*, *Phys. Rev.* **D27**, 285 (1983). For ep scattering at $Q^2 \simeq 10 \text{ GeV}^2/\text{c}^2$, the result is $\sigma_L/\sigma_T = 0.22 \pm 0.10$.

[10] A summary is given by P. Söding and G. Wolf, *Ann. Rev. Nucl. Part. Sci.* **31**, 231 (1981). See also the papers quoted in Problem 1-5.

[11] M. Gell-Mann, *Phys. Rev.* **92**, 833 (1953); T. Nakano and K. Nishijima, *Prog. Theor. Phys. (Kyoto)* **10**, 581 (1955).

[12] These predictions are in fact shared by a model of integrally charged colored quarks due originally to M. Y. Han and Y. Nambu, *Phys. Rev.* **139**, B1005 (1965), and it is notoriously difficult to draw distinctions, as emphasized by M. Chanowitz, *Particles and Fields—1975*, edited by H. J. Lubatti and P. M. Mockett, University of Washington, Seattle, 1975, p. 448; and by H. J. Lipkin, *Nucl. Phys.* **B155**, 104 (1979). Properties of the $\eta(549)$ and $\eta'(958)$ strongly favor the fractional charge assignment: M. Chanowitz, *Phys. Rev. Lett.* **44**, 59 (1980). Another experimental test will be developed in Section 8.6 and Problem 8–13.

[13] See the data compiled by I. Mannelli, *Proceedings 1981 International Symposium on Lepton and Photon Interactions at High Energies*, edited by W. Pfeil, Physikalisches Institut Universität Bonn, Bonn, 1981, p. 730.

[14] See, for example, the comparison in Fig. 9 of G. Smadja, *Proceedings 1981 International Symposium on Lepton and Photon Interactions at High Energies*, edited by W. Pfeil, Physikalisches Institut Universität Bonn, Bonn, 1981, p. 444.

[15] O. W. Greenberg, *Phys. Rev. Lett.* **13**, 598 (1964).

[16] Refined theoretical estimates for the leptonic branching ratio (17.75%) have been given by F. J. Gilman and D. H. Miller, *Phys. Rev.* **D17**, 1846 (1978) and by N. Kawamoto and A. I. Sanda, *Phys. Lett.* **76B**, 446 (1978).

[17] A clear and thorough discussion of the calculation of the π^0-lifetime has been given by C. H. Llewellyn Smith, *Quantum Flavordynamics, Quantum Chromodynamics, and Unified Theories*, edited by K. T. Mahanthappa and J. Randa, Plenum, New York, 1980, p. 59.

[18] The current "standard model" of weak and electromagnetic interactions, based on spontaneously broken $SU(2)_L \otimes U(1)_Y$ gauge symmetry, was proposed by S. Weinberg, *Phys. Rev. Lett.* **19**, 1264 (1967) and by A. Salam, *Elementary Particle Theory: Relativistic Groups and Analyticity* (8th Nobel Symposium), edited by N. Svartholm, Almqvist and Wiksell, Stockholm, 1968, p. 367. For a meticulous intellectual history see M. Veltman, *Proceedings of the 6th International Symposium on Electron and Photon Interactions at High Energies*, edited by H. Rollnik and W. Pfeil, North-Holland, Amsterdam, 1974, p. 429. A later historical perspective is that of S. Coleman, *Science* **206**, 1290 (1979).

[19] N. Cabibbo, *Phys. Rev. Lett.* **10**, 531 (1963).

[20] S. L. Glashow, J. Iliopoulos, and L. Maiani, *Phys. Rev.* **D2**, 1285 (1970).

[21] B. J. Bjørken and S. L. Glashow, *Phys. Lett.* **11**, 255 (1964).

[22] J. J. Aubert *et al.*, *Phys. Rev. Lett.* **33**, 1404 (1974); J.-E. Augustin *et al.*, *ibid.* **33**, 1406 (1974).

[23] Definitive evidence for charmed mesons was presented by G. Goldhaber *et al.*, *Phys. Rev. Lett.* **37**, 255 (1976), and by I. Peruzzi *et al.*, *ibid.* **37**, 569 (1976). The first example of a charmed baryon decay was that seen by E. G. Cazzoli *et al.*, *ibid.* **34**, 1125 (1975).

[24] The first thorough discussions of this phenomenon are to be found in the reports by H. Newman (Mark-J Collaboration), Ch. Berger (PLUTO Collaboration), G. Wolf (TASSO Collaboration), and S. Orito (JADE Collaboration), in *Proceedings of the 1979 International Symposium on Lepton and Photon Interactions at High Energies*, edited by T. B. W. Kirk and H. D. I. Abarbanel, Fermilab, Batavia, Illinois, 1979, pp. 3, 19, 34, 52.

CHAPTER 2

LAGRANGIAN FORMALISM AND CONSERVATION LAWS

There are many ways of formulating the relativistic quantum field theory of interacting particles, each with its own set of advantages and shortcomings or inconveniences. The Lagrangian formalism has a number of attributes that make it particularly felicitous for our rather utilitarian purposes. Not to be neglected among its assets is the fact that it is a familiar construct in classical mechanics, and that many of its practical advantages can be understood already at the classical level. The Lagrangian approach is characterized by a simplicity in that field theory may be regarded as the limit of a system with n degrees of freedom as n tends toward infinity. Perhaps more to the point, it provides a formalism in which relativistic covariance is manifest, because the four coordinates of space–time enter symmetrically. This is a decided, though by no means indispensable, advantage for the construction of a relativistic theory. Lagrangian field theory is also particularly suited to the systematic discussion of invariance principles and the conservation laws to which they are related. In addition, a variational principle provides a direct link between the Lagrangian and the equations of motion. The foregoing properties are of particular value for the development of gauge theories, in which the interactions arise as consequences of what will be called local gauge symmetries. Finally, the path that leads from a Lagrangian to a quantum field theory by the method of canonical quantization is one that is extremely well traveled. This makes it possible to conduct much of the discussion of the formulation of gauge theories at what is essentially the classical level, and to make the leap to quantum field theory simply by using standard results, without repeating developments that are to be found in every field theory textbook. Thus equipped with the Feynman rules for a theory, we may proceed to calculate the consequences.

Three things are done in this brief chapter. First, we shall summarize the basic elements of the Lagrangian formulation of classical mechanics and of field theory, and the derivation of the equations of motion. Second, we shall evoke the intimate connection between continuous symmetries of the Lagrangian and constants of the motion, which is embodied in Noether's theorem. Third, in the course of these discussions, we shall take the opportunity to recall several elementary free-particle Lagrangians and the associated equations of motion. These will recur in more interesting physical contexts throughout this volume.

2.1 Hamilton's Principle

The Lagrangian function L may be regarded as the fundamental object in classical mechanics.[1] From it is constructed the classical action

$$S \equiv \int_{t_1}^{t_2} dt\, L(q, \dot{q}), \tag{2.1.1}$$

where $q(t)$ is the generalized coordinate and $\dot{q}(t) \equiv dq/dt$ the generalized velocity. The equations of motion follow from Hamilton's principle of least action, according to which the variation

$$\delta S = \delta \int_{t_1}^{t_2} dt\, L(q, \dot{q}) = 0, \tag{2.1.2}$$

subject to the constraint that variations of the generalized coordinates vanish at the endpoints t_1 and t_2. Thus the physical path is the particular trajectory joining $q_1 \equiv q(t_1)$ and $q_2 \equiv q(t_2)$ along which the action is stationary. An important generalization to quantum mechanics as a weighted sum over paths has been developed by Feynman.[2]

In a large number of problems of physical interest, the Lagrangian depends only upon the generalized coordinates and their first derivatives. In this case, satisfaction of Hamilton's principle is guaranteed by the Euler–Lagrange equations in the form

$$\frac{\partial L}{\partial q} = \frac{d}{dt}\left(\frac{\partial L}{\partial \dot{q}}\right). \tag{2.1.3}$$

To show this, let us note that the variation in the action is given by

$$\delta S = \int_{t_1}^{t_2} dt\left[\frac{\partial L}{\partial q(t)}\delta q(t) + \frac{\partial L}{\partial \dot{q}(t)}\delta \dot{q}(t)\right]. \tag{2.1.4}$$

Since $\delta \dot{q}(t) = (d/dt)\,\delta q(t)$, the second term may be integrated by parts:

$$\int_{t_1}^{t_2} dt\, \frac{\partial L}{\partial \dot{q}} \cdot \frac{d}{dt}\,\delta q = \frac{\partial L}{\partial \dot{q}}\,\delta q\Big|_{t_1}^{t_2} - \int_{t_1}^{t_2} dt\left(\frac{d}{dt}\frac{\partial L}{\partial \dot{q}}\right)\delta q. \tag{2.1.5}$$

Because the endpoints are constrained,

$$\delta q(t_1) = 0 = \delta q(t_2), \tag{2.1.6}$$

the first term in (2.1.5) is identically zero and the variation of the action becomes

$$\delta S = \int_{t_1}^{t_2} \left(\frac{\partial L}{\partial q} - \frac{d}{dt} \frac{\partial L}{\partial \dot{q}} \right) \delta q, \tag{2.1.7}$$

which vanishes provided that (2.1.3) holds. For the case of several (separately varied) generalized coordinates the same arithmetic leads to a set of Euler–Lagrange equations of the form (2.1.3), one for each coordinate.

The most familiar cases in classical mechanics are those for which the Lagrangian is simply the difference of kinetic and potential energies. Indeed, if we write in one dimension

$$L = \tfrac{1}{2}m\dot{x}^2 - V(x), \tag{2.1.8}$$

the Euler–Lagrange equation gives

$$-dV/dx = m\ddot{x}. \tag{2.1.9}$$

Upon identifying the applied force as

$$F \equiv -dV/dx, \tag{2.1.10}$$

we simply recover Newton's equation of motion.

The formalism can easily be generalized to include a Lagrangian that depends explicitly on the time t or on some other parameter. This is appropriate for an "open" system, which exchanges energy and momentum with its surroundings. A frequently encountered case is that of an external driving force $F(t)$, represented in the Lagrangian by a term $qF(t)$. The mathematical formalism can also be extended to treat Lagrangians that contain higher than first-order derivatives of the generalized coordinates. The ensuing Euler–Lagrange equations of motion are then higher than second order, but a "local" description in terms of local differential equations remains possible so long as the Lagrangian contains only derivatives of finite order. These generalizations will not be required for the study of gauge theories.

With the Lagrangian description of classical mechanics in hand, the transition to Lagrangian field theory, guided by the requirement of relativistic invariance, is straightforward. Construction of a field theory begins from the Lagrangian density $\mathscr{L}(\phi(x), \partial_\mu \phi(x))$, a functional of the field $\phi(x)$ and its four-gradient $\partial_\mu \phi(x) = \partial \phi(x)/\partial x^\mu$. The field may be regarded as a separate, generalized coordinate at each value of its argument, the space–time coordinate x. Thus arises the description of field theory as a closed system with an infinite number of degrees of freedom.

The classical action is now defined as

$$S \equiv \int_{t_1}^{t_2} dt \int d^3\mathbf{x}\, \mathscr{L}(\phi, \partial_\mu \phi), \tag{2.1.11}$$

where the spatial integral of the Lagrangian density takes the part of the Lagrangian in classical mechanics:

$$L \equiv \int d^3\mathbf{x}\, \mathscr{L}(\phi, \partial_\mu \phi). \tag{2.1.12}$$

By the same reasoning followed in the classical problem, the action is required to be stationary,

$$\delta \int_{t_1}^{t_2} d^4x\, \mathscr{L}(\phi, \partial_\mu \phi) = 0, \tag{2.1.13}$$

subject as always to the constraint that the variations in the fields vanish at the endpoints characterized by t_1 and t_2. The requirement of least action is now ensured by the Euler–Lagrange equations for the Lagrangian density,

$$\frac{\partial \mathscr{L}}{\partial \phi(x)} = \partial_\mu \frac{\partial \mathscr{L}}{\partial(\partial_\mu \phi(x))}, \tag{2.1.14}$$

which lead in turn to explicit equations of motion for the fields. Evidently these equations will be covariant if the Lagrangian density itself transforms as a Lorentz scalar. Moreover, the equations of motion are unchanged if a total divergence is added to the Lagrangian density. For problems involving several fields ϕ_i, the variational principle applied separately to each field leads to a set of equations (2.1.14), precisely as for mechanical problems with several generalized coordinates.

In what follows, it will frequently be convenient to refer in context to the Lagrangian density $\mathscr{L}$ by the shorter term Lagrangian.

2.2   Free Field Theory Examples

In many situations of physical interest, it is possible to establish, or to divine, the equations of motion for the fields directly. When this can be done, the Lagrangian is an afterthought that facilitates a systematic quantization or serves as a useful expedient for the investigation of invariance principles. For our purposes of constructing gauge theories from physical symmetries, it will often be the Lagrangian that comes first and the equations of motion that appear as consequences. Even when we proceed in this manner, however, the construction of the Lagrangian will be guided by the desire to reproduce certain characteristics of simpler equations of motion. It is therefore invaluable to be conversant with some of the most commonly occurring Lagrangians.

What is in many ways the simplest field theory is that of a field $\phi(x)$ that transforms as a Lorentz scalar or pseudoscalar. The Lagrangian

$$\mathcal{L} = \tfrac{1}{2}[(\partial^\mu\phi)(\partial_\mu\phi) - m^2\phi^2] \tag{2.2.1}$$

leads, via the Euler–Lagrange equation (2.1.14), to the Klein–Gordon equation,

$$(\Box + m^2)\phi(x) = 0, \tag{2.2.2}$$

for a spinless free particle of mass m.

The complex scalar field

$$\phi(x) = \frac{a(x) + ib(x)}{\sqrt{2}}, \tag{2.2.3}$$

which may be expressed in terms of the two independent real functions $a(x)$ and $b(x)$, is appropriate for the description of particles that carry an additive quantum number, such as electric charge. For most purposes it is convenient to write the Lagrangian in terms of $\phi(x)$ and the complex-conjugate field $\phi^*(x) = (a(x) - ib(x))/\sqrt{2}$, instead of $a(x)$ and $b(x)$. In special circumstances, still other parametrizations are more economical. The Lagrangian of the complex field is written by analogy with that of the real scalar field (2.2.1) as

$$\begin{aligned} \mathcal{L} &= \tfrac{1}{2}[(\partial^\mu a)^2 - m^2 a^2] + \tfrac{1}{2}[(\partial^\mu b)^2 - m^2 b^2] \\ &= (\partial^\mu\phi)^*(\partial_\mu\phi) - m^2\phi^*\phi \\ &= |\partial^\mu\phi|^2 - m^2|\phi|^2. \end{aligned} \tag{2.2.4}$$

Independent variations of ϕ and ϕ^* (or equivalently of a and b) then yield as equations of motion the two Klein–Gordon equations

$$(\Box + m^2)\phi(x) = 0, \qquad (\Box + m^2)\phi^*(x) = 0. \tag{2.2.5}$$

It will be seen later that in the presence of the appropriate interactions the fields ϕ and ϕ^* can be identified with particles of opposite charge.

The Dirac equation

$$(i\gamma^\mu\partial_\mu - m)\psi(x) = 0 \tag{2.2.6}$$

for a free fermion follows from the Lagrangian

$$\mathcal{L} = \bar{\psi}(x)(i\gamma^\mu\partial_\mu - m)\psi(x) = 0, \tag{2.2.7}$$

where the Dirac conjugate field is as usual defined as $\bar{\psi}(x) = \psi^\dagger(x)\gamma^0$. (Notations and conventions are reviewed in Appendix A.)

Electromagnetic interactions will be discussed at some length in Chapter 3, when the subject of gauge invariance is studied in detail. For the present, let us simply recall that a convenient formulation of the Maxwell theory may be obtained by expressing the electric and magnetic field strengths $\mathbf{E}$ and $\mathbf{B}$ in terms of the four-vector potential $A_\mu(x)$. In the absence of sources, an appropriate Lagrangian for the free electromagnetic field is

$$\mathscr{L} = -\tfrac{1}{4}(\partial_\nu A_\mu - \partial_\mu A_\nu)(\partial^\nu A^\mu - \partial^\mu A^\nu), \tag{2.2.8}$$

from which Maxwell's equations may be readily verified. The factor of $1/4$ leads to the conventional normalization of the energy density.

Finally, for the case of a massive vector field with mass M, the Lagrangian is

$$\mathscr{L} = -\tfrac{1}{4}(\partial_\nu A_\mu - \partial_\mu A_\nu)(\partial^\nu A^\mu - \partial^\mu A^\nu) + \tfrac{1}{2}M^2 A^\mu A_\mu. \tag{2.2.9}$$

The difference in sign between the mass term for the vector field and that for the scalar field in (2.2.1) is simply a consequence of the fact that the physical degrees of freedom are represented by the spacelike components of A_μ, which contribute with a minus sign to $A^\mu A_\mu$. The equation of motion, which follows from (2.2.9) upon regarding each component of A_μ as an independent field, is the Proca equation,

$$\partial_\nu(\partial^\nu A^\mu - \partial^\mu A^\nu) + M^2 A^\mu = 0. \tag{2.2.10}$$

Taking the divergence of this equation, we find that

$$M^2(\partial \cdot A) = 0, \tag{2.2.11}$$

which implies that the vector field is divergenceless,

$$\partial \cdot A = 0, \tag{2.2.12}$$

because $M^2 \neq 0$. The vanishing of the divergence $(\partial \cdot A)$ corresponds to the covariant elimination of one of the four apparent degrees of freedom of the vector field A_μ. It also permits the Proca equation to be written in a more familiar form,

$$(\Box + M^2)A^\mu = 0. \tag{2.2.13}$$

2.3   Symmetries and Conservation Laws

Much of the rest of this volume will be devoted to recognizing symmetries in Nature and deducing their implications for the fundamental particles and their interactions. Invariance principles and the associated symmetry transformations are of many types: continuous or discrete, geometrical or internal, and, as the following chapter will make apparent, global or local. They play many important roles in physics including that of restricting—and thus guiding—the formulation of theories, as we have already tacitly acknowledged in requiring the Lagrangian to be a Lorentz scalar.

When the equations of motion are derived from a variational principle, a general and systematic procedure is available for establishing conservation theorems and constants of the motion as a consequence of invariance properties. Thus, conservation laws and selection rules observed in Nature may be imposed as symmetries of the Lagrangian, restricting or prescribing its form. The general framework for this program is provided by Noether's theorem,[3] which correlates a conservation law with every continuous symmetry transformation under which the Lagrangian is invariant in form. Let us illustrate the utility of this theorem by considering two examples, one of a space–time, or geometrical, invariance, and the other of an internal symmetry.

First consider, as an example of geometrical transformations of the space–time variables, translations of the form

$$x_\mu \to x'_\mu = x_\mu + a_\mu, \tag{2.3.1}$$

where the infinitesimal displacement a_μ is independent of the coordinate x_μ. A Lagrangian that is invariant in form under a transformation of this type will therefore change by an amount

$$\delta \mathscr{L} = \mathscr{L}[x'] - \mathscr{L}[x] = a^\mu \, d\mathscr{L}/dx^\mu. \tag{2.3.2}$$

For a Lagrangian with no explicit dependence on the coordinates, we may equivalently compute the change as

$$\delta \mathscr{L} = \frac{\partial \mathscr{L}}{\partial \phi} \delta \phi + \frac{\partial \mathscr{L}}{\partial(\partial_\mu \phi)} \delta(\partial_\mu \phi), \tag{2.3.3}$$

where

$$\delta \phi = \phi(x') - \phi(x) = a^\mu \partial_\mu \phi(x) \tag{2.3.4}$$

and

$$\delta(\partial_\mu \phi) = \partial_\mu \phi(x') - \partial_\mu \phi(x) = a^\nu \partial_\nu \partial_\mu \phi(x). \tag{2.3.5}$$

Consequently, using the Euler–Lagrange equations to eliminate $\partial \mathscr{L}/\partial \phi$, we find

$$\delta \mathscr{L} = \left[\partial_\nu \frac{\partial \mathscr{L}}{\partial(\partial_\nu \phi)} \right] a^\mu \partial_\mu \phi + \frac{\partial \mathscr{L}}{\partial(\partial_\nu \phi)} a^\mu \partial_\mu \partial_\nu \phi$$

$$= \partial_\nu \frac{\partial \mathscr{L}}{\partial(\partial_\nu \phi)} a^\mu \partial_\mu \phi. \tag{2.3.6}$$

Equating the two expressions for δ yields

$$a_\mu \partial_\nu \left[\frac{\partial \mathscr{L}}{\partial(\partial_\nu \phi)} \partial^\mu \phi - g^{\mu\nu} \mathscr{L} \right] = 0, \tag{2.3.7}$$

which is to be satisfied for arbitrary infinitesimal displacements a_μ. We therefore conclude that the stress–energy–momentum flow characterized by the tensor

$$\Theta^{\mu\nu} \equiv \frac{\partial \mathscr{L}}{\partial(\partial_\nu \phi)} \partial^\mu \phi - g^{\mu\nu} \mathscr{L} \tag{2.3.8}$$

satisfies the local conservation law

$$\partial_\mu \Theta^{\mu\nu} = 0. \tag{2.3.9}$$

It is easily verified that Θ^{00} is the Hamiltonian density

$$\mathcal{H} = \frac{\partial \mathcal{L}}{\partial(\partial_0 \phi)} \partial^0 \phi - \mathcal{L} \qquad (2.3.10)$$

so that the total energy

$$H \equiv \int d^3 x \Theta^{00} \qquad (2.3.11)$$

is a constant of the motion, and that the Θ^{0v} correspond to momentum densities. Thus has translation invariance led to four-momentum conservation.

It is also instructive, and more central to our later purposes, to consider by a simple example the consequences of an internal symmetry. If the neutron and proton are taken as elementary particles with a common mass m, the Lagrangian for free nucleons may be written in an obvious notation as

$$\mathcal{L} = \bar{p}(i\gamma^\mu \partial_\mu - m)p + \bar{n}(i\gamma^\mu \partial_\mu - m)n. \qquad (2.3.12)$$

In terms of the composite spinor

$$\psi \equiv \begin{pmatrix} p \\ n \end{pmatrix}, \qquad (2.3.13)$$

the Lagrangian may be rewritten more compactly as

$$\mathcal{L} = \bar{\psi}(i\gamma^\mu \partial_\mu - m)\psi. \qquad (2.3.14)$$

It is evidently invariant under global isospin rotations,

$$\psi \to \exp\frac{i\boldsymbol{\tau}\cdot\boldsymbol{\alpha}}{2}\psi, \qquad (2.3.15)$$

where $\boldsymbol{\tau} = (\tau_1, \tau_2, \tau_3)$ consists of the usual 2×2 Pauli isospin matrices, and $\boldsymbol{\alpha} = (\alpha_1, \alpha_2, \alpha_3)$ is an arbitrary constant (three-vector) parameter of the transformation. A global transformation is one that subjects the spinor ψ to the same rotation everywhere in space–time. Thus the parameter $\boldsymbol{\alpha}$ is independent of the coordinate x.

A continuous transformation such as an isospin rotation can be built up out of infinitesimal rotations. It is therefore interesting to consider the effect on the Lagrangian of an infinitesimal transformation

$$\psi(x) \to \psi(x) + \frac{i}{2}\boldsymbol{\alpha}\cdot\boldsymbol{\tau}\psi(x), \qquad (2.3.16)$$

which leads to the infinitesimal variations

$$\delta\psi = \frac{i}{2}\boldsymbol{\alpha}\cdot\boldsymbol{\tau}\psi,$$
$$\qquad (2.3.17)$$
$$\delta(\partial_\mu\psi) = \frac{i}{2}\boldsymbol{\alpha}\cdot\boldsymbol{\tau}(\partial_\mu\psi).$$

If the Lagrangian is to be invariant under such a transformation, we require that

$$\delta \mathscr{L} = 0. \tag{2.3.18}$$

Explicit computation yields

$$\delta \mathscr{L} = \frac{\partial \mathscr{L}}{\partial \psi} \delta \psi + \frac{\partial \mathscr{L}}{\partial (\partial_\mu \psi)} \delta(\partial_\mu \psi) + \frac{\partial \mathscr{L}}{\partial \bar{\psi}} \delta(\bar{\psi}) + \frac{\partial \mathscr{L}}{\partial (\partial_\mu \bar{\psi})} \delta(\partial_\mu \bar{\psi})$$

$$= \left[\partial_\mu \frac{\partial \mathscr{L}}{\partial (\partial_\mu \psi)} \right] \frac{i}{2} \boldsymbol{\alpha} \cdot \boldsymbol{\tau} \psi + \frac{\partial \mathscr{L}}{\partial (\partial_\mu \psi)} \frac{i}{2} \boldsymbol{\alpha} \cdot \boldsymbol{\tau} (\partial_\mu \psi)$$

$$= \partial_\mu \boldsymbol{\alpha} \cdot \left[\frac{i}{2} \frac{\partial \mathscr{L}}{\partial (\partial_\mu \psi)} \boldsymbol{\tau} \psi \right], \tag{2.3.19}$$

where the second line follows from the equations of motion. The quantity in square brackets may be identified as a conserved current (density),

$$\mathbf{J}^\mu = \frac{i}{2} \frac{\partial \mathscr{L}}{\partial (\partial_\mu \psi)} \boldsymbol{\tau} \psi, \tag{2.3.20}$$

which satisfies the continuity equation

$$\partial_\mu \mathbf{J}^\mu = 0. \tag{2.3.21}$$

For the specific case at hand of the free nucleon Lagrangian, the explicit form of the conserved current is

$$\mathbf{J}^\mu = \bar{\psi} \gamma^\mu \frac{\boldsymbol{\tau}}{2} \psi \tag{2.3.22}$$

which is immediately recognizable as the isospin current, in analogy with the familiar electromagnetic current for Dirac particles. It is a standard exercise to show that these classical results carry over to quantum field theory.

In the following chapter we shall pursue further the consequences of internal symmetries and encounter the possibility of deriving interactions from symmetries. There we shall develop in detail the idea of local gauge invariance and study its implications for a theory of electrodynamics.

Problems

2-1. Verify explicitly that changing the Lagrangian density by a total divergence leaves the Euler–Lagrange equations unchanged.

2-2. Using the freedom to add a total divergence to the Lagrangian, show that the Dirac Lagrangian for a free massless fermion may be written in the form

$$\mathscr{L}' = \tfrac{1}{2} \bar{\psi} i \gamma^\mu \partial_\mu \psi - \tfrac{1}{2} (\partial_\mu \bar{\psi}) i \gamma^\mu \psi.$$

2-3. Beginning from the Lagrangian for the free electromagnetic field (2.2.8), derive the equations of motion by varying independently each component of the electromagnetic field A_μ.

2-4. Use the requirement that the Lagrangian be invariant under a continuous symmetry to deduce the conserved quantity that corresponds to a particular transformation. Show that invariance under (*i*) translations in space, (*ii*) translations in time, (*iii*) spatial rotations implies conservation of (*i*) momentum, (*ii*) energy, (*iii*) angular momentum.

For Further Reading

LAGRANGIAN MECHANICS AND ELECTRODYNAMICS. The Lagrangian description of classical phenomena is developed in many places, including

H. Goldstein, *Classical Mechanics*, second edition, Addison-Wesley, Reading, Massachusetts, 1980.

L. Landau and E. M. Lifshitz, *Mechanics*, translated by J. B. Sykes and J. S. Bell, Addison-Wesley, Reading, Massachusetts, 1960.

L. Landau and E. M. Lifshitz, *The Classical Theory of Fields*, second edition, translated by M. Hamermesh, Addison-Wesley, Reading, Massachusetts, 1962.

D. E. Soper, *Classical Field Theory*, Wiley–Interscience, New York, 1976.

A particularly detailed discussion of the Lagrangian description of the classical electromagnetic field appears in

F. Rohrlich, *Classical Charged Particles*, Addison-Wesley, Reading, Massachusetts, 1965.

LAGRANGIAN FORMULATION OF QUANTUM FIELD THEORY. Elaboration of the material of this chapter may be found in many of the standard field theory textbooks, among which the following are especially accessible:

J. D. Bjorken and S. D. Drell, *Relativistic Quantum Fields*, McGraw-Hill, New York, 1965.

N. N. Bogoliubov and D. V. Shirkov, *Introduction to the Theory of Quantized Fields*, translated by G. M. Volkoff, Interscience, New York, 1959.

C. Itzykson and J.-B. Zuber, *Quantum Field Theory*, McGraw-Hill, New York, 1980.

F. Mandl, *Introduction to Quantum Field Theory*, Interscience, New York, 1959.

J. J. Sakurai, *Advanced Quantum Mechanics*, Addison-Wesley, Reading, Massachusetts, 1967.

CONSERVATION LAWS. The classic reference on Noether's theorem and Hamilton's principle is

E. L. Hill, *Rev. Mod. Phys.* **23**, 253 (1953).

Some useful explanatory remarks are offered by

T. H. Boyer, *Am. J. Phys.* **34**, 475 (1966).

The physical basis of numerous invariance principles is admirably explained in

J. J. Sakurai, *Invariance Principles and Elementary Particles*, Princeton University Press, Princeton, New Jersey, 1964.

PATH INTEGRALS. The path-integral formulation of nonrelativistic quantum mechanics is developed thoroughly in

R. P. Feynman and A. R. Hibbs, *Quantum Mechanics and Path Integrals*, McGraw-Hill, New York, 1965.

The path-integral approach to quantum field theory is pursued in two recent books:

L. D. Faddeev and A. A. Slavnov, *Gauge Fields, Introduction to Quantum Theory*, Benjamin, Reading, Massachusetts, 1980.

P. Ramond, *Field Theory, A Modern Primer*, Benjamin, Reading, Massachusetts, 1981.

See also

E. S. Abers and B. W. Lee, *Phys. Rep.* **9C**, 1 (1973).

C. DeWitt Morette, A. Maheshwari, and B. Nelson, *Phys. Rep.* **50**, 255 (1979).
M. S. Marinov, *Phys. Rep.* **60**, 1 (1980).
J. C. Taylor, *Gauge Theories of Weak Interactions*, Cambridge University Press, Cambridge, 1976.
VARIATIONAL PRINCIPLES. The development of field theory from an action principle is expounded by
R. E. Peierls, *Proc. Roy. Soc. (London)* **A214**, 143 (1952).
J. Schwinger, *Phys. Rev.* **82**, 914 (1951), **91**, 91 (1953).

References

[1] An instructive intellectual history of classical mechanics appears in R. B. Lindsay and H. Margenau, *Foundations of Physics*, Dover, New York, 1957.

[2] The method of path integrals has its roots in the work of P. A. M. Dirac, *Phys. Zeits. Sowjetunion* **3**, 64 (1933) and was brought to its modern form by R. P. Feynman, *Rev. Mod. Phys.* **20**, 367 (1948); *Phys. Rev.* **80**, 440 (1950). These papers are reprinted in *Quantum Electrodynamics*, edited by J. Schwinger, Dover, New York, 1958.

[3] E. Noether, *Nachr. Kgl. Ges. Wiss. Göttingen* p. 235 (1918).

CHAPTER 3

THE IDEA OF GAUGE INVARIANCE

3.1 Historical Preliminaries

We turn now to a discussion of the theory of electrodynamics, which is both the simplest gauge theory and the most familiar. The foundations for our present understanding of the subject were laid down by Maxwell in 1864 in his equations unifying the electric and magnetic interactions. The electromagnetic potential that one is led to introduce in order to generate fields that comply with Maxwell's equations by construction is not uniquely defined. The resulting freedom to choose many potentials that describe the same electromagnetic fields has come to be called *gauge invariance*. We shall see that the gauge invariance of electromagnetism can be phrased in terms of a continuous symmetry of the Lagrangian, which leads, through Noether's theorem, to the conservation of electric charge and to other important consequences. Although it is clearly possible to regard gauge invariance as simply an outcome of Maxwell's unification, one may wonder whether a greater importance might not attach to the symmetry itself and thus be led to investigate the degree to which Maxwell's equations might be seen to follow from the symmetry. Indeed, the idea of gauge invariance as a dynamical principle arose from efforts by Hermann Weyl[1] to find a geometric basis for both gravitation and electromagnetism. Weyl's attempts to unify the fundamental interactions of his day through the requirement of invariance under a space–time-dependent change of scale were unsuccessful. His terminology, *Eichinvarianz* (*Eich* = gauge or standard of calibration), has nevertheless survived, and his original program is worth recalling.

Consider the change in a function $f(x)$ between the point x_μ and the point $x_\mu + dx_\mu$. In a space with uniform scale, it is simply

$$f(x + dx) = f(x) + \partial^\mu f(x)\, dx_\mu. \qquad (3.1.1)$$

37

But if in addition the scale, or unit of measure, for f changes by a factor $(1 + S^\mu dx_\mu)$ in going from x_μ to $x_\mu + dx_\mu$, the value of the function becomes

$$f(x + dx) = (f(x) + \partial^\mu f(x) dx_\mu)(1 + S^\nu dx_\nu)$$
$$= f(x) + (\partial^\mu f(x) + f(x)S^\mu) dx_\mu + O(dx)^2. \qquad (3.1.2)$$

To first order in the infinitesimal translation dx_μ, the increment in the function f is therefore

$$\Delta f = (\partial^\mu + S^\mu)f \, dx_\mu. \qquad (3.1.3)$$

Weyl wished to base a theory upon the modified differential operator $(\partial^\mu + S^\mu)$ and to identify the four-vector potential $A^\mu(x)$ of electromagnetism with a space–time-dependent generator of scale changes, $S^\mu(x)$. Thus would electromagnetism find a basis in geometry.

Let us see why the proposed identification is incorrect. Recall from elementary quantum mechanics that the classical four-momentum

$$p^\mu = (E; p_x, p_y, p_z) \qquad (3.1.4)$$

goes over to the quantum-mechanical operator

$$p^\mu = i\partial^\mu = (i\partial^0; -i\nabla). \qquad (3.1.5)$$

For a particle with electric charge e, the canonical replacement is

$$(p^\mu - eA^\mu) \to i(\partial^\mu + ieA^\mu), \qquad (3.1.6)$$

in natural units with $\hbar = c = 1$. This suggests that Weyl's program could be implemented successfully if one identified

$$S^\mu = ieA^\mu \qquad (3.1.7)$$

and required invariance of the laws of physics under a change of phase

$$(1 + ieA^\mu dx_\mu) \simeq \exp(ieA^\mu dx_\mu) \qquad (3.1.8)$$

rather than under a change of scale. Following work by Fock[2] and London,[3] Weyl[4] began in 1929 to study invariance under this phase rotation, but retained the terminology "gauge invariance."[5]

In the course of this chapter, we shall seek to develop an increasingly precise understanding of gauge invariance and its consequences. We begin by reviewing the manifestations of gauge invariance in the classical electro-dynamics of Maxwell's equations. Next we consider the implications of phase invariance in quantum mechanics and see for the first time how imposition of a local symmetry requires the existence of interactions. We then digress briefly on the importance of the potential and above all of the phase factor (3.1.8) in quantum theory. There follows a systematic investigation of phase invariance in Lagrangian field theory. Finally we close the chapter by deducing the Feynman rules for scalar and spinor electrodynamics from the Lagrangians for those theories.

3.2   Gauge Invariance in Classical Electrodynamics

We have already remarked on two of the motivations for examining gauge invariance in detail: the hope of finding an explanation, or at least a deeper understanding, of the conservation of electric charge, and the desire (represented in the first instance by Weyl's attempts) to derive electro-dynamics from some basic principle. As a necessary preliminary to these studies, let us review the manifestations of gauge arbitrariness in classical electrodynamics.

Maxwell's equation for magnetic charge,

$$\nabla \cdot \mathbf{B} = 0, \tag{3.2.1}$$

invites us to write the magnetic field as

$$\mathbf{B} = \nabla \times \mathbf{A}, \tag{3.2.2}$$

where $\mathbf{A}$ is called the vector potential. This identification ensures that $\mathbf{B}$ will be divergenceless, by virtue of the identity

$$\nabla \cdot (\nabla \times \mathbf{A}) = 0. \tag{3.2.3}$$

If we add an arbitrary gradient to the vector potential

$$\mathbf{A} \rightarrow \mathbf{A} + \nabla \Lambda, \tag{3.2.4}$$

the magnetic field is unchanged, because

$$\mathbf{B} = \nabla \times (\mathbf{A} + \nabla \Lambda) = \nabla \times \mathbf{A}. \tag{3.2.5}$$

In similar fashion, the curl equation (Faraday–Lenz) for the electric field,

$$\nabla \times \mathbf{E} = -\partial \mathbf{B}/\partial t, \tag{3.2.6}$$

which can be rewritten as

$$\nabla \times (\mathbf{E} + \partial \mathbf{A}/\partial t) = 0, \tag{3.2.7}$$

suggests the identification

$$\mathbf{E} + \partial \mathbf{A}/\partial t = -\nabla V, \tag{3.2.8}$$

where V is known as the scalar potential. In order that the electric field remain invariant under the shift (3.2.4), we must also require

$$V \rightarrow V - \partial \Lambda/\partial t. \tag{3.2.9}$$

All this can be expressed compactly in covariant notation. The electromagnetic field-strength tensor

$$F^{\mu\nu} = -F^{\nu\mu} = \partial^\nu A^\mu - \partial^\mu A^\nu = \begin{array}{c} \mu\ \nu \rightarrow \\ \downarrow \end{array} \begin{pmatrix} 0 & E_1 & E_2 & E_3 \\ -E_1 & 0 & B_3 & -B_2 \\ -E_2 & -B_3 & 0 & B_1 \\ -E_3 & B_2 & -B_1 & 0 \end{pmatrix}, \quad (3.2.10)$$

built up from the four-vector potential

$$A^\mu = (V; \mathbf{A}), \quad (3.2.11)$$

is unchanged by the "gauge transformation"

$$A^\mu \rightarrow A^\mu - \partial^\mu \Lambda, \quad (3.2.12)$$

where $\Lambda(x)$ is an arbitrary function of the coordinate. The fact that many different four-vector potentials yield the same electromagnetic fields, and thus describe the same physics, is a manifestation of the gauge invariance of classical electrodynamics.

The Maxwell equations (3.2.1) and (3.2.6), which motivated the introduction of a potential, may be rewritten in covariant form as

$$\partial^\lambda F^{\mu\nu} + \partial^\mu F^{\nu\lambda} + \partial^\nu F^{\lambda\mu} = 0. \quad (3.2.13)$$

A more compact expression follows upon introduction of the dual field-strength tensor:

$$*F^{\mu\nu} = -\tfrac{1}{2}\varepsilon^{\mu\nu\rho\sigma} F_{\rho\sigma} = \begin{pmatrix} 0 & B_1 & B_2 & B_3 \\ -B_1 & 0 & -E_3 & E_2 \\ -B_2 & E_3 & 0 & -E_1 \\ -B_3 & -E_2 & E_1 & 0 \end{pmatrix}, \quad (3.2.14)$$

which may be obtained formally from $F^{\mu\nu}$ by replacing $\mathbf{E} \rightarrow \mathbf{B}$ and $\mathbf{B} \rightarrow -\mathbf{E}$. Here we adopt the convention that Levi-Civita's antisymmetric symbol $\varepsilon^{\mu\nu\rho\sigma}$ is equal to ∓ 1 for even or odd permutations of $(0, 1, 2, 3)$ and that $\varepsilon^{\mu\nu\rho\sigma} = -\varepsilon_{\mu\nu\rho\sigma}$. Evidently the inverse relation is

$$F^{\mu\nu} = \tfrac{1}{2}\varepsilon^{\mu\nu\rho\sigma} *F_{\rho\sigma}. \quad (3.2.15)$$

Taking the dual of equation (3.2.13) we find that the Maxwell equations (3.2.1) and (3.2.6) may be written as

$$\partial_\mu *F^{\mu\nu} = 0. \quad (3.2.16)$$

The remaining Maxwell equations,

$$\nabla \cdot \mathbf{E} = \rho = -\nabla \cdot \dot{\mathbf{A}} - \nabla^2 V \quad (3.2.17)$$

and

$$\nabla \times \mathbf{B} = \mathbf{J} + \dot{\mathbf{E}} = \mathbf{J} - \ddot{\mathbf{A}} - \nabla \dot{V}$$
$$\parallel$$
$$\nabla \times (\nabla \times \mathbf{A}) = -\nabla^2 \mathbf{A} + \nabla(\nabla \cdot \mathbf{A}),$$

(3.2.18)

(where $\dot{\mathbf{A}} \equiv \partial \mathbf{A}/\partial t$, etc.) correspond, in covariant notation, to

$$\partial_\mu F^{\mu\nu} = -J^\nu,$$

(3.2.19)

with the electromagnetic current given by

$$J^\nu = (\rho; \mathbf{J}).$$

(3.2.20)

Two consequences are immediately apparent. First, the electromagnetic current is conserved:

$$\partial_\nu J^\nu = -\partial_\nu \partial_\mu F^{\mu\nu} = 0.$$

(3.2.21)

Second, the wave equation (3.2.19) may be expanded as

$$\Box A^\nu - \partial^\nu(\partial_\mu A^\mu) = J^\nu,$$

(3.2.22)

which becomes, in the absence of sources and in Lorentz gauge ($\partial \cdot A = 0$),

$$\Box A^\nu = 0.$$

(3.2.23)

Each component of the vector potential, to be identified with the photon field, thus satisfies a Klein–Gordon equation for a massless particle.

We see in these familiar results a relationship between gauge invariance, current conservation, and massless vector fields. Let us now attempt to understand these connections more completely.

3.3   Phase Invariance in Quantum Mechanics

Suppose that we knew the Schrödinger equation, but not the laws of electrodynamics. Would it be possible to derive (in other words, to guess) Maxwell's equations from a gauge principle? The answer is yes! It is worthwhile to trace the steps in the argument in detail.

A quantum-mechanical state is described by a complex Schrödinger wave function $\psi(x)$. Quantum-mechanical observables involve inner products of the form

$$\langle \mathcal{O} \rangle = \int \psi^* \mathcal{O} \psi,$$

(3.3.1)

which are unchanged under a global phase rotation:

$$\psi(x) \rightarrow e^{i\theta} \psi(x).$$

(3.3.2)

In other words, the absolute phase of the wave function cannot be measured and is a matter of convention. *Relative* phases between wave functions, as measured in interference experiments, are unaffected by such a global rotation.

This raises the question: Are we free to choose one phase convention in Paris and another in Batavia? Differently stated, can quantum mechanics be formulated to be invariant under local (position-dependent) phase rotations

$$\psi(x) \to \psi'(x) = e^{i\alpha(x)}\psi(x)? \tag{3.3.3}$$

We shall see that this can be accomplished, but at the price of introducing an interaction that will be constructed to be electromagnetism.

The quantum-mechanical equations of motion, such as the Schrödinger equation, always involve derivatives of the wave function ψ, as do many observables. Under local phase rotations, these transform as[6]

$$\partial_\mu\psi(x) \to \partial_\mu\psi'(x) = e^{i\alpha(x)}[\partial_\mu\psi(x) + i(\partial_\mu\alpha(x))\psi(x)], \tag{3.3.4}$$

which involves more than a mere phase change. The additional gradient-of-phase term spoils local phase invariance. Local phase invariance may be achieved, however, if the equations of motion and the observables involving derivatives are modified by the introduction of the electromagnetic field $A_\mu(x)$. If the gradient ∂_μ is everywhere replaced by the *gauge-covariant derivative*

$$\mathcal{D}_\mu \equiv \partial_\mu + ieA_\mu, \tag{3.3.5}$$

where e is the charge in natural units of the particle described by $\psi(x)$ and the field $A_\mu(x)$ transforms under phase rotations (3.3.3) as

$$A_\mu(x) \to A'_\mu(x) \equiv A_\mu(x) - (1/e)\partial_\mu\alpha(x), \tag{3.3.6}$$

it is easily verified that under local phase rotations

$$\mathcal{D}_\mu\psi(x) \to e^{i\alpha(x)}\mathcal{D}_\mu\psi(x). \tag{3.3.7}$$

Consequently quantities such as $\psi^*\mathcal{D}_\mu\psi$ are invariant under local phase transformations. The required transformation law (3.3.6) for the four-vector A_μ is precisely the form (3.2.12) of a gauge transformation in electrodynamics. Moreover, the covariant derivative defined in (3.3.5) corresponds to the familiar replacement $p \to p - eA$ already noted in (3.1.6). Thus the form of the coupling $(\mathcal{D}_\mu\psi)$ between the electromagnetic field and matter is suggested, if not uniquely dictated, by local phase invariance.

This example has shown the possibility of using local gauge invariance as a dynamical principle, as promised in the discussion of Weyl's program. It should be remarked that the idea of modifying the equations of motion to accommodate an invariance principle is not without a successful precedent. It was precisely to accommodate local charge conservation that Maxwell modified Ampère's law by the addition of a "displacement current" $\partial\mathbf{E}/\partial t$. Before proceeding to a more systematic application of the gauge principle, some commentary is in order on the import of the vector potential in quantum mechanics.

3.4 Significance of Potentials in Quantum Theory

In classical electrodynamics, the field strengths $\mathbf{E}$ and $\mathbf{B}$ are regarded as the basic physical quantities and, as we have seen in Section 3.2, the potential $A^\mu = (V; \mathbf{A})$ is introduced as a convenient calculational device. Because of the gauge ambiguity of electrodynamics, the potential corresponding to a given configuration of the fields is not uniquely defined. It may therefore appear that the potential is no more than an auxiliary mathematical quantity with no independent physical significance. This is decidedly not the case in quantum theory, as the analysis of Aharonov and Bohm[7] has made clear.

To see the effect of potentials in the absence of fields, consider a non-relativistic charged particle moving through a static vector potential, which corresponds to a vanishing magnetic field. If the wave function $\psi^0(\mathbf{x}, t)$ is a solution of the Schrödinger equation in the absence of the vector potential,

$$\frac{-\hbar^2}{2m} \nabla^2 \psi^0 = i\hbar \frac{\partial \psi^0}{\partial t}, \tag{3.4.1}$$

then the solution in the presence of the vector potential will be

$$\psi(\mathbf{x}, t) = \psi^0(\mathbf{x}, t) e^{iS/\hbar}, \tag{3.4.2}$$

with

$$S = e \int d\mathbf{x} \cdot \mathbf{A}. \tag{3.4.3}$$

The new solution follows from the Schrödinger equation

$$(-i\hbar\nabla - e\mathbf{A})^2 \psi / 2m = i\hbar \frac{\partial \psi}{\partial t}, \tag{3.4.4}$$

and is a special case of the familiar fact that the phase shift experienced by a particle is given by $(i/\hbar)$ times the change in its classical action. That the new solution differs from the old simply by a phase factor implies that there is no change in any physical result, as expected.

Now suppose that a single coherent beam of charged particles is split into two parts, each passing on opposite sides of a solenoid, but excluded from it, as shown in Fig. 3-1. After having passed the solenoid, the beams are recombined and the resulting interference pattern is observed.

Now let

$$\psi^0(\mathbf{x}, t) = \psi_1^0(\mathbf{x}, t) + \psi_2^0(\mathbf{x}, t) \tag{3.4.5}$$

represent the wave function in the absence of a vector potential, where ψ_1^0 and ψ_2^0 denote the components of the beam that pass above and below the solenoid. When an electric current flows through the solenoid, it creates a

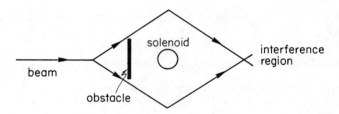

F$_{\text{IG}}$. 3-1. Schematic experiment to demonstrate interference with a time-independent vector potential.

magnetic field $\mathbf{B}$, which is essentially confined within the solenoid. In the experimental arrangement described, the beams pass only through field-free regions. The vector potential $\mathbf{A}$ cannot, however, be zero everywhere outside the solenoid, because the total flux through every loop containing the solenoid is a constant given by

$$\Phi = \int d\boldsymbol{\sigma} \cdot \mathbf{B} = \int d\mathbf{x} \cdot \mathbf{A}. \tag{3.4.6}$$

By analogy with the single-beam configuration, the perturbed wave function is then

$$\psi = \psi_1^0 e^{iS_1/\hbar} + \psi_2^0 e^{iS_2/\hbar}, \tag{3.4.7}$$

where

$$S_i = e \int_{\text{path } i} d\mathbf{x} \cdot \mathbf{A}. \tag{3.4.8}$$

Evidently the interference of the two components of the recombined beam will depend upon the phase difference $(S_1 - S_2)/\hbar$. Consequently there is a physical effect of the vector potential in spite of the fact that the beams have experienced no forces due to electromagnetic fields.

Some elementary remarks may help to reconcile this remarkable result with classical experience. First, because the effect of the potentials appears in an interference phenomenon, it is essentially quantum-mechanical in nature. Second, the phase difference

$$(S_1 - S_2)/\hbar = (e/\hbar) \oint d\mathbf{x} \cdot \mathbf{A}(\mathbf{x}) \tag{3.4.9}$$

can be written in the form of an integral around a closed path. This form emphasizes that shifting the vector potential by a gradient, the familiar gauge freedom of classical electrodynamics, has no effect upon the result. Finally, let us remark that the covariant generalization of (3.4.9) is simply

$$(-e/\hbar) \oint dx_\mu A^\mu. \tag{3.4.10}$$

The experiment suggested by this analysis was performed by Chambers[8] soon after the Aharonov–Bohm proposal. The observation that interference fringes shift as the enclosed flux is varied confirms that electromagnetic effects do occur in regions free of electric and magnetic fields. Thus the field-strength tensor $F^{\mu\nu}$ is insufficient to determine all electromagnetic effects in quantum mechanics. A knowledge of the phase factor

$$\exp\left[(-ie/\hbar)\oint dx_\mu A^\mu\right],\tag{3.4.11}$$

whose resemblance to (3.1.8) should not be overlooked, does make possible comprehensive predictions.

This brief digression has served several purposes: First, to call attention to the Aharonov–Bohm effect, which is interesting in its own right. Second, to dispel the conviction, born of classical experience, that the electromagnetic vector potential is a mathematical artifice, without physical significance. Third, to introduce albeit sketchily the important role played in quantum theory by path-dependent phase factors.

3.5   Phase Invariance in Field Theory

In the preceding sections we have seen something of the connection between electromagnetic gauge invariance and the conservation of charge, and have found that by generalizing the global phase invariance of quantum mechanics to a local phase symmetry one may be led from a theory describing free particles to one in which the particles experience electromagnetic inter-actions. With this as background, it is now time to bring the logical structure into clearer focus by making a more systematic investigation of phase invariance, in the framework of Lagrangian field theory. To avoid becoming lost in formalism, and because we shall take this opportunity to derive some specific results that will be of use later, we proceed by example.

Consider the Lagrangian for the free complex scalar field,

$$\mathcal{L} = |\partial^\mu\phi|^2 - m^2|\phi|^2,\tag{3.5.1}$$

from which the Euler–Lagrange equations lead to the Klein–Gordon equations

$$(\Box + m^2)\phi(x) = 0, \qquad (\Box + m^2)\phi^*(x) = 0.\tag{3.5.2}$$

A global transformation on these fields,

$$\phi(x) \to e^{iq\alpha}\phi(x), \qquad \phi^*(x) \to e^{-iq\alpha}\phi^*(x),\tag{3.5.3}$$

leads to infinitesimal variations

$$\delta\phi = iq(\delta\alpha)\phi, \qquad \delta\phi^* = -iq(\delta\alpha)\phi^*;\tag{3.5.4}$$
$$\delta(\partial_\mu\phi) = iq(\delta\alpha)\partial_\mu\phi, \qquad \delta(\partial_\mu\phi^*) = -iq(\delta\alpha)\partial_\mu\phi^*.\tag{3.5.5}$$

The statement of global phase invariance is that such transformations leave the Lagrangian unchanged:

$$\delta\mathscr{L} = 0. \tag{3.5.6}$$

Explicit computation yields

$$\delta\mathscr{L} = \frac{\partial\mathscr{L}}{\partial\phi}\delta\phi + \frac{\partial\mathscr{L}}{\partial(\partial_\mu\phi)}\delta(\partial_\mu\phi) + \frac{\partial\mathscr{L}}{\partial\phi^*}\delta\phi^* + \frac{\partial\mathscr{L}}{\partial(\partial_\mu\phi^*)}\delta(\partial_\mu\phi^*)$$

$$= \left[\partial_\mu\frac{\partial\mathscr{L}}{\partial(\partial_\mu\phi)}\right]iq(\delta\alpha)\phi + \frac{\partial\mathscr{L}}{\partial(\partial_\mu\phi)}iq(\delta\alpha)\partial_\mu\phi - (\phi \to \phi^*)$$

$$= iq(\delta\alpha)\partial_\mu\left[\frac{\partial\mathscr{L}}{\partial(\partial_\mu\phi)}\phi - \frac{\partial\mathscr{L}}{\partial(\partial_\mu\phi^*)}\phi^*\right] \equiv 0, \tag{3.5.7}$$

where the equations of motion have been used in passing to the second line. Evidently we may identify a conserved Noether current,

$$j^\mu = -iq\left[\frac{\partial\mathscr{L}}{\partial(\partial_\mu\phi)}\phi - \frac{\partial\mathscr{L}}{\partial(\partial_\mu\phi^*)}\phi^*\right]$$

$$= iq[\phi^*\partial^\mu\phi - (\partial^\mu\phi^*)\phi] \equiv iq\phi^*\overset{\leftrightarrow}{\partial}^\mu\phi, \tag{3.5.8}$$

which satisfies

$$\partial_\mu j^\mu = 0. \tag{3.5.9}$$

With the identification of q as the electric charge, (3.5.8) is recognizable at once as the electromagnetic current of the charged scalar field. The connection between global phase invariance and current conservation is thus made explicit. Noether's theorem guarantees that this relation is in fact a general one.

What are the consequences of imposing invariance under *local* phase rotations, which transform the fields as

$$\phi(x) \to e^{iq\alpha(x)}\phi(x)? \tag{3.5.10}$$

Terms in the Lagrangian that depend only on the fields are left invariant, just as before, so there are no additional consequences beyond those of global gauge invariance. However, as we have seen in the quantum-mechanical discussion of Section 3.3, gradient terms transform as

$$\partial_\mu\phi(x) \to e^{iq\alpha(x)}[\partial_\mu\phi(x) + iq(\partial_\mu\alpha(x))\phi(x)], \tag{3.5.11}$$

which necessitate the introduction of the gauge-covariant derivative

$$\mathscr{D}_\mu \equiv \partial_\mu + iqA_\mu(x). \tag{3.5.12}$$

Objects such as $\mathscr{D}_\mu\phi$ will then simply undergo the same phase rotation as the fields, namely

$$\mathscr{D}_\mu\phi \to e^{iq\alpha(x)}\mathscr{D}_\mu\phi, \tag{3.5.13}$$

provided that the vector field A_μ transforms as

$$A_\mu(x) \rightarrow A_\mu(x) - \partial_\mu \alpha(x). \qquad (3.5.14)$$

It was remarked in Section 3.3 that the replacement $\partial_\mu \rightarrow \mathcal{D}_\mu$ prescribes the form of the interaction between the gauge field A_μ and matter. To see how this comes about in Lagrangian field theory, let us look explicitly at the Dirac equation. The free-particle Lagrangian

$$\mathcal{L}_{\text{free}} = \bar{\psi}(i\gamma^\mu \partial_\mu - m)\psi \qquad (3.5.15)$$

is replaced by the locally gauge-invariant expression

$$\begin{aligned}
\mathcal{L} &= \bar{\psi}(i\gamma^\mu \mathcal{D}_\mu - m)\psi \\
&= \bar{\psi}(i\gamma^\mu \partial_\mu - m)\psi - q A_\mu \bar{\psi}\gamma^\mu \psi \\
&= \mathcal{L}_{\text{free}} - J^\mu A_\mu,
\end{aligned} \qquad (3.5.16)$$

where the (conserved) electromagnetic current has the familiar form

$$J^\mu = q\bar{\psi}\gamma^\mu \psi. \qquad (3.5.17)$$

Precisely this form of the current follows from the requirement of global gauge invariance. It is easy to verify that the Lagrangian (3.5.16) is indeed invariant under the combined transformations (3.5.10) and (3.5.14).

To arrive at the complete Lagrangian for quantum electrodynamics it remains only to add a kinetic energy term for the vector field, which describes the propagation of free photons. The Lagrangian (2.2.8) leads to Maxwell's equations and is manifestly invariant under local gauge transformations (3.5.14). Assembling all the pieces we therefore have

$$\mathcal{L}_{\text{QED}} = \mathcal{L}_{\text{free}} - J^\mu A_\mu - \tfrac{1}{4} F_{\mu\nu} F^{\mu\nu}, \qquad (3.5.18)$$

which is indeed the usual QED Lagrangian.[9]

A photon mass term would have the form

$$\mathcal{L}_\gamma = \tfrac{1}{2} m^2 A^\mu A_\mu, \qquad (3.5.19)$$

which obviously violates local gauge invariance because

$$A^\mu A_\mu \rightarrow (A^\mu - \partial^\mu \alpha)(A_\mu - \partial_\mu \alpha) \neq A^\mu A_\mu. \qquad (3.5.20)$$

Thus we find that local gauge invariance has led us to the existence of a massless photon. (Although this conclusion is not inescapable, it can be avoided only at a price, as will be explained in Chapter 5.) The best direct limit on the photon mass comes from measurements of the magnetic field of Jupiter by the Pioneer 10 spacecraft.[10] The upper limit at 90% confidence level is

$$M_\gamma < 4.5 \times 10^{-16} \text{ eV/c}^2, \qquad (3.5.21)$$

corresponding to a modified Coulomb potential of the form

$$V \sim \exp(-r/r_0)/r, \qquad (3.5.22)$$

with $r_0 > 4.4 \times 10^5$ km ($= 4.4 \times 10^{23}$ fm). Subsequent space probes may be expected to improve this sensitivity further.

We have now seen how global phase invariance leads to the existence of a conserved charge. The more constraining requirement of local phase invariance necessitates the introduction of a massless gauge field and restricts the possible interactions of radiation with matter. For the Dirac equation, the interaction term was simply of the form $-J^\mu A_\mu$, with J^μ the conserved current of the free-fermion Lagrangian. In general, some care is required in making this identification because the structure of the current may be altered by the interactions. Such a modification occurs, for example, in the other case we have examined, that of the complex scalar field.

Replacement of the gradient ∂_μ by the gauge-covariant derivative $\mathscr{D}_\mu$ in the Lagrangian (3.5.1) yields the locally gauge-invariant form

$$\begin{aligned}
\mathscr{L} &= |\mathscr{D}^\mu \phi|^2 - m^2 |\phi|^2 \\
&= |\partial^\mu \phi|^2 - m^2 |\phi|^2 - iqA_\mu \phi^* \overleftrightarrow{\partial}^\mu \phi + q^2 A^2 \phi^* \phi \\
&= |\partial^\mu \phi|^2 - m^2 |\phi|^2 - j^\mu A_\mu + q^2 A^2 \phi^* \phi, \qquad (3.5.23)
\end{aligned}$$

where j^μ is the conserved current (3.5.8) for the free-particle Lagrangian. The final term in (3.5.23), which corresponds to a "contact interaction," can be identified as a $J \cdot A$ contribution when Noether's theorem is used to define a conserved current for the full Lagrangian, including interactions.

3.6   Feynman Rules for Electromagnetism

The passage from Lagrangian to Feynman rules is documented in many places. Here we shall merely outline the final procedure without derivation, assuming the reader to be as familiar with the methods as he cares to be.

The recipe for extracting vertex factors from the interaction Lagrangian $\mathscr{L}_{\text{int}}$ is uncomplicated. In the quantity $i\mathscr{L}_{\text{int}}$, replace all the field operators by free-particle wave functions, and evaluate the resulting expressions in momentum space. Omit all factors that correspond to the external lines. What remains is the vertex factor for the interaction.

Consider first the interaction in spinor electrodynamics. We have from (3.5.16)

$$i\mathscr{L}_{\text{int}} = -iqA_\mu \bar{\psi}\gamma^\mu \psi, \quad = \quad j^\mu A_\mu \qquad (3.6.1)$$

into which we insert

$$\begin{aligned}
\psi &= u_1 e^{-ip_1 \cdot x} \\
\bar{\psi} &= \bar{u}_2 e^{ip_2 \cdot x} \\
A_\mu &= \varepsilon_\mu^*(k) e^{ik \cdot x}
\end{aligned} \qquad (3.6.2)$$

so that

$$i\mathscr{L}_{\text{int}} = -iq\varepsilon_\mu^*(k)\bar{u}_2\gamma^\mu u_1 e^{i(p_2+k-p_1)\cdot x}, \tag{3.6.3}$$

from which, after discarding the external factors, we retain

$$-iq\gamma^\mu. \tag{3.6.4}$$

For scalar electrodynamics, the interaction Lagrangian of (3.5.23) contains two terms. The first, the trilinear term in $\mathscr{L}_{\text{int}}$, yields

$$i\mathscr{L}_{\text{int}}[3] = qA_\mu[\phi^*\partial^\mu\phi - (\partial^\mu\phi^*)\phi]. \tag{3.6.5}$$

Upon substitution of

$$\begin{aligned}
\phi &= S_1 e^{-ip_1\cdot x} \\
\phi^* &= S_2 e^{ip_2\cdot x}
\end{aligned} \tag{3.6.6}$$

and the usual form for the photon, we find

$$i\mathscr{L}_{\text{int}}[3] \rightarrow \varepsilon_\mu^*(k)S_1 S_2 e^{i(p_2+k-p_1)\cdot x}q(-ip_1-ip_2)^\mu, \tag{3.6.7}$$

which yields a vertex factor

$$-iq(p_1+p_2)^\mu. \tag{3.6.8}$$

Finally, for the quadrilinear term

$$i\mathscr{L}_{\text{int}}[4] = iq^2 A_\mu A_\nu g^{\mu\nu}\phi^*\phi, \tag{3.6.9}$$

we write

$$i\mathscr{L}_{\text{int}}[4] \rightarrow iq^2 g^{\mu\nu}(\varepsilon_\mu^*(k_1)\varepsilon_\nu^*(k_2) + \varepsilon_\mu^*(k_2)\varepsilon_\nu^*(k_1))e^{i(p_2+k_1+k_2-p_1)\cdot x}, \tag{3.6.10}$$

from which the vertex factor is

$$2iq^2 g^{\mu\nu}. \tag{3.6.11}$$

The momentum–space propagators may be obtained with the aid of a similar mnemonic device. The general form of the wave equations is typified by that for the Klein–Gordon field,

$$(\Box + m^2)\phi(x) = J(x), \tag{3.6.12}$$

where the source term J may itself depend on the fields. The propagator is simply related to the inverse of the momentum–space representation of the operator on the left-hand side. Thus for the Klein–Gordon field we have

$$G(p^2) = \frac{i}{p^2 - m^2 + i\varepsilon}, \tag{3.6.13}$$

while for the fermion field in the Dirac equation

$$(\not{p} - m)\psi(x) = J(x), \tag{3.6.14}$$

the propagator is

$$G(p) = \frac{i}{\not{p} - m + i\varepsilon} = \frac{i(\not{p} + m)}{p^2 - m^2 + i\varepsilon}. \tag{3.6.15}$$

To obtain the photon propagator, a choice of gauge is required. This is because the equation of motion

$$\square A^\nu - \partial^\nu(\partial^\mu A_\mu) = J^\nu \tag{3.2.22}$$

does not uniquely determine the potential A^ν in terms of the conserved current J^ν. To resolve the gauge ambiguity in the most pedestrian fashion we may choose the gauge $(\partial \cdot A) = 0$, in which case (3.2.22) collapses to a set of massless Klein–Gordon equations, which lead to the propagator

$$G(k^2) = \frac{-ig^{\mu\nu}}{k^2 + i\varepsilon}. \tag{3.6.16}$$

Other gauge-fixing procedures will be discussed briefly in Chapter 7. The gauge invariance of the theory guarantees that observable quantities will be independent of the choice of gauge.

The analysis presented in this chapter has shown the theory of electro-magnetism to be the gauge theory associated with the phase transformations that form the Abelian group $U(1)$. We shall next investigate the generalization of these ideas to non-Abelian groups. The resulting theories are known as non-Abelian gauge theories, or Yang–Mills theories.

Problems

3-1. Using the Feynman rules given in Appendix B.5, compute the differential cross section $d\sigma/d\Omega$ and the total (integrated) cross section $\sigma \equiv \int d\Omega(d\sigma/d\Omega)$ for the reaction $e^-\sigma^+ \rightarrow e^-\sigma^+$, where σ^+ is a charged scalar particle. Express the differential cross section in terms of (a) kinematic invariants, (b) c.m. variables, and (c) laboratory variables corresponding to the scalar target at rest. How would the results differ for $e^-\sigma^-$ scattering?

3-2. Making use of the Dirac equation, show that the most general parity-conserving form for the electromagnetic current of the proton is

$$J_\mu \sim \bar{u}(p')[\Gamma_1(q^2)\gamma_\mu + \Gamma_2(q^2)i\sigma_{\mu\nu}q^\nu + \Gamma_3(q^2)q_\mu]u(p),$$

where $\sigma_{\mu\nu} \equiv (i/2)[\gamma_\mu, \gamma_\nu]$ and $q \equiv p' - p$. What are the consequences of current conservation, $\partial^\mu J_\mu = 0$?

3-3. Calculate the differential cross section in the laboratory frame for elastic electron–proton scattering (a) for a structureless proton (i.e., a Dirac particle); (b) for a real proton (using the results of Problem 3-2). How do these results differ from the cross section for $e\sigma$ scattering?

3-4.   Reformulate Maxwell's equations, taking into account the possibility that magnetic monopoles exist. Show that classical electrodynamics is invariant under the transformation

$$\mathbf{E} \to \mathbf{E} \cos \theta + \mathbf{B} \sin \theta,$$
$$\mathbf{B} \to -\mathbf{E} \sin \theta + \mathbf{B} \cos \theta,$$
$$q \to q \cos \theta + g \sin \theta,$$
$$g \to -q \sin \theta + g \cos \theta,$$

where q and g are electric and magnetic charges, respectively. Show that, if the ratio g/q has the same value for all sources, the magnetic charge can be rotated away, and the theory expressed in terms of electric charges only. What is the value of the new effective electric charge? Analyze the gauge invariance of the modified classical electrodynamics. [Reference: G. Wentzel, *Prog. Theoret. Phys. Suppl.* **37**–**38**, 163 (1966).]

3-5.   Construct the $O(e^4)$ amplitude for the reaction $\gamma\gamma \to \gamma\gamma$ in QED, which represents the sum of six Feynman diagrams. Show that your final result is gauge-invariant (in the sense that the amplitude vanishes upon replacement of $\varepsilon_\mu(k)$ by k_μ) and finite, whereas the contribution of each diagram is separately gauge-dependent and divergent.

3-6.   (a) Consider a nonrelativistic charged particle moving along the axis of a cylindrical Faraday cage connected to an external generator, which causes the potential $V(t)$ on the cage to vary with time only when the particle is well within the cage. Show that, if the wave function $\psi^0(\mathbf{x}, t)$ is a solution of the Schrödinger equation for $V(t) \equiv 0$, the solution when the generator is operating will be $\psi(\mathbf{x}, t) = \psi^0(\mathbf{x}, t)e^{iS/\hbar}$, where $S = -\int^t dt' eV(t')$.

(b) Now suppose that a single coherent beam of charged particles is split into two parts, each of which is allowed to pass through its own long cylindrical cage of the kind just described. On emerging from the Faraday cages the beams are recombined and the resulting interference pattern is observed. The beam is chopped into bunches that are long compared with the wavelength of an individual particle but short compared with the Faraday cages. The potentials on the two cages vary independently, but are nonzero only when a bunch is well within the tubes. This ensures that the beam traverses a time-varying potential without experiencing electric or magnetic forces. Describe how the interference pattern depends upon the applied voltages. [Reference: Aharonov and Bohm, ref. 7.]

3-7.   If baryon number is absolutely conserved, the conservation law may be a consequence of a global phase symmetry like that of electromagnetism, with the electric charge replaced by baryon number. How would Newton's law of gravitation be modified if the baryonic phase symmetry were a *local* gauge invariance? In view of the close equality of inertial and gravitational masses imposed by the Eötvös experiment [P. G. Roll, R. Krotkov, and

R. H. Dicke, *Ann. Phys.* (*NY*) **26**, 442 (1967)], what can be said about the strength of a hypothetical gauge interaction coupled to the baryon current? [Reference: T. D. Lee and C. N. Yang, *Phys. Rev.* **98**, 150 (1955).]

For Further Reading

CLASSICAL ELECTROMAGNETISM. The path to Maxwell's equations is described in the standard textbooks,

J. D. Jackson, *Classical Electrodynamics*, second edition, Wiley, New York, 1975.

W. K. H. Panofsky and M. Phillips, *Classical Electricity and Magnetism*, second edition, Addison-Wesley, Reading, Massachusetts, 1962.

and in the historical introduction to

M. Born and E. Wolf, *Principles of Optics*, fourth edition, Pergamon, New York, 1970.

The definitive account is that of

E. T. Whitaker, *A History of Theories of Aether and Electricity*, 2 volumes, Nelson, London, Vol. 1: *The Classical Theories*, 1910, revised and enlarged 1951, Vol. 2: *The Modern Theories 1900–1926*, 1953; reprinted by Harper Torchbooks, New York, 1960.

GAUGE INVARIANCE. The history of the concept of gauge invariance has been reviewed by

C. N. Yang, *Proceedings of the 6th Hawaii Topical Conference in Particle Physics*, edited by P. N. Dobson, Jr. *et al.*, University of Hawaii Press, Honolulu, 1976, p. 489.

C. N. Yang, *Five Decades of Weak Interaction Theory*, edited by N.-P. Chang, *Ann. NY Acad. Sci.* **294**, 86 (1977).

For additional remarks on the connection between global gauge invariance and current conservation, see

E. P. Wigner, "Invariance in Physical Theory," in *Symmetries and Reflections*, Indiana University Press, Bloomington, 1967, p. 3.

AHARONOV–BOHM EFFECT. This is an apparently inexhaustible subject, in part because the conclusion that physics can be affected by potentials is still resisted in some quarters and in part because the analysis of specific experimental arrangements is indeed subtle. Among discussions of matters of principle, see

W. H. Furry and N. F. Ramsey, *Phys. Rev.* **118**, 623 (1960).

Y. Aharonov and D. Bohm, *Phys. Rev.* **123**, 1511 (1961).

R. P. Feynman, R. B. Leighton, and M. Sands, *The Feynman Lectures in Physics*, Vol. II, Addison-Wesley, Reading, Massachusetts, 1964, Section 15-5.

M. Peshkin, *Phys. Rep.* **80**, 375 (1981).

Specific experimental results are reported or analyzed in

F. G. Werner and D. R. Brill, *Phys. Rev. Lett.* **4**, 344 (1960).

R. C. Jaklevic, J. J. Lambe, A. H. Silver, and J. E. Mercereau, *Phys. Rev. Lett.* **12**, 274 (1964).

G. Matteucci and G. Pozzi, *Am. J. Phys.* **46**, 619 (1978).

D. M. Greenberger and A. W. Overhauser, *Rev. Mod. Phys.* **51**, 43 (1979).

J. J. Sakurai, *Phys. Rev.* **D21**, 2993 (1980).

ELECTROMAGNETISM AS A GAUGE THEORY. Lucid introductions may be found in

E. S. Abers and B. W. Lee, *Phys. Rep.* **9C**, 1 (1973).

I. J. R. Aitchison and A. J. G. Hey, *Gauge Theories in Particle Physics*, Adam Hilger, Bristol, 1982.

PHOTON MASS. The fascinating history of measurements of the photon mass is reviewed in many places, including

A. S. Goldhaber and M. M. Nieto, *Rev. Mod. Phys.* **43**, 277 (1971); *Sci. Am.* **234**, 86 (May, 1976).

I. Yu. Kobzarev and L. B. Okun, *Usp. Fiz. Nauk* **95**, 131 (1968) [English translation: *Sov. Phys.-Uspekhi* **11**, 338 (1968)]. J. D. Jackson, *op. cit.*, Section 1.2.

The upper bound on the mass of the photon can be improved by the analysis of galactic magnetic fields. The most restrictive limit, $M_\gamma < 3 \cdot 10^{-28}$ eV/c^2, has been inferred from the stability of the Magellanic clouds, as discussed by

G. V. Chibisov, *Usp. Fiz. Nauk* **119**, 551 (1976) [English translation: *Sov. Phys.-Uspekhi* **19**, 624 (1976)].

INTEGRAL FORMULATION OF GAUGE THEORIES. The idea that electrodynamics can be derived from a path-dependent, or nonintegrable, phase, as evoked by the discussion of the Aharonov–Bohm effect, goes back to the celebrated "monopole paper" by

P. A. M. Dirac, *Proc. Royal Soc.* (*London*) **A133**, 60 (1931).

This idea has been implemented by

S. Mandelstam, *Ann. Phys.* (*NY*) **19**, 1 (1962).

C. N. Yang, *Phys. Rev. Lett.* **33**, 445 (1974).

T. T. Wu and C. N. Yang, *Phys. Rev.* **D12**, 3845 (1975)

The later work makes contact with the mathematical concepts of the theory of fiber bundles, which arose in the study of abstract problems in geometry. A brief nontechnical introduction to the mathematics has been given by

I. M. Singer, *Phys. Today* **35**, 41 (March, 1982).

Extensive references to the mathematical literature may be found in

G. H. Thomas, *Riv. del Nuovo Cim.* **3**, No. 4 (1980).

References

[1] H. Weyl, *Space–Time–Matter*, translated by H. L. Brose, Dover, New York, 1951, Chapter IV, Section 35, p. 282. The German original dates from 1921.

[2] V. Fock, *Z. Phys.* **39**, 226 (1927).

[3] F. London, *Z. Phys.* **42**, 375 (1927).

[4] H. Weyl, *Z. Phys.* **56**, 330 (1929).

[5] For a summary of these early developments, see W. Pauli, *Handbuch der Physik*, edited by S. Flügge, Springer-Verlag, Berlin-Göttingen-Heidelberg, 1958, Vol. V/1, p. 1, which is largely identical with the 1933 edition.

[6] Covariant notation is adopted here in anticipation of the more general discussion to follow. Space and time components may be distinguished if an explicit analysis of the Schrödinger equation is desired.

[7] Y. Aharonov and D. Bohm, *Phys. Rev.* **115**, 485 (1959). Similar obervations were made earlier, in the context of electron microscopy, by W. Ehrenberg and R. E. Siday, *Proc. Phys. Soc.* (*London*) **62B**, 8 (1949).

[8] R. G. Chambers, *Phys. Rev. Lett.* **5**, 3 (1960).

[9] This is by no means the most general gauge-invariant Lagrangian that may be constructed. Although it does not arise from the minimal substitution $\partial_\mu \to \mathscr{D}_\mu$ a magnetic moment interaction involving the spin tensor $\sigma_{\mu\nu}$ is compatible with local gauge invariance, as Problem 3-2 will show. This is but a single example. What does seem to distinguish (3.5.18), apart from its agreement with experiment, is that it defines a renormalizable theory.

[10] L. Davis, A. S. Goldhaber, and M. M. Nieto, *Phys. Rev. Lett.* **35**, 1402 (1975).

CHAPTER 4

NON-ABELIAN GAUGE THEORIES

In this chapter we undertake the extension of our ideas about local gauge invariance to gauge groups more complicated than the group of phase rotations. We shall find that it is possible to enforce local gauge invariance by following essentially the same strategy as succeeded for electrodynamics. The principal difference, apart from algebraic complexity, will be the appearance of interactions among the gauge bosons as a consequence of the non-Abelian nature of the gauge symmetry. As before, we proceed by example, developing the $SU(2)$-isospin gauge theory put forward by Yang and Mills[1] and by Shaw.[2] The generalization to other gauge groups proceeds without complication.

4.1 Motivation

The near degeneracy of the neutron and proton masses, the charge- independence of nuclear forces, and many subsequent observations support the notion of isospin conservation in the strong interactions. What is meant by isospin conservation is that the laws of physics should be invariant under rotations in isospin space, and that the proton and neutron should appear symmetrically in all equations. This means that if electromagnetism can be neglected, the isospin orientation is of no significance. The distinction between proton and neutron thus becomes entirely a matter of arbitrary convention. In such a world, the existence of two distinct kinds of nucleons could be inferred from the properties of the ground state of the ^{4}He nucleus, in much the same manner as we have deduced from the spin-3/2 baryons Δ^{++} or Ω^- the need for three colors of quarks.

This sort of reasoning lay behind the introduction in Section 2.3 of the free-nucleon Lagrangian

$$\mathscr{L}_0 = \bar{\psi}(i\gamma^\mu \partial_\mu - m)\psi, \tag{2.3.14}$$

written in terms of the composite fermion fields

$$\psi = \begin{pmatrix} p \\ n \end{pmatrix}. \tag{2.3.13}$$

As the earlier discussion showed, the Lagrangian (2.3.14) has an invariance under global isospin rotations, and the isospin current is conserved. Thus one has complete freedom in naming the proton and neutron (in the absence of electromagnetism), but only at a single point in space–time. Once freely chosen, the convention must be respected everywhere throughout space–time.

This single restriction may seem, as it did to Yang and Mills,[1] at odds with the idea of local field theory. Furthermore, we have just seen in Chapter 3 that electromagnetism possesses a *local* gauge invariance, and that by imposing that local symmetry on a free-particle Lagrangian it is possible to construct an interesting (and indeed correct) theory of electrodynamics. In analogy with electromagnetism we are led to ask whether we can require that the freedom to name the two states of the nucleon be available independently at every space–time point. Can we, in other words, turn the global $SU(2)$ invariance of the free field theory into a mathematically consistent local $SU(2)$ invariance? If so, what are the physical consequences?

4.2 Construction

The formulation of the theory proceeds just as in the Abelian case. If under a local gauge transformation the field transforms as

$$\psi(x) \to \psi'(x) = G(x)\psi(x), \tag{4.2.1}$$

with

$$G(x) \equiv \exp\left(\frac{i}{2}\tau \cdot \alpha(x)\right), \tag{4.2.2}$$

then the gradient transforms as

$$\partial_\mu \psi \to G(\partial_\mu \psi) + (\partial_\mu G)\psi. \tag{4.2.3}$$

To ensure the local gauge invariance of the theory we first introduce a gauge-covariant derivative

$$\mathscr{D}_\mu \equiv I\partial_\mu + igB_\mu, \tag{4.2.4}$$

where

$$I = \begin{pmatrix} 1 & 0 \\ 0 & 1 \end{pmatrix} \tag{4.2.5}$$

serves as a reminder that the operators are 2×2 matrices in isospin space and g will be seen to play the role of a strong-interaction coupling constant. The object B_μ is the 2×2 matrix defined by

$$B_\mu = \tfrac{1}{2}\boldsymbol{\tau} \cdot \mathbf{b}_\mu = \tfrac{1}{2}\tau^a b_\mu^a = \tfrac{1}{2}\begin{pmatrix} b_3 & b_1 - ib_2 \\ b_1 + ib_2 & -b_3 \end{pmatrix}, \qquad (4.2.6)$$

where the three gauge fields are $\mathbf{b}_\mu = (b_1, b_2, b_3)$, bold-face quantities denote isovectors, and the isospin index a runs from 1 to 3.

The point of introducing the gauge fields and the gauge-covariant derivative is to obtain a generalization of the gradient that transforms as

$$\mathscr{D}_\mu \psi \to \mathscr{D}'_\mu \psi' = G(\mathscr{D}_\mu \psi). \qquad (4.2.7)$$

Requiring this to be so will show us how B_μ must behave under gauge transformations. By explicit computation we have

$$\begin{aligned} \mathscr{D}'_\mu \psi' &= (\partial_\mu + igB'_\mu)\psi' \\ &= G(\partial_\mu \psi) + (\partial_\mu G)\psi + igB'_\mu(G\psi) \\ &\equiv G(\partial_\mu + igB_\mu)\psi \\ &= G(\partial_\mu \psi) + igG(B_\mu \psi), \end{aligned} \qquad (4.2.8)$$

which may be solved to yield the condition

$$igB'_\mu(G\psi) = igG(B_\mu \psi) - (\partial_\mu G)\psi, \qquad (4.2.9)$$

which must hold for arbitrary values of the nucleon field ψ. Regarding the transformation law as an operator equation and multiplying on the right by G^{-1}, we obtain

$$\begin{aligned} B'_\mu &= GB_\mu G^{-1} + \frac{i}{g}(\partial_\mu G)G^{-1} \\ &= G\left[B_\mu + \frac{i}{g}G^{-1}(\partial_\mu G)\right]G^{-1}. \end{aligned} \qquad (4.2.10)$$

Although this transformation law may appear formidable at first sight, it has a rather simple interpretation. Recall that in the case of electromagnetism the local gauge transformation was the phase rotation

$$G_{\mathrm{EM}} = e^{iq\alpha(x)}, \qquad (4.2.11)$$

where $\alpha(x)$ is a numerical parameter. A transcription of the general transformation law (4.2.10) is therefore

$$\begin{aligned} A'_\mu &= G_{\mathrm{EM}}A_\mu G_{\mathrm{EM}}^{-1} + \frac{i}{q}(\partial_\mu G_{\mathrm{EM}})G_{\mathrm{EM}}^{-1} \\ &= A_\mu - \partial_\mu \alpha, \end{aligned} \qquad (4.2.12)$$

just as in (3.5.14). For the case of isospin gauge symmetry the meaning of (4.2.10) is that B_μ is transformed by an isospin rotation plus a gradient term. To see this explicitly, consider an infinitesimal gauge transformation, represented by

$$G = 1 + \frac{i}{2}\tau \cdot \boldsymbol{\alpha}, \qquad |\alpha_i| \ll 1. \tag{4.2.13}$$

From the transformation law (4.2.10), we have

$$B'_\mu = B_\mu + \frac{i}{2}\boldsymbol{\alpha} \cdot \tau B_\mu - \frac{i}{2}B_\mu \boldsymbol{\alpha} \cdot \tau - \frac{1}{2g}\partial_\mu(\boldsymbol{\alpha} \cdot \tau) + O(\alpha^2) \tag{4.2.14}$$

which, in view of the definition (4.2.6), is equivalent to

$$\tau \cdot \mathbf{b}'_\mu = \tau \cdot \mathbf{b}_\mu + \frac{i}{2}(\tau \cdot \boldsymbol{\alpha}\tau \cdot \mathbf{b}_\mu - \tau \cdot \mathbf{b}_\mu\tau \cdot \boldsymbol{\alpha}) - \frac{1}{g}\partial_\mu(\boldsymbol{\alpha} \cdot \tau). \tag{4.2.15}$$

The middle term may be simplified at once with the aid of the familiar Pauli-matrix identity

$$\tau \cdot \mathbf{a}\tau \cdot \mathbf{b} = \mathbf{a} \cdot \mathbf{b} + i\tau \cdot \mathbf{a} \times \mathbf{b}, \tag{4.2.16}$$

but some insight and the prospect of easy generalization are gained by proceeding more formally. Write the middle term in component form as

$$T_2 \equiv \frac{i}{2}\alpha^j b_\mu^k(\tau^j\tau^k - \tau^k\tau^j) = \frac{i}{2}\alpha^j b_\mu^k[\tau^j, \tau^k]. \tag{4.2.17}$$

For the isospin group $SU(2)$, the commutator is given by

$$[\tau^j, \tau^k] = 2i\varepsilon_{jkl}\tau^l, \tag{4.2.18}$$

so that

$$T_2 = -\varepsilon_{jkl}\alpha^j b_\mu^k\tau^l = -\boldsymbol{\alpha} \times \mathbf{b}_\mu \cdot \tau, \tag{4.2.19}$$

and

$$\tau \cdot \mathbf{b}'_\mu = \tau \cdot \mathbf{b}_\mu - \boldsymbol{\alpha} \times \mathbf{b}_\mu \cdot \tau - \frac{1}{g}\partial_\mu(\boldsymbol{\alpha} \cdot \tau). \tag{4.2.20}$$

Since the three isospin components of the gauge field are linearly independent, we have as the transformation law for infinitesimal gauge transformations

$$b_\mu'^l = b_\mu^l - \varepsilon_{jkl}\alpha^j b^k - \frac{1}{g}\partial_\mu\alpha^l,$$

$$\mathbf{b}'_\mu = \mathbf{b}_\mu - \boldsymbol{\alpha} \times \mathbf{b}_\mu - \frac{1}{g}\partial_\mu\boldsymbol{\alpha}, \tag{4.2.21}$$

which has the claimed structure. The result in component form, by the way, shows that the transformation rule depends on the structure constants ε_{jkl} and not on the representation of the isospin group. We learn from the intermediate steps (4.2.17,18) that the isospin rotation, which is the new feature compared with electromagnetism, arises from the noncommutativity of the gauge transformations, or in other words from the non-Abelian nature of the symmetry group.

To this point in the construction of the isospin gauge theory of nucleons, we have a Lagrangian given by

$$
\begin{aligned}
\mathscr{L} &= \bar{\psi}(i\gamma^\mu \mathscr{D}_\mu - m)\psi \\
&= \mathscr{L}_0 - g\bar{\psi}\gamma^\mu B_\mu \psi \\
&= \mathscr{L}_0 - \frac{g}{2}\mathbf{b}_\mu \cdot \bar{\psi}\gamma^\mu \boldsymbol{\tau}\psi,
\end{aligned}
\tag{4.2.22}
$$

namely a free Dirac Lagrangian plus an interaction term that couples the isovector gauge fields to the conserved isospin current of the nucleons. The structure of the interaction between the gauge fields and matter is precisely analogous to that found in the case of QED. To proceed further, we must construct a field-strength tensor and hence a kinetic term for the gauge fields. Although elegant means are available for motivating the correct form, some understanding is to be gained from a pedestrian approach. In analogy with electromagnetism, we seek a field-strength tensor

$$
F_{\mu\nu} \equiv \tfrac{1}{2}\mathbf{F}_{\mu\nu} \cdot \boldsymbol{\tau} = \tfrac{1}{2}F^a_{\mu\nu}\tau^a
\tag{4.2.23}
$$

from which to construct a gauge-invariant kinetic term

$$
\mathscr{L}_{\text{gauge}} = -\tfrac{1}{4}\mathbf{F}_{\mu\nu} \cdot \mathbf{F}^{\mu\nu} = -\tfrac{1}{2}\operatorname{tr}(F_{\mu\nu}F^{\mu\nu}),
\tag{4.2.24}
$$

where the last equality follows from the Pauli-matrix identity

$$
\operatorname{tr}(\tau^a \tau^b) = 2\delta^{ab}.
\tag{4.2.25}
$$

Thus, we wish to find a field-strength tensor that transforms under local gauge transformations G as

$$
F'_{\mu\nu} = GF_{\mu\nu}G^{-1}.
\tag{4.2.26}
$$

Note first that a transcription of the QED form is not satisfactory:

$$
\begin{aligned}
\partial_\nu B'_\mu - \partial_\mu B'_\nu &= \partial_\nu\left[GB_\mu G^{-1} + \frac{i}{g}(\partial_\mu G)G^{-1}\right] - \partial_\mu\left[GB_\nu G^{-1} + \frac{i}{g}(\partial_\nu G)G^{-1}\right] \\
&= G(\partial_\nu B_\mu - \partial_\mu B_\nu)G^{-1} + [(\partial_\nu G)B_\mu - (\partial_\mu G)B_\nu]G^{-1} \\
&\quad + G[B_\mu(\partial_\nu G^{-1}) - B_\nu(\partial_\mu G^{-1})] \\
&\quad + \frac{i}{g}[(\partial_\mu G)(\partial_\nu G^{-1}) - (\partial_\nu G)(\partial_\mu G^{-1})] \\
&\neq G(\partial_\nu B_\mu - \partial_\mu B_\nu)G^{-1}.
\end{aligned}
\tag{4.2.27}
$$

This result may be cast in slightly more symmetric form by recalling that

$$G^{-1}G = GG^{-1} = 1, \tag{4.2.28}$$

so that

$$\partial_\mu(G^{-1}G) = 0 = \partial_\mu(GG^{-1}), \tag{4.2.29}$$

whence

$$(\partial_\mu G^{-1})G = -G^{-1}(\partial_\mu G). \tag{4.2.30}$$

Judicious use of (4.2.30) in (4.2.27) then yields

$$\partial_\nu B'_\mu - \partial_\mu B'_\nu = G(\partial_\nu B_\mu - \partial_\mu B_\nu)G^{-1}$$
$$+ G\{[G^{-1}(\partial_\nu G), B_\mu] - [G^{-1}(\partial_\mu G), B_\nu]\}G^{-1}$$
$$+ \frac{1}{ig}G[(\partial_\nu G^{-1})(\partial_\mu G) - (\partial_\mu G^{-1})(\partial_\nu G)]G^{-1}, \tag{4.2.31}$$

showing manifestly that the additional terms arise from the non-Abelian group structure. It is natural to add a term to $\partial_\nu B_\mu - \partial_\mu B_\nu$ to recover the desired transformation properties.

For inspiration, we observe that the electromagnetic field-strength tensor can also be written in the form

$$F_{\mu\nu} = \frac{1}{iq}[\mathscr{D}_\nu, \mathscr{D}_\mu], \tag{4.2.32}$$

with

$$\mathscr{D}_\mu = \partial_\mu + iqA_\mu, \tag{3.5.12}$$

because

$$F_{\mu\nu} = \frac{1}{iq}[(\partial_\nu + iqA_\nu),(\partial_\mu + iqA_\mu)]$$
$$= \partial_\nu A_\mu - \partial_\mu A_\nu + iq[A_\nu, A_\mu], \tag{4.2.33}$$

where the commutator vanishes in an Abelian theory. This suggests that for the $SU(2)$ gauge theory a candidate field-strength tensor is the form

$$F_{\mu\nu} = \frac{1}{ig}[\mathscr{D}_\nu, \mathscr{D}_\mu] = \partial_\nu B_\mu - \partial_\mu B_\nu + ig[B_\nu, B_\mu]. \tag{4.2.34}$$

The commutator term transforms as

$$ig[B'_\nu, B'_\mu] = ig\left[\left(GB_\nu G^{-1} + \frac{i}{g}(\partial_\nu G)G^{-1}\right),\left(GB_\mu G^{-1} + \frac{i}{g}(\partial_\mu G)G^{-1}\right)\right]$$
$$= igG[B_\nu, B_\mu]G^{-1}$$
$$- G\{[G^{-1}(\partial_\nu G), B_\mu] - [G^{-1}(\partial_\mu G), B_\nu]\}G^{-1}$$
$$- \frac{1}{ig}G\{(\partial_\nu G^{-1})(\partial_\mu G) - (\partial_\mu G^{-1})(\partial_\nu G)\}G^{-1}, \tag{4.2.35}$$

where equation (4.2.30) has again been used to bring the result to a symmetric form. The terms after the first are precisely what is required to cancel the extra terms in (4.2.31). Thus we find that the field-strength tensor given by (4.2.34) has the desired behavior (4.2.26) under local gauge transformations. The Yang–Mills Lagrangian,

$$\mathscr{L}_{YM} = \bar{\psi}(i\gamma^\mu \mathscr{D}_\mu - m)\psi - \tfrac{1}{2}\,\mathrm{tr}\, F_{\mu\nu}F^{\mu\nu} \qquad (4.2.36)$$

is therefore invariant under local gauge transformations. Whereas a mass term $M^2 B_\mu B^\mu$ for the gauge fields is incompatible with local gauge invariance, as in electromagnetism, a common nonzero mass for the nucleons is clearly permitted.

It is of some interest to display the components of the field-strength tensor, as this will be of value for the generalization to other gauge groups. Using the definitions (4.2.6) and (4.2.23) and the commutation relations (4.2.18), it is straightforward to show that

$$F^l_{\mu\nu} = \partial_\nu b^l_\mu - \partial_\mu b^l_\nu + g\varepsilon_{jkl} b^j_\mu b^k_\nu. \qquad (4.2.37)$$

For a gauge group other than $SU(2)$, the Levi-Civita symbol will be replaced by the antisymmetric structure constants f_{jkl}.

4.3   Some Physical Consequences

It is now appropriate to consider briefly the differences between Yang–Mills theory and quantum electrodynamics, and to investigate some of the experimental consequences of the Yang–Mills Lagrangian (4.2.36). First let us note that sourceless QED is a free (or noninteracting) field theory. Only bilinear combinations of the photon gauge field A_μ occur in the QED Lagrangian (3.5.18), and thus the only object for which Feynman rules are required is the photon propagator, as indicated in Fig. 4-1. The Yang–Mills theory has a richer structure, however. Even in the absence of fermion sources there will be interactions as a consequence of the nonlinear term in $F_{\mu\nu}$. Trilinear and quadrilinear terms thus appear in $F_{\mu\nu}F^{\mu\nu}$, as well as the familiar bilinear term. Thus, in addition to the gauge field propagator the theory contains the three- and four-gauge-boson vertices displayed in Fig. 4-2. These additional interactions have numerous important physical consequences, which will command our attention later on. For the moment,

QED:

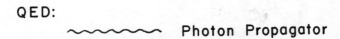

Photon Propagator

FIG. 4-1. The photon propagator of quantum electrodynamics.

SU (2):

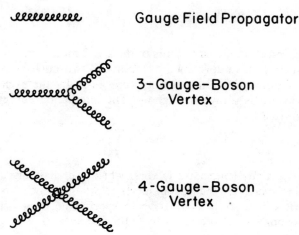

Gauge Field Propagator

3−Gauge−Boson
Vertex

4−Gauge−Boson
Vertex

FIG. 4-2. Gauge boson propagator and self-interactions in Yang–Mills theory.

it suffices to remark that they exist, and that they owe their existence to the non-Abelian structure of the gauge group.

Since the construction of the Yang–Mills theory was motivated by the search for an isospin-conserving theory of nuclear interactions, let us analyze some consequences of the theory for the interactions among fermions. The interaction term

$$\mathscr{L}_{\text{int}} = -\frac{g}{2}b_\mu^a\bar{\psi}\gamma^\mu\tau^a\psi \tag{4.3.1}$$

leads by the usual procedure to the Feynman rule for the nucleon–nucleon–gauge boson vertex. For the transition depicted in Fig. 4-3 of a nucleon with isospin label $\alpha\,(=p,n$ or $1,2)$ into a nucleon with isospin label β and a gauge boson with Lorentz index μ and isospin label $a\,(=1,2,3)$, the vertex factor is

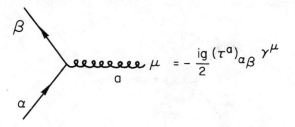

$$= -\frac{ig}{2}(\tau^a)_{\alpha\beta}\,\gamma^\mu$$

FIG. 4-3. Feynman rule for the nucleon–gauge field interaction in Yang–Mills theory.

simply

$$-\frac{ig}{2}(\tau^a)_{\alpha\beta}\gamma^\mu, \tag{4.3.2}$$

where α and β label components of the 2×2 matrix τ^a.

What does this imply for the stability of two-nucleon systems? Suppose the two-body nuclear interaction were given by the exchange of a single gauge boson, as shown in Fig. 4-4. The interaction energy would then be characterized by

$$\mathscr{E} = \frac{g^2}{4} \sum_a \tau^a_{\alpha\beta}\tau^a_{\gamma\delta} \tag{4.3.3}$$

times essentially kinematic factors, which we shall discuss momentarily. Instead of specifying the isospin labels $\alpha\beta\gamma\delta$ of the nucleons before and after interaction, it is more instructive to specify the two-nucleon state by its total isospin, 0 or 1, and to evaluate the average

$$\langle \tau^{(1)} \cdot \tau^{(2)} \rangle = \frac{\langle \tau^2 \rangle - \langle \tau^{(1)2} \rangle - \langle \tau^{(2)2} \rangle}{2}, \tag{4.3.4}$$

where $\tau^{(1,2)}$ refer to individual nucleons and τ refers to the composite. For nucleons of isospin $1/2$, we then find

$$\begin{aligned} \mathscr{E}(I = 0) &= -\tfrac{3}{4} \\ \mathscr{E}(I = 1) &= +\tfrac{1}{4}, \end{aligned} \tag{4.3.5}$$

times common factors. The interaction is attractive in the isoscalar channel and repulsive in the isovector channel. If the Yang–Mills theory were an otherwise plausible theory of nucleon structure, this analysis would suggest an explanation for the fact that the only bound two-nucleon state is the (isoscalar) deuteron.

The applicability of lowest-order perturbation theory to a problem in strong-interaction dynamics is open to challenge because the coupling constant that would play the role of an expansion parameter is of order 1 at least; the nuclear forces are strong. However, it is not much of an over-

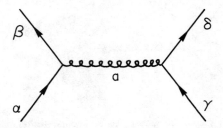

FIG. 4-4. Model for the nucleon–nucleon interaction in $SU(2)$ gauge theory.

statement to say that perturbation theory is all we know how to do. It may
also be reasonable to hope that the systematics signaled by the Born term
faithfully represent the pattern displayed by the full theory. At any rate, this
hope is a recurrent one. The elementary calculation just concluded has been
applied to other questions, such as the bound-state problem in quantum
chromodynamics, under the name of a "maximally attractive channel analysis." Nevertheless, the legitimate concern over the utility of perturbation
theory is beside the point in the situation at hand, because there is a direct
contradiction between the Yang–Mills theory and experimental nuclear
physics. As a consequence of local gauge invariance, the Yang–Mills quanta
b_μ are massless vector bosons, which implies that the force mediated by them
should be of infinite range, like the Coulomb interaction. On the contrary, a
successful phenomenological description[3] of the nuclear force involves the
exchange of massive particles: the pseudoscalar pion and the vector mesons
ρ^+, ρ^0, ρ^-, and ω. Thus the theory cannot serve the purpose for which it was
conceived.

4.4   Assessment

In this chapter and the preceding one, the examples of electromagnetism and
the Yang–Mills theory have shown how gauge principles may be used to
guide the construction of theories. Global gauge invariance implies the
existence of a conserved current, according to Noether's theorem. Local
gauge invariance requires the introduction of massless vector gauge bosons,
prescribes (or, more properly, restricts) the form of the interactions of gauge
bosons with sources, and generates interactions among the gauge bosons if
the symmetry is non-Abelian.

It is appealing to try to make use of the observed symmetries of Nature as
gauge symmetries, to exploit the potent idea that symmetries define interactions. This is indeed the course we shall follow in later applications.
However, the example of the Yang–Mills theory shows that, whereas
mathematical consistency is readily achieved, experimental success is not
assured in advance for every would-be gauge group. Long-range nuclear
forces mediated by massless vector quanta are not observed. Therefore the
Yang–Mills theory, based on the idea of isospin invariance, or flavor symmetry, as the strong-interaction gauge group is incorrect, or at least
incomplete.

If we are reluctant to abandon such an elegant theory, several courses are
open: the choice of a new gauge group, the reinterpretation of the theory,
and the evasion of massless gauge bosons. We shall make use of all three.
By finding a different hadronic symmetry to which the mathematical structure
of the Yang–Mills construction may be applied, we shall be led to quantum
chromodynamics. It is also natural to reinterpret the theory constructed in

this chapter as a theory of weak interactions, based upon the "weak-isospin" symmetry apparent in nuclear β-decay, but this attempt would also founder on the prediction of massless gauge fields. Before we proceed to specific applications to the fundamental interactions, it will be important to understand how to circumvent the prediction of massless gauge bosons, while preserving the local gauge invariance of the Lagrangian as a desirable restricting principle. This is the subject of the following chapter.

Problems

4-1.   Derive the Yang–Mills Lagrangian for a scalar field theory in which the three real scalar fields correspond to the triplet representation of $SU(2)$. The free-particle Lagrangian is

$$\mathcal{L} = \tfrac{1}{2}[(\partial_\mu \phi)^2 - m^2 \phi^2],$$

with

$$\phi = \begin{pmatrix} \phi_1 \\ \phi_2 \\ \phi_3 \end{pmatrix}.$$

4-2.   (a) By making the minimal substitution $\partial_\mu \rightarrow \mathcal{D}_\mu$ in the free-particle Lagrangian, construct a theory of the electrodynamics of a massive spin-one boson $V^\pm$ and deduce the Feynman rules for the theory.

(b) Now compute the differential cross section $d\sigma/d\Omega$ and the total (integrated) cross section $\sigma \equiv \int d\Omega (d\sigma/d\Omega)$ for the reaction $e^+ e^- \rightarrow V^+ V^-$, for various helicities of the produced particles. Work in the c.m. frame, and in the high-energy limit (where all the masses may be neglected). Assume the colliding beams are unpolarized. Compare the results with the cross sections for $e^+ e^- \rightarrow \sigma^+ \sigma^-$ and $e^+ e^- \rightarrow \mu^+ \mu^-$ derived in Problems 1-4 and 1-5. [Reference for Feynman rules: J. D. Bjorken and S. D. Drell, *Relativistic Quantum Mechanics*, McGraw-Hill, New York, 1964, Appendix B.]

4-3.   Show that the transformations (4.2.10) of the gauge fields form a group—that is, that under successive transformations on the matter fields given by $\psi \rightarrow \psi' = G_1 \psi \rightarrow \psi'' = G_2 \psi'$ the transformation of the gauge field is characterized by $G = G_2 G_1$.

4-4.   Apply the analysis of nucleon-nucleon forces of Section 4.3 to the nucleon–antinucleon case. Compare the Yang–Mills "predictions" with the spectrum of nonstrange mesons.

For Further Reading

NON-ABELIAN GAUGE THEORIES. A general discussion, similar in tone to the one in this chapter, may be found in

E. S. Abers and B. W. Lee, *Phys. Rep.* **9C**, 1 (1973).

A more abstract geometrical approach, which usefully complements this chapter, is followed by

C. Itzykson and J.-B. Zuber, *Quantum Field Theory*, McGraw-Hill, New York, 1980, Chapter 12.

A construction that does not make use of the Lagrangian formalism is given by

I. J. R. Aitchison and A. J. G. Hey, *Gauge Theories in Particle Physics*, Adam Hilger, Bristol, 1982, Section 8.5.

GENERAL GAUGE GROUPS. The Yang–Mills strategy of constructing a theory from a local symmetry group was elaborated by

R. Utiyama, *Phys. Rev.* **101**, 1597 (1956).

How to build a theory on a general gauge group is explained by

S. L. Glashow and M. Gell-Mann, *Ann. Phys. (NY)* **15**, 437 (1961).

GAUGE THEORIES OF THE STRONG INTERACTIONS. Following the work of Yang and Mills, there have been many attempts to construct phenomenologically acceptable strong-interaction theories based upon flavor symmetry groups. Among these, see

Y. Fujii, *Prog. Theor. Phys. (Kyoto)* **21**, 232 (1959).

J. J. Sakurai, *Ann. Phys. (NY)* **11**, 1 (1960).

Y. Ne'eman, *Nucl. Phys.* **26**, 222 (1961).

The paper by Sakurai is quite remarkable, more for what is said than for what is done. The one by Ne'eman is better known for the proposal that the flavor symmetry be $SU(3)$.

References

[1] C. N. Yang and R. L. Mills, *Phys. Rev.* **96**, 191 (1954).

[2] R. Shaw, "The Problem of Particle Types and Other Contributions to the Theory of Elementary Particles," Cambridge University Thesis, 1955.

[3] General features of the nucleon–nucleon interaction are described in A. Bohr and B. Mottelson, *Nuclear Structure*, Benjamin, Reading, Massachusetts, 1969, Vol. I, Section 2-5.

CHAPTER 5

HIDDEN SYMMETRIES

Much importance has been attached to symmetry principles in the preceding chapters. We have seen the connection between exact symmetries and conservation laws and have found that the requirement of local gauge invariance can serve as a dynamical principle to guide the construction of interacting field theories. Although a certain economy and mathematical elegance has thus been achieved, the results of the program to this point are unsatisfactory in several important respects. First, the gauge principle has led us to theories in which all the interactions are mediated by massless vector bosons, whereas only a single massless vector boson, the photon, is apparent in Nature. Second, the algorithm for constructing gauge theories that we have developed applies only to exact symmetries, whereas Nature exhibits numerous symmetries that are only approximate. Third, there are many situations in physics in which the exact symmetry of an interaction is concealed by circumstances. The canonical example, which will be elaborated below, is that of the Heisenberg ferromagnet, an infinite crystalline array of spin-1/2 magnetic dipoles. Below the Curie temperature the ground state is a completely ordered configuration in which all dipoles are aligned in some arbitrary direction, belying the rotation invariance of the underlying interaction. It is thus of interest to learn how to deal with symmetries that are not exact or not manifest, perhaps in the hope of evading the conclusion that interactions must be mediated by massless gauge bosons.

In this chapter we shall therefore analyze more thoroughly than we have before the various types of symmetries. The distinction between internal symmetries and the geometrical symmetries that involve coordinate transformations (such as the Poincaré invariance of relativistic theories) has already been drawn in Chapter 2. There, too, it was observed that continuous symmetries imply local conservation laws through Noether's theorem, whereas discrete symmetries do not. A related difference between discrete and continuous symmetries will be uncovered in the discussion that follows. Most of our attention, however, will be concentrated on the distinction between exact and approximate symmetries. Among approximate symmetries, several different realizations are possible. The Lagrangian may display an imperfect or explicitly broken symmetry, or it may happen that the Lagrangian is symmetric but the physical vacuum does not respect the symmetry. In the latter case, the symmetry of the Lagrangian is said to be spontaneously broken. The various possibilities will be illustrated in what follows.

Our principal concerns in this chapter will be the conditions under which a symmetry is spontaneously broken, and the consequences of spontaneous symmetry breakdown. We shall find that if the Lagrangian of a theory is invariant under an exact continuous symmetry that is not a symmetry of the physical vacuum, one or more massless spin-zero particles, known as Goldstone bosons, must occur. From the point of view of unobserved massless particles, this would seem to double our trouble: Gauge theories lead to unwanted massless vector bosons, and the spontaneous breakdown of a continuous symmetry implies the existence of unwanted spinless particles. However, if the spontaneously broken symmetry is a local gauge symmetry, a miraculous interplay between the would-be Goldstone boson and the normally massless gauge bosons endows the gauge bosons with mass and removes the Goldstone boson from the spectrum. The Higgs mechanism, by which this interplay occurs, is a central ingredient in our current understanding of the gauge bosons of the weak interactions.

The results of this chapter all can be codified and presented in a formal and even axiomatic manner. This was of some importance for the development of the subject because the precise statement of a theorem makes it possible to understand both its generality and its limitations. Our purposes will be better served, however, by considering a number of specific examples.

5.1   The Idea of Spontaneously Broken Symmetries

The physical world manifests a number of apparently exact conservation laws, which we believe reflect the operation of exact symmetries of Nature. These include the conservation of energy and momentum, of angular momentum, and of electric charge. In the language of Lagrangian field theory, exact symmetry is characterized by two conditions. The Lagrangian (density) is invariant under the symmetry in question,

$$\delta \mathscr{L} = 0, \tag{5.1.1}$$

and the unique physical vacuum is invariant under the symmetry transformations. From these requirements a standard analysis demonstrates the
mass degeneracy of particle multiplets.

Many of the useful internal symmetries, such as the flavor symmetries of
isospin and $SU(3)$ and the conservation of strangeness and charm, hold only
approximately. It is usual to treat these approximate symmetries by writing
the Lagrangian as

$$\mathscr{L} = \mathscr{L}_{\text{symmetric}} + \varepsilon\mathscr{L}_{\text{symmetry breaking}}. \tag{5.1.2}$$

This form is particularly useful if the symmetry-breaking term is small, in
some sense, and can be treated as a perturbation upon the symmetric interaction. The perturbation lifts the degeneracy of particle multiplets, with the
resulting intermultiplet splitting being a function of the parameter ε, which
vanishes as $\varepsilon \to 0$. A familiar example is

$$\mathscr{L} = \mathscr{L}_{\text{strong}} + \mathscr{L}_{\text{EM}}, \tag{5.1.3}$$

in which the strong-interaction Lagrangian is isospin-invariant and the
responsibility for isospin violations resides in the electromagnetic term
$\mathscr{L}_{\text{EM}}$. This characterization of explicit symmetry breaking lies behind the
conventional view of a hierarchy of strong, electromagnetic, and weak
interactions, which derives much of its utility from the fact that the dominant
interaction respects the largest group of symmetries.

We saw by example in Chapter 2 how continuous symmetries of the
Lagrangian lead to exact conservation laws. Approximate conservation laws
may arise if, as in the preceding paragraph, the Lagrangian is imperfectly
symmetric. It may also happen that the Lagrangian $\mathscr{L}$ is exactly invariant
under some symmetry, so that (5.1.1) holds, but that the dynamics determined
by $\mathscr{L}$ imply a degenerate set of vacuum states, which are not invariant under
the symmetry. This leads to exact local conservation laws (if not to conserved
charges), but conceals the symmetry of the theory—for example, by breaking
the mass degeneracy of particle multiplets. Finally, one may imagine a hybrid
situation in which an interaction that gives rise to a spontaneously broken
symmetry is accompanied by an explicit symmetry-breaking interaction.

Each of these four situations can be illustrated by the infinite ferromagnet,
to which we have already alluded. The nearest-neighbor interaction between
spins, or magnetic dipole moments, is invariant under the group of spatial
rotations $SO(3)$. In the disordered, or paramagnetic, phase, which exists above
the Curie temperature T_{C}, the medium displays an exact symmetry in the
absence of an external field. The spontaneous magnetization of the system
is zero, and there is no preferred direction in space, so the $SO(3)$ invariance
is manifest. A privileged direction may be selected by imposing an external
magnetic field, which tends to align the spins in the material. The $SO(3)$

symmetry is thus broken down to an axial $SO(2)$ symmetry of rotations about the external field direction. The full symmetry is restored when the external field is turned off.

For temperatures below T_C, when the system is in the ferromagnetic or ordered phase, the situation is rather different. In the absence of an impressed field, the configurations of lowest energy have a nonzero spontaneous magnetization because the nearest-neighbor force favors the parallel alignment of spins. In these circumstances the $SO(3)$ symmetry is said to be spontaneously broken down to $SO(2)$. The fact that the direction of the spontaneous magnetization is random and the fact that the measurable properties of the infinite ferromagnet do not depend upon its orientation are the vestiges of the original $SO(3)$ symmetry. The ground state is thus infinitely degenerate. A particular direction for the spontaneous magnetization may be chosen by imposing an external field which breaks the $SO(3)$ symmetry explicitly. In contrast to the paramagnetic case, however, the spontaneous magnetization does not return to zero when the applied field is turned off. For the rotational invariance to be broken spontaneously, it is crucial that the ferromagnet be infinite in extent, so that rotation from one degenerate ground state to another would require the impossible task of rotating an infinite number of elementary dipoles. This is the picturesque analog of the statement in field theory that the degenerate vacua lie in distinct Hilbert spaces.

After this intuitive discussion, we are prepared to see how spontaneous symmetry breaking may arise in a simple mathematical model. Consider a Lagrangian for a self-interacting real scalar field ϕ, which may be written in the form

$$\mathcal{L} = \tfrac{1}{2}(\partial_\mu \phi)(\partial^\mu \phi) - V(\phi). \tag{5.1.4}$$

How does the nature of the vacuum, and thus of the particle spectrum, depend upon the effective potential $V(\phi)$? If the potential is an even functional of the scalar field ϕ,

$$V(\phi) = V(-\phi), \tag{5.1.5}$$

then the Lagrangian (5.1.4) is invariant under the parity transformation

$$\phi \to -\phi. \tag{5.1.6}$$

To explore the possibilities, it is convenient to consider an explicit potential

$$V(\phi) = \tfrac{1}{2}\mu^2\phi^2 + \tfrac{1}{4}|\lambda|\phi^4. \tag{5.1.7}$$

The quadratic term has the familiar appearance of the mass term of scalar field theory. The positive coefficient of the quartic term is chosen to ensure stability against unbounded oscillations. Higher powers than the fourth are omitted in order that the theory be renormalizable.

Two cases, which correspond to manifest or spontaneously broken symmetry, can now be distinguished. If the parameter $\mu^2 > 0$, the potential (5.1.7) has a unique minimum at $\phi = 0$, as shown in Fig. 5-1(a), which corresponds to the vacuum state. This identification is perhaps most easily made in the Hamiltonian formalism. The Hamiltonian density is given by

$$\mathcal{H} = \pi\dot{\phi} - \mathcal{L}, \qquad (5.1.8)$$

where

$$\dot{\phi} = \partial_0\phi \qquad (5.1.9)$$

and the canonical momentum is

$$\pi \equiv \partial\mathcal{L}/\partial\dot{\phi}. \qquad (5.1.10)$$

In the case at hand we therefore have

$$\mathcal{H} = \tfrac{1}{2}[(\partial_0\phi)^2 + (\nabla\phi)^2] + V(\phi). \qquad (5.1.11)$$

The state of lowest energy is thus seen to be one for which the value of the field ϕ is a constant, which we denote by $\langle\phi\rangle_0$. The value of this constant is determined by the dynamics of the theory; it corresponds to the absolute minimum (or minima) of the potential $V(\phi)$. We shall refer to $\langle\phi\rangle_0$ as the vacuum expectation value of the field ϕ. If the parameter μ^2 is positive, the minimum of the potential (5.1.7) is at

$$\langle\phi\rangle_0 = 0. \qquad (5.1.12)$$

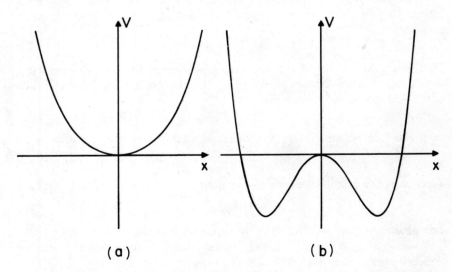

(a)   (b)

FIG. 5-1. (a) Ordinary effective potential with a unique minimum at $\phi = 0$. (b) Potential with a degenerate vacuum, corresponding to a spontaneously broken symmetry.

The approximate form of the Lagrangian appropriate to the study of small oscillations around this minimum is then

$$\mathcal{L}_{\text{s.o.}} = \tfrac{1}{2}[(\partial_\mu \phi)(\partial^\mu \phi) - \mu^2 \phi^2], \tag{5.1.13}$$

which is that of a free particle with mass $= \mu$.

If $\mu^2 < 0$, the situation is that of a spontaneously broken symmetry. The potential

$$V(\phi) = -\tfrac{1}{2}|\mu^2|\phi^2 + \tfrac{1}{4}|\lambda|\phi^4, \tag{5.1.14}$$

shown in Fig. 5-1(b), has minima at

$$\langle\phi\rangle_0 = \pm\sqrt{-\mu^2/|\lambda|} \equiv \pm v, \tag{5.1.15}$$

which correspond to two degenerate lowest-energy states, either of which may be chosen to be the vacuum. Because of the parity invariance of the Lagrangian, the ensuing physical consequences must be independent of this choice. Whatever is the choice, however, the symmetry of the theory is spontaneously broken; the parity transformation (5.1.6) is then an invariance of the Lagrangian, but not of the vacuum state. Let us choose

$$\langle\phi\rangle_0 = +v, \tag{5.1.16}$$

and define a shifted field

$$\phi' \equiv \phi - \langle\phi\rangle_0 = \phi - v, \tag{5.1.17}$$

so that the vacuum state corresponds to

$$\langle\phi'\rangle_0 = 0. \tag{5.1.18}$$

In terms of the shifted field the Lagrangian is

$$\mathcal{L} = \frac{1}{2}(\partial_\mu \phi')(\partial^\mu \phi') - |\mu^2|\left(\frac{\phi'^4}{4v^2} + \frac{\phi'^3}{v} + \phi'^2 - \frac{v^2}{4}\right), \tag{5.1.19}$$

which has no manifest symmetry properties with respect to the shifted field ϕ'. For small oscillations about the vacuum, we have

$$\mathcal{L}_{\text{s.o.}} = \tfrac{1}{2}[(\partial_\mu \phi')(\partial^\mu \phi') - 2|\mu^2|\phi'^2] \tag{5.1.20}$$

(plus an irrelevant constant), which describes the oscillation of a particle with (mass)$^2 = 2|\mu^2| = -2\mu^2 > 0$.

This simple example has illustrated how spontaneous symmetry breaking occurs when a symmetry of the Lagrangian is not respected by the vacuum state, defined as the state of lowest energy. The same methods of choosing a vacuum from among a degenerate set of vacua and discovering the particle spectra apply equally well to more complicated and more interesting physical situations, to which we now turn.

5.2 Spontaneous Breaking of Continuous Symmetries

The leap to the spontaneous breaking of continuous symmetries is easily made by considering a model first investigated by Goldstone,[1] based on the Lagrangian for two scalar fields ϕ_1 and ϕ_2,

$$\mathscr{L} = \tfrac{1}{2}[(\partial_\mu\phi_1)(\partial^\mu\phi_1) + (\partial_\mu\phi_2)(\partial^\mu\phi_2)] - V(\phi_1^2 + \phi_2^2). \qquad (5.2.1)$$

The Lagrangian is invariant under the group $SO(2)$ of rotations in the plane

$$\phi \equiv \begin{pmatrix} \phi_1 \\ \phi_2 \end{pmatrix} \rightarrow \begin{pmatrix} \cos\theta & \sin\theta \\ -\sin\theta & \cos\theta \end{pmatrix}\begin{pmatrix} \phi_1 \\ \phi_2 \end{pmatrix}. \qquad (5.2.2)$$

As we have done before, we consider the effective potential

$$V(\phi^2) = \tfrac{1}{2}\mu^2\phi^2 + \tfrac{1}{4}|\lambda|(\phi^2)^2, \qquad (5.2.3)$$

where $\phi^2 = \phi_1^2 + \phi_2^2$, and distinguish two cases.

A positive value of the parameter $\mu^2 > 0$ corresponds to the ordinary case of exact symmetry. The unique vacuum occurs at

$$\langle\phi\rangle = \begin{pmatrix} 0 \\ 0 \end{pmatrix}, \qquad (5.2.4)$$

and so for small oscillations the Lagrangian takes the form

$$\mathscr{L}_{\text{s.o.}} = \tfrac{1}{2}[(\partial_\mu\phi_1)(\partial^\mu\phi_1) - \mu^2\phi_1^2] + \tfrac{1}{2}[(\partial_\mu\phi_2)(\partial^\mu\phi_2) - \mu^2\phi_2^2]. \quad (5.2.5)$$

This is none other than the Lagrangian (2.2.4) for a pair of scalar particles with common mass μ. Thus the introduction of a symmetric interaction preserves the degenerate multiplet structure of the free theory, in accord with our simplest expectations.

The choice $\mu^2 < 0$ leads to a spontaneous breakdown of the $SO(2)$ symmetry. The absolute minimum of the potential now occurs for

$$\langle\phi\rangle_0^2 = -\mu^2/|\lambda| \equiv v^2, \qquad (5.2.6)$$

which corresponds to a continuum of distinct vacuum states, degenerate in energy. The degeneracy is of course a consequence of the $SO(2)$ symmetry of the effective potential. Let us select as the physical vacuum state the configuration

$$\langle\phi\rangle_0 = \begin{pmatrix} v \\ 0 \end{pmatrix}, \qquad (5.2.7)$$

as we may always do with a suitable definition of coordinates. Expanding about the vacuum configuration by defining

$$\phi' \equiv \phi - \langle\phi\rangle_0 \equiv \begin{pmatrix} \eta \\ \zeta \end{pmatrix}, \qquad (5.2.8)$$

we obtain the Lagrangian for small oscillations

$$\mathscr{L}_{\text{s.o.}} = \tfrac{1}{2}[(\partial_\mu\eta)(\partial^\mu\eta) + 2\mu^2\eta^2] + \tfrac{1}{2}[(\partial_\mu\zeta)(\partial^\mu\zeta)], \qquad (5.2.9)$$

plus an irrelevant constant. There are two particles in the spectrum. The η-particle, associated with radial oscillations, has $(\text{mass})^2 = 2\mu^2 > 0$, just as we found for the particle in the case of a spontaneously broken parity invariance. The ζ-particle, however, is massless. The mass of the η-particle may be viewed as a consequence of the restoring force of the potential against radial oscillations. In contrast, the masslessness of ζ is a consequence of the $SO(2)$ invariance of the Lagrangian, which means that there is no restoring force against angular oscillations. The splitting of the spectrum and the appearance of the massless particle are known as the Goldstone phenomenon. The massless particles, which are referred to as Goldstone bosons, are the zero-energy excitations that connect possible vacua. In the general case, one massless spin-zero particle will occur for each broken generator of the original symmetry group—that is, for each generator that connects distinct vacuum states. This is illustrated by further examples in Problems 5-1 and 5-2.

The splitting of the spectrum and the appearance of massless spinless bosons is a quite general consequence of the spontaneous breakdown of a continuous symmetry. This statement has attained the status of a theorem.[2] In any field theory that obeys the "usual axioms," including locality, Lorentz invariance, and positive-definite norm on the Hilbert space, if an exact continuous symmetry of the Lagrangian is not a symmetry of the physical vacuum, then the theory must contain a massless spin-zero particle (or particles) whose quantum numbers are those of the broken group generator (or generators).

When coupled with the nonobservation of massless scalars or pseudo-scalars (except perhaps for the pion), this strong statement would seem to preclude the use of spontaneous symmetry breaking in realistic theories. However, as Higgs[3] was first to notice, gauge theories do not satisfy the assumptions on which the theorem is based, although they are respectable field theories. To quantize electrodynamics, for example, one must choose between the covariant Gupta–Bleuler formalism with its unphysical indefinite metric states or quantization in a physical gauge wherein manifest covariance is lost. With no further ado, let us next investigate the spontaneous breaking of a gauge symmetry.

5.3 The Higgs Mechanism

We consider locally gauge-invariant Lagrangians that give rise to spontaneously broken symmetries. There will arise an unexpected cooperation between the massless gauge fields (such as made the Yang–Mills theory an unacceptable description of the strong interactions of nucleons) and the massless Goldstone bosons that have been seen to accompany spontaneous symmetry breaking. The simplest example of the Higgs phenomenon, as this interplay is known, is provided by the Abelian Higgs model, which is nothing but the locally gauge-invariant extension of the Goldstone model discussed in the previous section: a $U(1)$-invariant theory that describes, in the absence of spontaneous symmetry breaking, the electrodynamics of charged scalars. The Lagrangian is simply

$$\mathscr{L} = |\mathscr{D}^\mu \phi|^2 - \mu^2|\phi|^2 - |\lambda|(\phi^*\phi)^2 - \tfrac{1}{4}F_{\mu\nu}F^{\mu\nu}, \tag{5.3.1}$$

where

$$\phi = \frac{\phi_1 \pm i\phi_2}{\sqrt{2}} \tag{5.3.2}$$

is a complex scalar field[4] and as usual

$$\mathscr{D}_\mu \equiv \partial_\mu + iqA_\mu \tag{5.3.3}$$

and

$$F_{\mu\nu} \equiv \partial_\nu A_\mu - \partial_\mu A_\nu. \tag{5.3.4}$$

The Lagrangian (5.3.1) is invariant under $U(1)$ rotations

$$\phi \to \phi' = e^{i\theta}\phi \tag{5.3.5}$$

and under the local gauge transformations

$$\begin{aligned}\phi(x) &\to \phi'(x) = e^{iq\alpha(x)}\phi(x),\\ A_\mu(x) &\to A'_\mu(x) = A_\mu(x) - \partial_\mu\alpha(x).\end{aligned} \tag{5.3.6}$$

As usual, there are two cases, depending upon the parameters of the effective potential.

For $\mu^2 > 0$, the potential has a unique minimum at $\phi = 0$ and the exact symmetry of the Lagrangian is preserved. The spectrum is simply that of ordinary QED of charged scalars, with a single massless photon A_μ and two scalar particles, ϕ and ϕ^*, with common mass μ.

The situation when $\mu^2 = -|\mu^2| < 0$ is that of spontaneously broken symmetry and requires a closer analysis. The potential has a continuum of absolute minima, corresponding to a continuum of degenerate vacua, at

$$\langle|\phi|^2\rangle_0 = -\mu^2/2|\lambda| \equiv v^2/2. \tag{5.3.7}$$

To explore the spectrum, we shift the fields in order to rewrite the Lagrangian (5.3.1) in terms of displacements from the physical vacuum. The latter may be chosen, without loss of generality, as

$$\langle\phi\rangle_0 = v/\sqrt{2}, \tag{5.3.8}$$

where $v > 0$ is a real number. We then define the shifted field

$$\phi' = \phi - \langle\phi\rangle_0, \qquad (5.3.9)$$

which is conveniently parametrized in terms of

$$\phi = e^{i\zeta/v}(v + \eta)/\sqrt{2}$$
$$\simeq (v + \eta + i\zeta)/\sqrt{2}. \qquad (5.3.10)$$

Then the Lagrangian appropriate for the study of small oscillations is

$$\mathscr{L}_{\text{s.o.}} = \frac{1}{2}[(\partial_\mu\eta)(\partial^\mu\eta) + 2\mu^2\eta^2] + \frac{1}{2}[(\partial^\mu\zeta)(\partial_\mu\zeta)]$$
$$- \frac{1}{4}F_{\mu\nu}F^{\mu\nu} + qvA_\mu(\partial^\mu\zeta) + \frac{q^2v^2}{2}A_\mu A^\mu + \cdots \qquad (5.3.11)$$

As we expect from our study of the Goldstone phenomenon, the η-field, which corresponds to radial oscillations, has a $(\text{mass})^2 = 2\mu^2 > 0$. The gauge field A_μ appears to have acquired a mass, but it is mixed up in the penultimate term with the seemingly massless ζ-field.

An astute choice of gauge will make it easier to sort out the spectrum of the spontaneously broken theory. To this end, it is convenient to rewrite the terms involving A_μ and ζ as

$$\frac{q^2v^2}{2}\left(A_\mu + \frac{1}{qv}\partial_\mu\zeta\right)\left(A^\mu + \frac{1}{qv}\partial^\mu\zeta\right), \qquad (5.3.12)$$

a form that pleads for the gauge transformation

$$A_\mu \to A'_\mu = A_\mu + \frac{1}{qv}\partial_\mu\zeta, \qquad (5.3.13)$$

which corresponds to the phase rotation on the scalar field

$$\phi \to \phi' = e^{-i\zeta(x)/v}\phi(x) = (v + \eta)/\sqrt{2}. \qquad (5.3.14)$$

Knowing that the Lagrangian is locally gauge-invariant, we may return to the original expression (5.3.1) to compute

$$\mathscr{L}_{\text{s.o.}} = \frac{1}{2}[(\partial_\mu\eta)(\partial^\mu\eta) + 2\mu^2\eta^2] - \frac{1}{4}F_{\mu\nu}F^{\mu\nu} + \frac{q^2v^2}{2}A'_\mu A'^\mu, \qquad (5.3.15)$$

plus an irrelevant constant. In this gauge the particle spectrum is manifest:

- an η-field, with $(\text{mass})^2 = -2\mu^2 > 0$;
- a massive vector field A'_μ, with mass $= qv$;
- no ζ-field.

By virtue of our choice of gauge, the ζ-particle has disappeared from the Lagrangian entirely. Where has it gone? The gauge transformation (5.3.13) shows that what was formerly the ζ-field is responsible for the longitudinal component of the massive vector field A'_μ. Before spontaneous symmetry breaking, the theory had four particle degrees of freedom: two scalars ϕ and ϕ^* plus two helicity states of the massless gauge field A_μ. After spontaneous symmetry breaking, we are left with one scalar particle η plus three helicity states of the massive gauge field A'_μ, for a total of four particle degrees of freedom. In common parlance it is said that the massless photon "ate" the massless Goldstone boson to become a massive vector boson. The remaining massive scalar (η) is known as the Higgs boson. The special gauge in which the particle spectrum became transparent is known as the unitary gauge or U-gauge, because only physical states appear in the Lagrangian. This is not to say that the unitarity of the S-matrix, or the complete set of Feynman rules for the theory, is obvious in this gauge.

This is a truly remarkable result, suggesting as it does the possibility of constructing spontaneously broken gauge theories in which the interactions are mediated by massive vector bosons, rather than the phenomenologically unacceptable massless vector bosons of the unbroken theories. In a sense, each of the massless particle diseases that we have encountered—gauge bosons and Goldstone bosons—has provided the cure for the other.

5.4   Spontaneous Breakdown of a Non-Abelian Symmetry

To approach the additional complications that attend the spontaneous breakdown of a non-Abelian gauge symmetry, we choose as a useful prototype an $SU(2)$ gauge theory and study scalar fields that make up the triplet (or isovector) representation

$$\phi = \begin{pmatrix} \phi_1 \\ \phi_2 \\ \phi_3 \end{pmatrix}. \tag{5.4.1}$$

Construction of the unbroken gauge theory was the subject of Problem 4-1. We require invariance under the gauge transformation

$$\phi \to \phi' = \exp(i\mathbf{T} \cdot \boldsymbol{\alpha})\phi, \tag{5.4.2}$$

where the exponential factor is a 3×3 matrix. The operator T_i generates isospin rotations about the i-axis and satisfies the usual $SU(2)$ algebra

$$[T^j, T^k] = i\varepsilon_{jkl}T^l. \tag{5.4.3}$$

The explicit matrix representation is

$$(T^j)_{kl} = -i\varepsilon_{jkl}. \tag{5.4.4}$$

As usual, the covariant derivative takes the form

$$\mathscr{D}_\mu = I\partial_\mu + ig\mathbf{T} \cdot \mathbf{b}_\mu \tag{5.4.5}$$

or, in isospin-component form,

$$(\mathscr{D}_\mu)_{kl} = \delta_{kl}\partial_\mu + g\varepsilon_{jkl}b_\mu^j. \tag{5.4.6}$$

In the presence of an effective potential, the Lagrangian is

$$\mathscr{L} = \tfrac{1}{2}(\mathscr{D}_\mu\boldsymbol{\phi}) \cdot (\mathscr{D}^\mu\boldsymbol{\phi}) - V(\boldsymbol{\phi} \cdot \boldsymbol{\phi}) - \tfrac{1}{4}\mathbf{F}_{\mu\nu} \cdot \mathbf{F}^{\mu\nu}. \tag{5.4.7}$$

When $\boldsymbol{\phi} = 0$ is a unique minimum of the effective potential V, the spectrum is that of an ordinary, isospin-conserving Yang–Mills field theory: three massive scalar mesons, each with mass μ, and three massless gauge bosons $\mathbf{b}_\mu$. The number of particle degrees of freedom is thus $3 \times 1 + 3 \times 2 = 9$.

Of more interest in the present context is the spontaneously broken case, in which the field configuration that minimizes $V(\boldsymbol{\phi}^2)$ may be chosen as

$$\langle\boldsymbol{\phi}\rangle_0 = \begin{pmatrix} 0 \\ 0 \\ v \end{pmatrix}. \tag{5.4.8}$$

We shift the scalar fields and expand about the minimum configuration, using

$$\boldsymbol{\phi} = \exp\left[\frac{i}{v}(\zeta_1 T_1 + \zeta_2 T_2)\right]\begin{pmatrix} 0 \\ 0 \\ v + \eta \end{pmatrix}. \tag{5.4.9}$$

It is unnecessary to retrace step by step the arithmetic of the Abelian Higgs model of Section 5.3. We may immediately exploit the gauge invariance of the theory by transforming to U-gauge, by letting

$$\boldsymbol{\phi} \to \boldsymbol{\phi}' = \exp\left[\frac{-i}{v}(\zeta_1 T_1 + \zeta_2 T_2)\right]\boldsymbol{\phi}$$

$$= \begin{pmatrix} 0 \\ 0 \\ v + \eta \end{pmatrix}, \tag{5.4.10}$$

with the implied transformation on the gauge fields. In the new gauge, the Lagrangian appropriate to the description of small oscillations about the minimum is

$$\mathscr{L}_{\text{s.o.}} = \frac{1}{2}\left[(\partial_\mu\eta)(\partial^\mu\eta) + 2\mu^2\eta^2\right] - \frac{1}{4}\mathbf{F}_{\mu\nu} \cdot \mathbf{F}^{\mu\nu}$$

$$+ \frac{g^2v^2}{2}(b_\mu^1 b^{1\mu} + b_\mu^2 b^{2\mu}) + \cdots \tag{5.4.11}$$

In this form the Lagrangian reveals that

- η has become a massive Higgs scalar, with $(\text{mass})^2 = -2\mu^2 > 0$;
- the would-be Goldstone bosons ζ_1 and ζ_2 have disappeared entirely; i.e., they have been "gauged away";
- the vector bosons b_μ^1 and b_μ^2 corresponding to the (broken symmetry) generators T_1 and T_2 acquire a common mass gv;
- the gauge boson b_μ^3 remains massless, reflecting the invariance of the vacuum under the generator T_3.

After spontaneous symmetry breaking, the number of particle degrees of freedom remains 9, now given by $1 \times 1 + 2 \times 3 + 1 \times 2$.

5.5   Prospects

The analysis of this chapter has made available to us a large variety of field theories that exhibit the spontaneous breaking of internal symmetries. The breaking of a discrete symmetry poses no particular problems. If the broken symmetry is continuous, however, the spontaneous breakdown of the symmetry is accompanied by the appearance of massless spin-zero particles known as Goldstone bosons. These appear as an impediment to the use of spontaneously broken symmetries to describe the physics of elementary particles, because no such massless scalars have yet been observed. If we go further and require the continuous symmetry to be a local gauge symmetry, the degrees of freedom associated with the Goldstone bosons are absorbed into the gauge bosons that correspond to the broken generators of the gauge group. Thus the disappearance of unwanted massless scalars and the acquisition of mass by the gauge bosons are coupled phenomena. The way is now open to the construction of theories in which interactions restricted by gauge principles are mediated by massive vector particles.

All the discussion of this chapter has been at the level of classical field theory. In particular situations, it will be important to verify that the elementary analysis of the minimum of the effective potential remains valid in the presence of quantum corrections. Generally speaking, this is the case unless the interactions represented by the effective potential are in some measure too strong. A brief discussion of one specific case will be presented in Section 6.5, where nontrivial constraints on the Higgs boson mass will be seen to ensue. It is also worthwhile to remark that the scalar fields by the interactions of which the symmetry is spontaneously broken need not be elementary, but might arise as dynamical bound states of elementary fermions. This is the case, for example, in the BCS theory of superconductivity,[5] where Cooper pairs of electrons are the objects of interest. This is a possibility to which we shall return.

We have developed the essential ideas of gauge theories and most of the tools needed to apply these ideas to the fundamental interactions. As a first application, we now turn to a unified description of the weak and electromagnetic interactions, the theory beyond QED that has been most extensively tested by experiment.

Problems

5-1. Analyze the spontaneous breakdown of a global $SU(2)$ symmetry. Consider the case of three real scalar fields ϕ_1, ϕ_2, ϕ_3, which constitute an $SU(2)$ triplet, denoted

$$\phi = \begin{pmatrix} \phi_1 \\ \phi_2 \\ \phi_3 \end{pmatrix}.$$

The Lagrangian density is

$$\mathscr{L} = \tfrac{1}{2}(\partial_\mu \phi) \cdot (\partial^\mu \phi) - V(\phi \cdot \phi),$$

where as usual

$$V = \tfrac{1}{2}\mu^2 \phi \cdot \phi + \tfrac{1}{4}|\lambda|(\phi \cdot \phi)^2.$$

Assume that for $\mu^2 < 0$ the potential has a minimum at

$$\langle \phi \rangle_0 = \begin{pmatrix} 0 \\ 0 \\ v \end{pmatrix}.$$

Then show that (a) the vacuum remains invariant under the action of the generator T_3, but not under T_1 or T_2; (b) the particles associated with T_1 and T_2 become massless (Goldstone) particles; (c) the particle associated with T_3 acquires a mass $= \sqrt{-2\mu^2}$.

5-2. Generalize the preceding example to a Lagrangian that describes the interactions of n scalar fields and is invariant under global transformations under the group O(n). After spontaneous symmetry breaking and the choice of a vacuum state, show that the vacuum is invariant under the group O($n - 1$). Verify that the number of Goldstone bosons corresponds to the number of broken generators of the original symmetry group—i.e., to the difference between the number of generators of O(n) and O($n - 1$).

5-3. The Ginzburg–Landau theory of superconductivity provides a phenomenological understanding of the Meissner effect: the observation that an external magnetic field does not penetrate the superconductor. Ginzburg and Landau introduce an "order parameter" ψ, such that $|\psi|^2$ is related to the density of superconducting electrons. In the absence of an impressed magnetic field, expand the free energy of the superconductor as

$$G_{\text{super}}(0) = G_{\text{normal}}(0) + \alpha|\psi|^2 + \beta|\psi|^4,$$

where α and β are phenomenological parameters.

(a) Minimize $G_{\text{super}}(0)$ with respect to the order parameter and discuss the circumstances under which spontaneous symmetry breaking occurs. Compute $\langle|\psi|^2\rangle_0$, the value at which $G_{\text{super}}(0)$ is minimized.

(b) In the presence of an external magnetic field B, a gauge-invariant expression for the free energy is

$$G_{\text{super}}(B) = G_{\text{super}}(0) + B^2/2 + (1/2m^*)\psi^*(-i\nabla - e^*\mathbf{A})^2\psi.$$

(The effective charge e^* turns out to be $2e$, because $|\psi|^2$ represents the density of Cooper pairs.) Derive the field equations that follow from minimizing $G_{\text{super}}(B)$ with respect to ψ and $\mathbf{A}$. Show that in the weak-field approximation ($\nabla\psi \simeq 0$, $\psi \simeq \langle\psi\rangle_0$) the photon acquires a mass within the superconductor. [Reference: V. L. Ginzburg and L. D. Landau, *Zh. Eksp. Teor. Fiz.* **20**, 1064 (1950).]

5-4.   Derive the equation of motion for the photon field A_v in the Abelian Higgs model and show that it amounts to a relativistic generalization of the Ginzburg–Landau description of a superconductor.

For Further Reading

SYMMETRY AND THE SPECTRUM. Implications of an exact quantum-mechanical symmetry are discussed in many places. Particularly clear presentations appear in

K. Gottfried, *Quantum Mechanics*, Vol. I: Fundamentals, Benjamin, Reading, Massachusetts, 1966, Chapter VI.

and in The Source,

E. P. Wigner, *Group Theory and Its Application to the Quantum Mechanics of Atomic Spectra*, translated by J. J. Griffin, Academic, New York, 1959.

SPONTANEOUSLY BROKEN GAUGE THEORIES. Fine introductory discussions at roughly the same technical level as this chapter are given by

E. S. Abers and B. W. Lee, *Phys. Rep.* **9C**, 1 (1973).

S. Coleman, "Secret Symmetry," in *Laws of Hadronic Matter*, 1973 Erice School, edited by A. Zichichi, Academic, New York and London, 1975, p. 139.

P. W. Higgs, "Spontaneous Symmetry Breaking," in *Phenomenology of Particles at High Energies*, 14th Scottish Universities Summer School in Physics, 1973, edited by R. L. Crawford and R. Jennings, Academic, New York and London, 1974, p. 529.

C. Itzykson and J.-B. Zuber, *Quantum Field Theory*, McGraw-Hill, New York, 1980, Chapters 11, 12.

A. D. Linde, *Rep. Prog. Phys.* **42**, 389 (1979).

J. C. Taylor, *Gauge Theories of the Weak Interactions*, Cambridge University Press, Cambridge, 1976, Chapters 5, 6.

A number of specific examples are treated by

H. Fritzsch and P. Minkowski, *Phys. Rep.* **73**, 67 (1981).

A comprehensive review at a slightly more formal level is presented by

J. Bernstein, *Rev. Mod. Phys.* **46**, 7 (1974).

Still more formal is the article by

L. O'Raifeartaigh, *Rep. Prog. Phys.* **42**, 259 (1979).

Special attention to the interplay between spontaneous symmetry breaking and conservation laws is given by

G. S. Guralnik, C. R. Hagen, and T. W. B. Kibble, in *Advances in Particle Physics*, edited by R. L. Cool and R. E. Marshak, Interscience, New York, 1968, Vol. 2, p. 567.

A non-field-theoretic introduction, with reference to the physical analogy of the superconductor, is presented in

I. J. R. Aitchison and A. J. G. Hey, *Gauge Theories in Particle Physics*, Adam Hilger, Bristol, 1982, Chapter 9.

SPONTANEOUS SYMMETRY BREAKING IN OTHER PHYSICAL CONTEXTS. The phenomenon of spontaneously broken symmetries is a ubiquitous one, present in crystallization, the onset of turbulence, the Meissner effect, and many other circumstances. An interesting tour is given by

P. W. Anderson, in *Gauge Theories and Modern Field Theory*, edited by R. Arnowitt and P. Nath, MIT Press, Cambridge, Massachusetts, 1976, p. 311; *Phys. Rev.* **130**, 439 (1963).

For more on the Heisenberg ferromagnet, see

D. C. Mattis, *The Theory of Magnetism*, Harper and Row, New York, 1965.

A brief account of spontaneous symmetry breaking in superconductors appears in

E. A. Lynton, *Superconductivity*, Methuen, London, 1962, Chapter 5.

J. R. Schrieffer, *Theory of Superconductivity*, Benjamin, New York, 1964.

An analogy between ferromagnetism and the laser effect is described by

M. Sargent III, M. O. Scully, and W. E. Lamb, Jr., *Laser Physics*, Addison-Wesley, Reading, Massachusetts, 1974, Section 21-3.

Spontaneous symmetry breaking in the Bogoliubov superfluid is reviewed by Higgs, *op. cit.*, and in

D. Forster, *Hydrodynamic Fluctuations, Broken Symmetry, and Correlation Functions*, Benjamin, Reading, Massachusetts, 1975.

SPONTANEOUSLY BROKEN SYMMETRIES IN PARTICLE PHYSICS. Attempts to account for the breaking of hadronic symmetries in terms of spontaneous symmetry breaking were initiated by

Y. Nambu, *Phys. Rev. Lett.* **4**, 380 (1960).

Y. Nambu and G. Jona-Lasinio, *Phys. Rev.* **122**, 345 (1961); **124**, 246 (1961).

GAUGE INVARIANCE AND MASS. That dynamical effects could circumvent the prediction of massless gauge bosons was first recognized by

J. Schwinger, *Phys. Rev.* **125**, 397 (1962).

who presented a concrete example of a massive photon in the exactly solvable model of the quantum electrodynamics of massless fermions in two dimensions, in

J. Schwinger, *Phys. Rev.* **128**, 2425 (1962).

For the three-dimensional case, see

J. F. Schonfeld, *Nucl. Phys.* **B185**, 157 (1981).

A recent review, emphasizing topological considerations, is given by

R. Jackiw, "Gauge Invariance and Mass, III," in *Asymptotic Realms of Physics*, edited by A. Guth, K. Huang, and R. L. Jaffe, MIT Press, Cambridge, Massachusetts, 1983.

An explicit demonstration that spontaneous symmetry breaking in the form of nonvanishing expectation values of a multiplet of scalar fields endows gauge bosons with mass was first given by

F. Englert and R. Brout, *Phys. Rev. Lett.* **13**, 321 (1964).

The gauge-invariant extension of the Goldstone model was studied nearly simultaneously by

P. W. Higgs, *Phys. Rev. Lett.* **13**, 508 (1964); *Phys. Rev.* **145**, 1156 (1966).

Similar conclusions were reached in a different context at nearly the same moment by

G. S. Guralnik, C. R. Hagen, and T. W. B. Kibble, *Phys. Rev. Lett.* **13**, 585 (1964).

The general group-theoretical analysis of this phenomenon is due to

T. W. B. Kibble, *Phys. Rev.* **155**, 1554 (1967).

UNITARY GAUGE. The demonstration that a U-gauge can be found for a general spontaneously broken gauge theory was given by

S. Weinberg, *Phys. Rev.* **D7**, 1068 (1973).

References

[1] J. Goldstone, *Nuovo Cim.* **19**, 154 (1961).

[2] J. Goldstone, A. Salam, and S. Weinberg, *Phys. Rev.* **127**, 965 (1962); S. Bludman and A. Klein, *ibid.* **131**, 2364 (1963); W. Gilbert, *Phys. Rev. Lett.* **12**, 713 (1964); R. F. Streater, *Proc. Roy. Soc. (London)* **A287**, 510 (1965); D. Kastler, D. W. Robinson, and A. Swieca, *Commun. Math. Phys.* **2**, 108 (1966).

[3] P. W. Higgs, *Phys. Lett.* **12**, 132 (1964).

[4] The analysis can be carried out equally well in terms of the two real fields ϕ_1 and ϕ_2, or in a two-dimensional vector notation with $\phi = \begin{pmatrix} \phi_1 \\ \phi_2 \end{pmatrix}$. Attention to factors of 2!

[5] J. Bardeen, L. N. Cooper, and J. R. Schrieffer, *Phys. Rev.* **106**, 162 (1962).

CHAPTER 6

ELECTROWEAK INTERACTIONS OF LEPTONS

In the preceding chapters we have developed a general strategy for the construction of exact or spontaneously broken gauge theories of the fundamental interactions. The point of departure for applications to particle physics is the recognition of a symmetry respected by the elementary fermions in the problem. What follows in the case of a spontaneously broken symmetry is an effort to conceal the exact symmetry in the same fashion that Nature has chosen. There is much art in the selection of the gauge symmetry and of a pattern of symmetry breaking. Both experimental results and theoretical requirements offer constraints. In the course of this chapter, we shall examine some of the ways in which the theory of weak and electromagnetic interactions has been shaped by internal and external pressures. Although the construction of an internally consistent theory that is in agreement with experiment is nontrivial, the present state of our knowledge leaves much room for arbitrariness. Much, if not all, of the remaining incompleteness of the theory has to do with conceptual questions that remain to be understood, and not simply with parameters that improved experiments will determine with greater precision. We know of nothing to indicate that a complete theory is unattainable.

To illustrate the construction of a realistic gauge theory, we shall treat the model of leptons first given explicitly by Weinberg and, in related form, by Salam.[1] We do not yet know how nearly this approaches an ultimate theory of the weak and electromagnetic interactions of leptons, but it has passed many experimental tests and makes a number of interesting further predictions. Crucial among these are the predicted properties of the intermediate vector bosons of the weak interactions, for which the tools for experimental tests are nearly at hand. If the so-called "standard model" should turn out to be incomplete or incorrect, it is nevertheless a supremely useful case study.

Before constructing the theory, it will be necessary to recount some basic elements of weak-interaction phenomenology that must be reproduced, and to see the shortcomings of earlier descriptions. To this end, we shall summarize the effective Lagrangian approach that began with Fermi and the intermediate-boson extension of that picture. We then present the partially unified $SU(2) \otimes U(1)$ gauge theory of weak and electromagnetic interactions and compare its implications with the earlier descriptions. A central element of the Weinberg–Salam model is the prediction of weak neutral-current phenomena. These will be discussed at some length. We next consider the Higgs boson of the theory, the existence of which is an ineluctable consequence of the mechanism of spontaneous symmetry breaking that we have developed. The properties of this scalar particle and prospects for its experimental detection are also discussed. Spontaneously broken gauge theories of the type we discuss here have the important asset of being renormalizable. Predictions are thus in principle calculable to all orders in perturbation theory. A proof of renormalizability is enormously complex and is beyond the scope of this book. We shall, however, look briefly at two aspects of the renormalization program: the definition of gauges better suited to higher-order calculations than the U-gauge, which was well-adapted to the examination of particle spectra, and the issue of anomalies. We shall find that the standard model, when restricted to the lepton sector, is nonrenormalizable, and that renormalizability can be achieved by introducing hadrons in a manner that will later be seen to have far-reaching implications. Throughout the chapter, many sample calculations are carried out in detail.

6.1   An Effective Lagrangian for the Weak Interactions

Our contemporary view of the weak interactions is the result of a long evolution of the theoretical picture of radioactive β-decay introduced by Fermi[2] in 1933. The history of the subject up to the invention of spontaneously broken gauge theories, and the low-energy phenomenology of nuclear β-decay, muon decay, etc., are well-documented in a number of excellent textbooks and monographs cited at the end of this chapter. We shall not reproduce this material but shall instead emphasize some aspects of the phenomenological theory that are not given prominence in the traditional textbook treatments. Our aims are to recall some calculational techniques, to point the way toward high-energy phenomena, and to expose the limitations of the historical approach.

We begin our exploration of weak-interaction phenomena with an investigation of neutrino–electron scattering. We shall briefly consider the implications of various possible space–time forms for the charged-current interaction and then examine in greater detail the properties of the vector-minus-axial-vector structure favored by experiment.

The most general matrix element for the charged-current $\nu_e e$ interaction, free of derivative couplings, is

$$\mathcal{M} = \sum \mathcal{M}_i = \sum C_i \bar{\nu} \mathcal{O}_i e \, \bar{e} \mathcal{O}^i (1 - \gamma_5) \nu, \qquad (6.1.1)$$

where the operators $\mathcal{O}_i$ are

$$
\left.
\begin{aligned}
\mathcal{O}_S &= 1 && \text{scalar} \\
\mathcal{O}_P &= \gamma_5 && \text{pseudoscalar} \\
\mathcal{O}_T &= \sigma_{\mu\nu} && \text{tensor} \\
\mathcal{O}_V &= \gamma_\mu && \text{vector} \\
\mathcal{O}_A &= \gamma_\mu \gamma_5 && \text{axial vector}
\end{aligned}
\right\}, \qquad (6.1.2)
$$

the summation runs over $i = $ S, P, T, V, A, and the explicit factor of $(1 - \gamma_5)$ appears in recognition of the fact that the neutrino sources of interest in Nature yield left-handed neutrinos. For calculations it will be convenient to employ the combinations $V \pm A$, which contribute incoherently to the absolute squares of matrix elements. Let us now compute the cross section that arises from each interaction in turn. Although much of this exercise is only of academic interest for the well-studied charged-current interactions, it will serve as an indication of what may be learned from scattering experiments. It will also be a useful reference when we turn to the study of neutral-current interactions.

Denote the incoming and outgoing four-momenta

$$
\left.
\begin{aligned}
\nu_{\text{in}}: &\quad q_1 \\
\nu_{\text{out}}: &\quad q_2 \\
e_{\text{in}}: &\quad p_1 \\
e_{\text{out}}: &\quad p_2
\end{aligned}
\right\}. \qquad (6.1.3)
$$

In the laboratory frame, depicted in Fig. 6-1(a), the four-momenta are explicitly given by

$$
\left.
\begin{aligned}
q_1^\mu &= (E; 0, 0, E) \\
p_1^\mu &= (m; 0, 0, 0) \\
p_2^\mu &= (E'; P' \sin\theta_L, 0, P' \cos\theta_L) \\
q_2^\mu &= (E + m - E'; -P' \sin\theta_L, 0, E - P' \cos\theta_L)
\end{aligned}
\right\}, \qquad (6.1.4)
$$

where the energy of the recoiling electron is

$$E' = \sqrt{P'^2 + m^2} \equiv yE, \qquad (6.1.5)$$

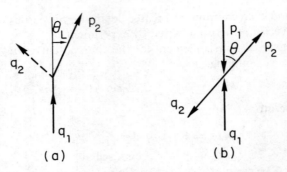

FIG. 6-1. Kinematics of neutrino–electron scattering in the (a) laboratory and (b) center of momentum frames.

with m the electron mass. With these definitions, the invariants of interest may be expressed as

$$\left. \begin{aligned} p_1 \cdot q_1 &= p_2 \cdot q_2 = mE \\ p_1 \cdot q_2 &= p_2 \cdot q_1 = m(E + m - E') \simeq mE(1 - y) \\ p_1 \cdot p_2 &= mE' = mEy \\ q_1 \cdot q_2 &= m(E' - m) \simeq mEy \end{aligned} \right\}. \tag{6.1.6}$$

In the c.m. frame, indicated in Fig. 6-1(b), the four-momenta are conveniently written in terms of

$$p^* = (s - m^2)/2\sqrt{s} = mE/\sqrt{s} \simeq \sqrt{s}/2, \tag{6.1.7}$$

and

$$\omega^* = \sqrt{p^{*2} + m^2} = (s + m^2)/2\sqrt{s}, \tag{6.1.8}$$

as

$$\left. \begin{aligned} q_1^\mu &= (p^*; 0, 0, p^*) \\ p_1^\mu &= (\omega^*; 0, 0, -p^*) \\ p_2^\mu &= (\omega^*; p^* \sin\theta, 0, p^* \cos\theta) \\ q_2^\mu &= (p^*; -p^* \sin\theta, 0, -p^* \cos\theta) \end{aligned} \right\}. \tag{6.1.9}$$

The invariants are then given by

$$\left. \begin{aligned} p_1 \cdot q_1 &= p_2 \cdot q_2 = p^*(p^* + \omega^*) \simeq 2p^{*2} \\ p_1 \cdot q_2 &= p_2 \cdot q_1 = p^*(\omega^* - p^* \cos\theta) \simeq p^{*2}(1 - \cos\theta) \\ p_1 \cdot p_2 &= \omega^{*2} + p^{*2} \cos\theta \simeq p^{*2}(1 + \cos\theta) \\ q_1 \cdot q_2 &= p^{*2}(1 + \cos\theta) \end{aligned} \right\}. \tag{6.1.10}$$

We now compute the differential and integral cross sections for the various interaction terms contained in (6.1.1).

(*i*) Scalar interaction: $v_e e \to v_e e$. The matrix element is

$$\mathcal{M} = \bar{u}(v, q_2)u(e, p_1)\bar{u}(e, p_2)(1 - \gamma_5)u(v, q_1). \tag{6.1.11}$$

Squaring the matrix element and summing over the spins of the outgoing particles, we have

$$
\begin{aligned}
|\mathcal{M}|^2 &= \text{tr}[u(e, p_1)\bar{u}(e, p_1)u(v, q_2)\bar{u}(v, q_2)] \\
&\quad \times \text{tr}[(1 - \gamma_5)u(v, q_1)\bar{u}(v, q_1)(1 + \gamma_5)u(e, p_2)\bar{u}(e, p_2)] \\
&= \text{tr}[(m + p\!\!\!/_1)q\!\!\!/_2]\,\text{tr}[(1 - \gamma_5)q\!\!\!/_1(1 + \gamma_5)(m + p\!\!\!/_2)]. \tag{6.1.12}
\end{aligned}
$$

The first trace is simply

$$\text{tr}[(m + p\!\!\!/_1)q\!\!\!/_2] = 4p_1 \cdot q_2, \tag{6.1.13}$$

by virtue of (A.3.3) and (A.3.5b). The second is conveniently rearranged as

$$
\begin{aligned}
\text{tr}[(1 - \gamma_5)q\!\!\!/_1(1 + \gamma_5)(m + p\!\!\!/_2)] &= 2\,\text{tr}[(1 - \gamma_5)q\!\!\!/_1(m + p\!\!\!/_2)] \\
&= 8q_1 \cdot p_2, \tag{6.1.14}
\end{aligned}
$$

where (A.3.3), (A.3.5b), (A.3.8), and (A.3.9) have been used. We therefore have $|\mathcal{M}|^2 = 32p_1 \cdot q_2 p_2 \cdot q_1$ or, averaging over the spin states of the incident electron,

$$\overline{|\mathcal{M}|^2} = 16p_1 \cdot q_2 p_2 \cdot q_1 = 16p^{*4}(1 - \cos\theta)^2. \tag{6.1.15}$$

The differential cross section in the c.m. frame is easily computed with the aid of (B.1.7) as

$$\frac{d\sigma}{d\Omega_{\text{c.m.}}} = \frac{\overline{|\mathcal{M}|^2}}{64\pi^2 s} = \frac{mE(1 - \cos\theta)^2}{32\pi^2}, \tag{6.1.16}$$

so that

$$\sigma = \int d\Omega\left(\frac{d\sigma}{d\Omega}\right) = \frac{mE}{16\pi}\int_{-1}^{1} dz(1 - z)^2 = \frac{mE}{6\pi}. \tag{6.1.17}$$

Comparing (6.1.6) and (6.1.10), we find that

$$(1 - \cos\theta) = 2(1 - y). \tag{6.1.18}$$

Consequently we may write the differential cross section in the useful form

$$\frac{d\sigma}{dy} = 4\pi\frac{d\sigma}{d\Omega_{\text{c.m.}}} = \frac{mE(1 - y)^2}{2\pi}, \tag{6.1.19}$$

which leads of course to the same integrated cross section,

$$\sigma = \int_0^1 dy(d\sigma/dy) = mE/6\pi. \tag{6.1.20}$$

(*ii*) Scalar interaction: $\bar{v}_e e \to \bar{v}_e e$. The matrix element is

$$\mathcal{M} = \bar{v}(v, q_1)u(e, p_1)\bar{u}(e, p_2)(1 - \gamma_5)v(v, q_2), \tag{6.1.21}$$

so that

$$\begin{aligned}
|\mathscr{M}|^2 &= \mathrm{tr}[u(e,p_1)\bar{u}(e,p_1)v(v,q_1)\bar{v}(v,q_1)]\\
&\quad \times \mathrm{tr}[(1-\gamma_5)v(v,q_2)\bar{v}(v,q_2)(1+\gamma_5)u(e,p_2)\bar{u}(e,p_2)]\\
&= \mathrm{tr}[(m+\not{p}_1)\not{q}_1]\mathrm{tr}[(1-\gamma_5)\not{q}_2(1+\gamma_5)(m+\not{p}_2)],
\end{aligned}\tag{6.1.22}$$

which may be obtained from the matrix-element-squared for $\nu_e e$ scattering by interchanging q_1 and q_2. As a result, we find

$$\overline{|\mathscr{M}|^2} = 16p_1 \cdot q_1 p_2 \cdot q_2,\tag{6.1.23}$$

so that

$$d\sigma/dy = mE/2\pi,\tag{6.1.24}$$

and

$$\sigma = mE/2\pi.\tag{6.1.25}$$

It is straightforward to verify that the pseudoscalar interaction yields the same cross sections.

(*iii*) Tensor coupling: $\nu_e e \to \nu_e e$. The matrix element is

$$\mathscr{M} = \bar{u}(v,q_2)\sigma_{\mu v}u(e,p_1)\bar{u}(e,p_2)\sigma^{\mu v}(1-\gamma_5)u(v,q_1),\tag{6.1.26}$$

so we are required to evaluate

$$|\mathscr{M}|^2 = \mathrm{tr}[\sigma_{\mu v}(m+\not{p}_1)\sigma_{\kappa\lambda}\not{q}_2]\mathrm{tr}[\sigma^{\mu v}(1-\gamma_5)\not{q}_1(1+\gamma_5)\sigma^{\kappa\lambda}(m+\not{p}_2)].\tag{6.1.27}$$

To evaluate these traces, it is convenient to exploit the identity

$$\sigma^{\mu v} = i(\gamma^\mu \gamma^v - g^{\mu v})\tag{A.2.10}$$

and to note that, by virtue of the anticommutation relations (A.2.1), a product of six γ-matrices may be simplified as

$$\begin{aligned}
\mathrm{tr}(\not{a}\not{b}\not{c}\not{d}\not{e}\not{f}) &= a\cdot b\,\mathrm{tr}(\not{c}\not{d}\not{e}\not{f}) - a\cdot c\,\mathrm{tr}(\not{b}\not{d}\not{e}\not{f}) + a\cdot d\,\mathrm{tr}(\not{b}\not{c}\not{e}\not{f})\\
&\quad - a\cdot e\,\mathrm{tr}(\not{b}\not{c}\not{d}f) + a\cdot f\,\mathrm{tr}(\not{b}\not{c}\not{d}\not{e}).
\end{aligned}\tag{6.1.28}$$

A slightly tedious calculation then yields

$$\begin{aligned}
|\mathscr{M}|^2 &= 256(2q_1 \cdot q_2 p_1 \cdot p_2 + 2q_1 \cdot p_1 q_2 \cdot p_2 - q_1 \cdot p_2 q_2 \cdot p_1)\\
&= 256(mE)^2(1+y)^2,
\end{aligned}\tag{6.1.29}$$

so that

$$\frac{d\sigma}{dy} = \frac{4mE}{\pi}(1+y)^2,\tag{6.1.30}$$

and

$$\sigma = 28mE/3\pi.\tag{6.1.31}$$

(*iv*) Tensor coupling: $\bar{v}_e e \to \bar{v}_e e$. Again the result may be obtained by interchanging q_1 and q_2. We have

$$\begin{aligned}
|\mathscr{M}|^2 &= 256(2q_1 \cdot q_2 p_1 \cdot p_2 + 2q_2 \cdot p_1 q_1 \cdot p_2 - q_2 \cdot p_2 q_1 \cdot p_1)\\
&= 256(mE)^2(1-2y)^2
\end{aligned}\tag{6.1.32}$$

from which

$$\frac{d\sigma}{dy} = \frac{4mE}{\pi}(1 - 2y)^2,$$

(6.1.33)

and

$$\sigma = 4mE/3\pi.$$

(6.1.34)

(v) S − T interference: $\nu_e e \to \nu_e e$. The matrix element is

$$\mathcal{M} = C_S \bar{u}(\nu, q_2) u(e, p_1) \bar{u}(e, p_2)(1 - \gamma_5) u(\nu, q_1)$$
$$+ C_T \bar{u}(\nu, q_2)\sigma_{\mu\nu} u(e, p_1) \bar{u}(e, p_2)\sigma^{\mu\nu}(1 - \gamma_5) u(\nu, q_1),$$

(6.1.35)

so the form of the interference terms is

$$\mathcal{I} = C_S C_T^* \, \text{tr}[\bar{u}(\nu, q_2) u(e, p_1) \bar{u}(e, p_1) \sigma_{\mu\nu} u(\nu, q_2)]$$
$$\times \text{tr}[\bar{u}(e, p_2)(1 - \gamma_5) u(\nu, q_1) \bar{u}(\nu, q_1)(1 + \gamma_5)\sigma^{\mu\nu} u(e, p_2)]$$
$$+ C_S^* C_T \, \text{tr}[\bar{u}(\nu, q_2)\sigma_{\mu\nu} u(e, p_1) \bar{u}(e, p_1) u(\nu, q_2)]$$
$$\times \text{tr}[\bar{u}(e, p_2)\sigma^{\mu\nu}(1 - \gamma_5) u(\nu, q_1) \bar{u}(\nu, q_1)(1 + \gamma_5) u(e, p_2)]$$
$$= \text{tr}[(m + \not{p}_1)\sigma_{\mu\nu}\not{q}_2] \text{tr}[(1 - \gamma_5)\not{q}_1(1 + \gamma_5)\sigma^{\mu\nu}(m + \not{p}_2)] C_S C_T^*$$
$$+ \text{tr}[\sigma_{\mu\nu}(m + \not{p}_1)\not{q}_2] \text{tr}[\sigma^{\mu\nu}(1 - \gamma_5)\not{q}_1(1 + \gamma_5)(m + \not{p}_2)] C_S^* C_T$$
$$= 2\,\text{tr}[\sigma_{\mu\nu}\not{q}_2(m + \not{p}_1)] \text{tr}[(1 + \gamma_5)\sigma^{\mu\nu}(m + \not{p}_2)\not{q}_1] C_S C_T^*$$
$$+ 2\,\text{tr}[\sigma_{\mu\nu}(m + \not{p}_1)\not{q}_2] \text{tr}[(1 - \gamma_5)\sigma^{\mu\nu}\not{q}_1(m + \not{p}_2)] C_S^* C_T.$$

(6.1.36)

The commutation relations show the two products of traces to be equal. Using again the identity (A. 2. 10), we easily compute

$$\mathcal{I} = -128(p_2 \cdot q_2 p_1 \cdot q_1 - q_1 \cdot q_2 p_1 \cdot p_2)\text{Re}(C_S^* C_T)$$
$$= -128(mE)^2(1 - y^2)\text{Re}(C_S^* C_T),$$

(6.1.37)

so that

$$\frac{d\sigma_{ST}}{dy} = -\frac{2mE}{\pi}(1 - y^2)\text{Re}(C_S^* C_T),$$

(6.1.38)

and

$$\sigma_{ST} = -\frac{4mE}{3\pi}\text{Re}(C_S^* C_T).$$

(6.1.39)

The cross section corresponding to an interference term need not, of course, be positive-definite. A parallel calculation shows that the same result is obtained for P − T interference.

(vi) S − T interference: $\bar{\nu}_e e \to \bar{\nu}_e e$. On interchanging q_1 and q_2 in (6.1.37) we find

$$\mathcal{I} = -128(p_2 \cdot q_1 p_1 \cdot q_2 - q_1 \cdot q_2 p_1 \cdot p_2)\text{Re}(C_S^* C_T)$$
$$= -128(mE)^2(1 - 2y)\text{Re}(C_S^* C_T),$$

(6.1.40)

from which

$$\frac{d\sigma_{ST}}{dy} = \frac{2mE}{\pi}(1 - 2y)\text{Re}(C_S^* C_T)$$

(6.1.41)

and

$$\sigma_{ST} = 0. \tag{6.1.42}$$

(*vii*) V − A interaction: $v_e e \to v_e e$. The matrix element is

$$\mathcal{M} = \bar{u}(v, q_2)\gamma_\mu(1 - \gamma_5)u(e, p_1)\bar{u}(e, p_2)\gamma^\mu(1 - \gamma_5)u(v, q_1). \tag{6.1.43}$$

Squaring the matrix element and summing over the spins of the outgoing particles, we obtain

$$\begin{aligned} |\mathcal{M}|^2 &= \text{tr}[\gamma_\mu(1 - \gamma_5)(m + \not{p}_1)(1 + \gamma_5)\gamma_\nu \not{q}_2] \\ &\quad \times \text{tr}[\gamma^\mu(1 - \gamma_5)\not{q}_1(1 + \gamma_5)\gamma^\nu(m + \not{p}_2)] \\ &\equiv A_{\mu\nu}B^{\mu\nu}. \end{aligned} \tag{6.1.44}$$

The first factor is then

$$\begin{aligned} A_{\mu\nu} &= 2\,\text{tr}[(1 + \gamma_5)\gamma_\nu \not{q}_2 \gamma_\mu(m + \not{p}_1)] \\ &= 8(q_{2\nu}p_{1\mu} - g_{\mu\nu}q_2 \cdot p_1 + q_{2\mu}p_{1\nu}) - 8i\varepsilon_{\mu\nu\rho\sigma}q_2^\rho p_1^\sigma, \end{aligned} \tag{6.1.45}$$

and the second is

$$\begin{aligned} B^{\mu\nu} &= 2\,\text{tr}[(1 - \gamma_5)\not{q}_1 \gamma^\nu(m + \not{p}_2)\gamma^\mu] \\ &= 8(q_1^\nu p_2^\mu - g^{\mu\nu}q_1 \cdot p_2 + q_1^\mu p_2^\nu) + 8i\varepsilon^{\mu\nu\kappa\lambda}q_{1\kappa}p_{2\lambda}. \end{aligned} \tag{6.1.46}$$

As a consequence,

$$\begin{aligned} |\mathcal{M}|^2 = A_{\mu\nu}B^{\mu\nu} &= 128(q_1 \cdot q_2 p_1 \cdot p_2 + q_1 \cdot p_1 q_2 \cdot p_2) \\ &\quad - 64i\varepsilon_{\mu\nu\rho\sigma}q_2^\rho p_1^\sigma(q_1^\nu p_2^\mu + q_1^\mu p_2^\nu) \\ &\quad + 64i\varepsilon^{\mu\nu\kappa\lambda}q_{1\kappa}p_{2\lambda}(q_{2\nu}p_{1\mu} + q_{2\mu}p_{1\nu}) \\ &\quad + 64\varepsilon_{\mu\nu\rho\sigma}\varepsilon^{\mu\nu\kappa\lambda}q_2^\rho p_1^\sigma q_{1\kappa}p_{2\lambda}. \end{aligned} \tag{6.1.47}$$

The second and third terms vanish because there are only three independent momenta in the problem. The last term may be evaluated with the aid of (A.3.15); it is simply

$$128(q_1 \cdot p_1 q_2 \cdot p_2 - q_1 \cdot q_2 p_1 \cdot p_2), \tag{6.1.48}$$

so that

$$\begin{aligned} |\mathcal{M}|^2 &= 256q_1 \cdot p_1 q_2 \cdot p_2 \\ &= 256(mE)^2, \end{aligned} \tag{6.1.49}$$

whereupon

$$d\sigma/dy = 4mE/\pi \tag{6.1.50}$$

and

$$\sigma = 4mE/\pi. \tag{6.1.51}$$

(*viii*) V − A interaction: $\bar{v}_e e \to \bar{v}_e e$. The replacement $q_1 \leftrightarrow q_2$ yields

$$\begin{aligned} |\mathcal{M}|^2 &= 256q_2 \cdot p_1 q_1 \cdot p_2 \\ &= 256(mE)^2(1 - y)^2, \end{aligned} \tag{6.1.52}$$

from which

$$\frac{d\sigma}{dy} = \frac{4mE}{\pi}(1 - y)^2 \qquad (6.1.53)$$

and

$$\sigma = 4mE/3\pi. \qquad (6.1.54)$$

(ix) V + A interaction: $v_e e \to v_e e$. We write the matrix element as

$$\mathcal{M} = \bar{u}(v, q_2)\gamma_\mu(1 + \gamma_5)u(e, p_1)\bar{u}(e, p_2)\gamma^\mu(1 - \gamma_5)u(v, q_1). \qquad (6.1.55)$$

The calculation differs from the V − A calculation only by a change in the sign of the $\varepsilon_{\mu\nu\rho\sigma}$ term. Thus we find that

$$\begin{aligned} |\mathcal{M}|^2 &= 256q_1 \cdot q_2 p_1 \cdot p_2 \\ &= 256(mE)^2 y^2, \end{aligned} \qquad (6.1.56)$$

which implies

$$\frac{d\sigma}{dy} = \frac{4mE}{\pi}y^2 \qquad (6.1.57)$$

and

$$\sigma = 4mE/3\pi.$$

(x) V + A interaction: $\bar{v}_e e \to \bar{v}_e e$. The previous results are unchanged by the interchange of q_1 and q_2, so we have at once

$$\frac{d\sigma}{dy} = \frac{4mE}{\pi}y^2 \qquad (6.1.59)$$

and

$$\sigma = 4mE/3\pi. \qquad (6.1.60)$$

The foregoing calculations of

$$\frac{d\sigma}{dy} = \frac{d\sigma}{dy}(\text{STP}) + \frac{d\sigma}{dy}(\text{V} - \text{A}) + \frac{d\sigma}{dy}(\text{V} + \text{A}) \qquad (6.1.61)$$

are summarized in Table 6.1. We see that energy-loss distributions (or

TABLE 6.1: DIFFERENTIAL CROSS SECTIONS FOR
CHARGED-CURRENT PROCESSES

Coupling	$\dfrac{\pi}{mE}\dfrac{d\sigma}{dy}(v_e e \to v_e e)$	$\dfrac{\pi}{mE}\dfrac{d\sigma}{dy}(\bar{v}_e e \to \bar{v}_e e)$
$|C_S|^2 + |C_P|^2$	$\frac{1}{2}(1 - y)^2$	$\frac{1}{2}$
$|C_T|^2$	$4(1 + y)^2$	$4(1 - 2y)^2$
$\text{Re}[(C_S^* + C_P^*)C_T]$	$-2(1 - y^2)$	$-2(1 - 2y)$
$|C_{V-A}|^2$	4	$4(1 - y)^2$
$|C_{V+A}|^2$	$4y^2$	$4y^2$

equivalently angular distributions) provide a means of probing the space–time structure of the interaction. It is, however, easy to verify the following "confusion theorem": for the pair of spin-averaged $(v, \bar{v})$ cross sections, any combination of $V - A$ and $V + A$ interactions can be reproduced by a suitably chosen combination of S, T, and P interactions. Spin dependences are in general different for the two sets of interactions.

A number of experiments in the late 1950s established the structure of the weak charged current as V − A. Let us now concentrate more closely on this form, in the context of high-energy ve scattering. It is conventional to write the effective Lagrangian of the leptonic weak interaction as a product of charged currents.

$$\mathcal{L}_{\text{eff}} = -\frac{G_F}{\sqrt{2}} \bar{v}\gamma_\mu(1 - \gamma_5)e \bar{e}\gamma^\mu(1 - \gamma_5)v, \tag{6.1.62}$$

where the Fermi constant has been measured[3] to be

$$G_F = (1.16632 \pm 0.00004) \times 10^{-5} \text{ GeV}^{-2}, \tag{6.1.63}$$

so that

$$G_F^2 = 5.29 \times 10^{-38} \text{ cm}^2/\text{GeV}^2. \tag{6.1.64}$$

The cross sections for $v_e e$ and $\bar{v}_e e$ scattering are then given, in the V − A model, by

$$\frac{d\sigma}{dy}(v_e e) = \frac{2mE}{\pi} G_F^2, \tag{6.1.65}$$

$$\sigma(v_e e) = \frac{2mE}{\pi} G_F^2 \simeq 1.72 \times 10^{-41} \text{ cm}^2 \left(\frac{E}{1 \text{ GeV}}\right), \tag{6.1.66}$$

$$\frac{d\sigma}{dy}(\bar{v}_e e) = \frac{2mE}{\pi} G_F^2(1 - y)^2, \tag{6.1.67}$$

$$\sigma(\bar{v}_e e) = \frac{2mEG_F^2}{3\pi} \simeq 0.574 \times 10^{-41} \text{ cm}^2 \left(\frac{E}{1 \text{ GeV}}\right). \tag{6.1.68}$$

The cross sections are extremely small, but not completely unattainable by experiment.

It is interesting to analyze the reason for the factor-of-3 difference between neutrino and antineutrino cross sections, or equivalently the difference in angular distributions plotted in Fig. 6-2. It is simply this: The operators $(1 \pm \gamma_5)$ are spin-projection operators in the limit of vanishing fermion mass. Thus, for a V − A interaction we may view all light fermions as left-handed (with helicity $-1/2$) and all light antifermions as right-handed (with helicity $+1/2$). Thus, as shown in Fig. 6-3(a), the initial state is a state of $J_z = 0$ for $v_e e$ scattering and $J_z = 1$ for $\bar{v}_e e$ scattering. The same holds for forward scattering (which we have labeled by $\cos\theta = -1$, $y = 0$). For backward

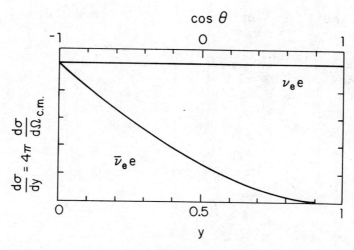

FIG. 6-2. Angular and energy-loss distributions of $\nu_e e$ and $\bar{\nu}_e e$ elastic scattering in the V–A model.

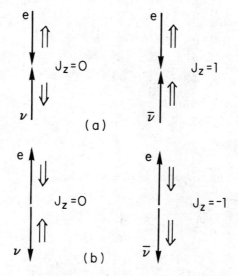

FIG. 6-3. Spin configurations for νe and $\bar{\nu} e$ scattering in the c.m. frame. (a) Initial state; (b) final state for backward scattering.

scattering (which we have described by $\cos \theta = +1$, $y = 1$), the situation is different, as shown in Fig. 6-3(b). The final state for $\nu_e e$ scattering is a state with total spin $J_z = 0$, as was the initial state, but the final state for $\bar{\nu}_e e$ scattering would correspond to $J_z = -1$, which would violate angular momentum conservation. Backward $\bar{\nu}_e e$ scattering is therefore forbidden, and the cross section must vanish at $y = 1$. This effect will recur many times in our study of the weak interactions.

One of the important observations of early experiments, which has been repeatedly confirmed and made more precise, is that the charged-current weak interactions are universal in strength: the same coupling constant applies to all leptonic and semileptonic interactions. As a consequence, the calculations we have just completed may be extended in straightforward fashion to the experimentally interesting case of inverse muon decay, the reaction

$$\nu_\mu(q_1)e(p_1) \to \mu(p_2)\nu_e(q_2). \tag{6.1.69}$$

Kinematic definitions are as in the case of $\nu_e e$ elastic scattering, except that the final muon energy is

$$E' = \sqrt{P'^2 + \mu^2}, \tag{6.1.70}$$

where μ is the muon mass. The invariants (6.1.6) become

$$\left.\begin{aligned}
q_1 \cdot p_1 &= mE \\
q_2 \cdot p_2 &= mE - (\mu^2 - m^2)/2 \\
p_1 \cdot p_2 &= mE' \\
p_1 \cdot q_2 &= m(E + m - E') \\
q_1 \cdot q_2 &= mE' - (\mu^2 + m^2)/2 \\
p_2 \cdot q_1 &= m(E - E') + (\mu^2 + m^2)/2
\end{aligned}\right\}. \tag{6.1.71}$$

The differential cross section is now

$$\frac{d\sigma}{d\Omega_{\text{c.m.}}} = \frac{\overline{|\mathscr{M}|^2}}{64\pi^2 s} \frac{p'_{\text{c.m.}}}{p_{\text{c.m.}}} = \frac{\overline{|\mathscr{M}|^2}}{64\pi^2 s}\left[1 - \frac{(\mu^2 - m^2)}{2mE}\right]. \tag{6.1.72}$$

Substitution of the matrix element (6.1.49) yields

$$\frac{d\sigma}{d\Omega_{\text{c.m.}}} = \frac{G_F^2 mE}{2\pi^2}\left[1 - \frac{(\mu^2 - m^2)}{2mE}\right]^2 \tag{6.1.73}$$

and therefore

$$\sigma(\nu_\mu e \to \mu\nu_e) = \frac{2G_F^2 mE}{\pi}\left[1 - \frac{(\mu^2 - m^2)}{2mE}\right]^2 \tag{6.1.74}$$

$$= \sigma_{\text{V}-\text{A}}(\nu_e e \to \nu_e e)\left[1 - \frac{(\mu^2 - m^2)}{2mE}\right]^2. \tag{6.1.75}$$

The expected total cross section is plotted in Fig. 6-4. Because of its high threshold energy ($E \simeq 10.9$ GeV), the reaction has been difficult to observe and has only recently been studied. The results[4] are in agreement with the

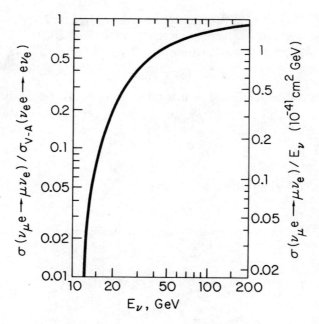

FIG. 6-4. Cross section for inverse muon decay, $v_\mu e \to \mu v_e$, as a function of beam energy.

theoretical expectations:

$$\frac{\sigma(v_\mu e \to \mu v_e)}{\sigma_{V-A}(v_e e \to v_e e)} = \begin{cases} 0.9 \pm 0.2, \text{ Gargamelle} \\ 0.98 \pm 0.18, \text{ CHARM} \end{cases} \qquad (6.1.76)$$

Inverse muon decay is a particularly interesting process for experimental study because it is purely leptonic and thus free from some of the ambiguities that will present themselves in the case of semileptonic interactions. The small cross section has been an impediment to detailed study. However, the reaction is also well suited to exposing the limitations of the effective Lagrangian approach. The angular distribution is, according to (6.1.73), isotropic in the c.m. system, which is to say that the scattering is purely s-wave. Partial-wave unitarity constrains the modulus of an inelastic partial-wave amplitude to be

$$|\mathscr{M}_J| < 1, \qquad (6.1.77)$$

where the partial-wave expansion for the scattering amplitude (here written for spinless external particles) is

$$f(\theta) = \left(2\frac{d\sigma}{d\Omega}\right)^{1/2} = \frac{1}{\sqrt{s}}\sum_{J=0}^{\infty}(2J+1)P_J(\cos\theta)\mathscr{M}_J, \qquad (6.1.78)$$

with $P_J(x)$ a Legendre polynomial. The explicit factor of 2 in (6.1.78) serves to undo the average over initial electron spins. The constraint (6.1.77) is equivalent to the familiar restriction

$$\sigma < \pi/p_{c.m.}^2 \tag{6.1.79}$$

for inelastic s-wave scattering. For inverse muon decay, we easily compute

$$\mathcal{M}_0 = \frac{G_F s}{\pi\sqrt{2}}\left[1 - \frac{(\mu^2 - m)^2}{s}\right]$$

$$\simeq G_F s/\pi\sqrt{2}, \tag{6.1.80}$$

which implies the unitarity constraint

$$G_F s/\pi\sqrt{2} < 1. \tag{6.1.81}$$

This means that the four-fermion theory can make sense only if

$$s < \pi\sqrt{2}/G_F, \tag{6.1.82}$$

which is to say that

$$\sqrt{s} < 617 \text{ GeV} \tag{6.1.83}$$

or

$$p_{c.m.} < 309 \text{ GeV}/c. \tag{6.1.84}$$

Although the energy at which unitarity is violated by the point-coupling theory is elevated, it is not astronomical. We must therefore take seriously the objection that the effective point-coupling theory is unacceptable, as we have formulated it. Suppose now that (6.1.62) is taken as a true Lagrangian, and not merely an effective one. Could the violation of unitarity that we have just exhibited be remedied in higher orders of perturbation theory? In second order we must compute the diagram shown in Fig. 6-5, for which

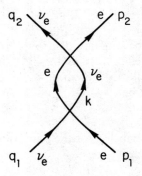

FIG. 6-5. Second-order contribution to $\nu_e e$ elastic scattering in the four-fermion (point-coupling) theory.

the matrix element is readily seen to be

$$\mathcal{M} \propto \int d^4k/k^2, \tag{6.1.85}$$

which is quadratically divergent. In fact, the divergence difficulty of the point-coupling theory grows more severe in each order of perturbation theory. In order to be saved—and this is evidently desirable because of its successes for low-energy phenomenology—the theory must be modified in a fundamental way. The simplest attempts to salvage the theory are the subject of the next section.

6.2   Intermediate Vector Bosons

Although Fermi's phenomenological interaction was inspired by the theory of electromagnetism, the analogy was not complete, and one may hope to obtain a more satisfactory theory by pushing the analogy further. An obvious device is to assume that the weak interaction, like quantum electrodynamics, is mediated by vector boson exchange. The weak intermediate boson must have the following three properties:

(*i*) It carries charge ± 1, because the familiar manifestations of the weak interactions (such as β-decay) are charge-changing;

(*ii*) It must be rather massive, to reproduce the short range of the weak force;

(*iii*) Its parity must be indefinite.

Furthermore, its couplings to the fermions are determined by the low-energy phenomenology. At each veW vertex will occur a factor

$$-i(G_F M_W^2/\sqrt{2})^{1/2}\gamma_\mu(1 - \gamma_5), \tag{6.2.1}$$

where M_W is the mass of the weak intermediate boson. The intermediate vector boson propagator has the standard form for a massive vector meson, derived in Problem 4-2,

$$\frac{-i(g_{\mu v} - k_\mu k_v/M_W^2)}{k^2 - M_W^2}. \tag{6.2.2}$$

According to this picture, inverse muon decay proceeds by the t-channel exchange diagram shown in Fig. 6-6. The Feynman amplitude is

$$\mathcal{M} = \frac{iG_F M_W^2}{\sqrt{2}} \bar{u}(v, q_2)\gamma^\mu(1 - \gamma_5)u(e, p_1)\frac{g_{\mu v} - k_\mu k_v/M_W^2}{k^2 - M_W^2}$$

$$\times \bar{u}(e, p_2)\gamma^v(1 - \gamma_5)u(v, q_1), \tag{6.2.3}$$

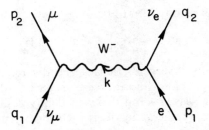

FIG. 6-6. Lowest-order contribution to inverse muon decay, in the intermediate boson model.

with $k = p_2 - q_1 = p_1 - q_2$. By substituting $k_\mu = (p_1 - q_2)_\mu$ and $k_\nu = (p_2 - q_1)_\nu$, and using the Dirac equation, we find that the contribution of the $k_\mu k_\nu$ term in the propagator is of order (m^2/M_W^2) and may be safely neglected. What remains is simply the amplitude we encountered in the point-coupling theory, times $M_W^2/(M_W^2 - k^2)$. The point-coupling result is recovered in the limit $M_W \to \infty$. Without further computation, we have

$$\frac{d\sigma}{d\Omega_{c.m.}}(\nu_\mu e \to \mu\nu_e) \simeq \frac{G_F^2 mE[1 - (\mu^2 - m^2)/2mE]^2}{2\pi^2[1 + mE(1 - \cos\theta)/M_W^2]^2}, \qquad (6.2.4)$$

where m and μ have been neglected with respect to M_W and E. The integrated cross section

$$\sigma(\nu_\mu e \to \mu\nu_e) \simeq \frac{2G_F^2 mE}{\pi} \frac{[1 - (\mu^2 - m^2)/2mE]^2}{(1 + 2mE/M_W^2)} \qquad (6.2.5)$$

reduces to the form (6.1.74) at modest energies, but approaches a constant value at high energies:

$$\lim_{E \to \infty} \sigma(\nu_\mu e \to \mu\nu_e) = G_F^2 M_W^2/\pi. \qquad (6.2.6)$$

This is a great improvement over the point-coupling theory, but a problem remains: the s-wave amplitude violates partial-wave unitarity. To see this we write

$$\mathcal{M}_0 = \frac{\sqrt{s}}{2} \int_{-1}^{1} d(\cos\theta) f(\theta)$$

$$= \frac{G_F mE}{\pi\sqrt{2}} \int_{-1}^{1} dz \left[1 + \frac{mE}{M_W^2}(1 - z)\right]^{-1}$$

$$= \frac{G_F M_W^2}{\pi\sqrt{2}} \log\left(1 + \frac{2mE}{M_W^2}\right)$$

$$= \frac{G_F M_W^2}{\pi\sqrt{2}} \log(1 + s/M_W^2). \qquad (6.2.7)$$

In this instance the unitarity constraint (6.1.77) is respected, so long as

$$s < M_W^2 \left[\exp\left(\frac{\pi\sqrt{2}}{G_F M_W^2} \right) - 1 \right]. \qquad (6.2.8)$$

Notice that in the point-coupling limit $M_W \to \infty$ we recover the result (6.1.82). These violations of unitarity arise at incredibly high energies (for $M_W = 100$ GeV/c^2, for example, unitarity is respected up to beam energies of 3.5×10^{23} GeV), and it might be hoped that the very mild violation of unitarity could be conquered in higher orders of perturbation theory. This is not the case. The $k_\mu k_\nu$ term in the W-boson propagator makes the theory nonrenormalizable by inducing new divergences in each order of perturbation theory. Equivalently, the introduction of the intermediate boson causes divergence problems in new physical processes, as we shall see presently, which make this *ad hoc* theory unacceptable. For the moment, however, let us continue to investigate the consequences of the W in leptonic interactions.

In the crossed reaction

$$\bar{\nu}_e e \to \bar{\nu}_\mu \mu, \qquad (6.2.9)$$

the intermediate boson appears as a direct-channel resonance, as shown in Fig. 6-7. The differential cross section is

$$\frac{d\sigma}{d\Omega}(\bar{\nu}_e e \to \bar{\nu}_\mu \mu) = \frac{G_F^2 m E}{8\pi^2} \frac{(1 - \cos\theta)^2}{(1 - 2mE/M_W^2)^2} \left[1 - \frac{(\mu^2 - m^2)}{2mE} \right]^2, \qquad (6.2.10)$$

which differs from the point-coupling cross section only by the factor $1/(1 - 2mE/M_W^2)^2$. The total cross section becomes

$$\sigma(\bar{\nu}_e e \to \bar{\nu}_\mu \mu) = \frac{2mEG_F^2[1 - (\mu^2 - m^2)/2mE]^2}{3\pi(1 - 2mE/M_W^2)^2}, \qquad (6.2.11)$$

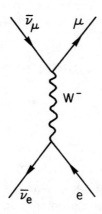

FIG. 6-7. Lowest-order contribution to the reaction $\bar{\nu}_e e \to \bar{\nu}_\mu \mu$, in the intermediate boson model.

which vanishes in the high-energy limit,

$$\lim_{E \to \infty} \sigma(\bar{v}_e e \to \bar{v}_\mu \mu) = G_F^2 M_W^4 / 6\pi m E. \qquad (6.2.12)$$

The apparent infinity in the cross section at $s = 2mE = M_W^2$ arises only because we have neglected the width of the unstable intermediate boson, an oversight we shall remedy at once. That aside, the unitarity problems for leptonic processes may be regarded as solved, at least in a practical sense, by the tree diagrams involving W exchange.

Nothing that we have said fixes the mass of the intermediate boson, which remains a parameter. (On present experimental evidence, to be reviewed in Section 7.3, $M_W \gtrsim 30 \text{ GeV}/c^2$.) Once the mass is specified, however, the decay characteristics of the intermediate vector boson are essentially determined by the requirement that the low-energy phenomenology of the $V - A$ theory be reproduced. In particular, we can predict definitely the rates for leptonic decays of the W.

In the rest frame of the W^-, the outgoing momenta for the decay

$$W^- \to e(p) + \bar{v}_e(q) \qquad (6.2.13)$$

are

$$\left. \begin{array}{l} p = (M_W/2)(1; \sin\theta, 0, \cos\theta) \\ q = (M_W/2)(1; -\sin\theta, 0, -\cos\theta) \end{array} \right\}, \qquad (6.2.14)$$

where the electron mass has been neglected. The matrix element for the decay is given by

$$\mathcal{M} = -i\left(\frac{G_F M_W^2}{\sqrt{2}}\right)^{1/2} \bar{u}(e, p)\gamma_\mu(1 - \gamma_5)v(v, q)\varepsilon^\mu, \qquad (6.2.15)$$

where $\varepsilon^\mu \equiv (0; \hat{\varepsilon})$ is the polarization vector of the decaying particle. Squaring the matrix element and summing over final-particle spins, we have, neglecting the electron mass,

$$|\mathcal{M}|^2 = \frac{G_F M_W^2}{\sqrt{2}} \text{tr}[\not{\varepsilon}(1 - \gamma_5)\not{q}(1 + \gamma_5)\not{\varepsilon}^*\not{p}]$$

$$= \frac{G_F M_W^2}{\sqrt{2}} 2\text{tr}[(1 + \gamma_5)\not{\varepsilon}\not{q}\not{\varepsilon}^*\not{p}]$$

$$= \frac{8G_F M_W^2}{\sqrt{2}}(\varepsilon \cdot q\varepsilon^* \cdot p - \varepsilon \cdot \varepsilon^* p \cdot q + \varepsilon \cdot p\varepsilon^* \cdot q + i\varepsilon_{\mu\nu\rho\sigma}\varepsilon^\mu q^\nu \varepsilon^{*\rho}p^\sigma). \quad (6.2.16)$$

The decay rate must be independent of the polarization of the decaying particle, so we choose first the simplest case of longitudinal polarization, helicity zero,

$$\varepsilon^\mu = (0; 0, 0, 1) = \varepsilon^{\mu*}, \qquad (6.2.17)$$

for which the $\varepsilon_{\mu\nu\rho\sigma}$ term vanishes. We then have at once

$$|\mathcal{M}|^2 = \frac{4G_F M_W^4}{\sqrt{2}} \sin^2 \theta \qquad (6.2.18)$$

so that, by virtue of (B.1.4), the differential decay rate

$$\frac{d\Gamma}{d\Omega}(W_0^- \to e^- \bar{\nu}_e) = \frac{|\mathcal{M}|^2}{64\pi^2 M_W} = \frac{G_F M_W^3}{16\pi^2 \sqrt{2}} \sin^2 \theta. \qquad (6.2.19)$$

The leptonic decay rate is therefore

$$\Gamma(W^- \to e^- \bar{\nu}_e) = \frac{G_F M_W^3}{6\pi \sqrt{2}} .$$

$$\simeq 437 \text{ MeV} \left(\frac{M_W}{100 \text{ GeV}/c^2} \right)^3. \qquad (6.2.20)$$

The difficulty of the calculation is not greatly increased for other polarizations. For helicity $= +1$,

$$\varepsilon^\mu = (0; -1, -i, 0)/\sqrt{2}, \qquad (6.2.21)$$

the square of the matrix element is

$$|\mathcal{M}|^2 = \frac{G_F M_W^4}{\sqrt{2}} 2[(1 + \cos^2 \theta) - 2 \cos \theta], \qquad (6.2.22)$$

where the final term is the contribution from $\varepsilon_{\mu\nu\rho\sigma}$. Consequently the differential decay rate is given by

$$\frac{d\Gamma}{d\Omega}(W_{+1}^- \to e^- \bar{\nu}_e) = \frac{G_F M_W^3}{32\pi^2 \sqrt{2}}(1 - \cos \theta)^2, \qquad (6.2.23)$$

which corresponds (as it must) to a total rate given by (6.2.20). The result for helicity $= -1$ is simply obtained by replacing $\varepsilon^\mu \leftrightarrow \varepsilon^{\mu *}$ in the expression for $|\mathcal{M}|^2$. This yields

$$\frac{d\Gamma}{d\Omega}(W_{-1}^- \to e^- \bar{\nu}_e) = \frac{G_F M_W^3}{32\pi^2 \sqrt{2}}(1 + \cos \theta)^2. \qquad (6.2.24)$$

The angular dependences merely reflect the helicity correlations of the $V - A$ theory. This is illustrated in Fig. 6-8. Since the electron emitted in W^- decay is left-handed, it is forbidden to travel parallel to the direction of W^- polarization. For the decay $W^+ \to e^+ \nu_e$ the situation is just reversed: the emitted positron, being right-handed, tends to follow the direction of the W^+ polarization. This effect is an example of C-violation in the weak

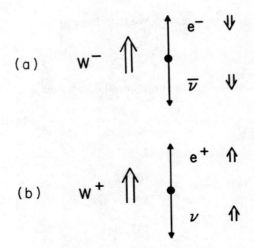

FIG. 6-8. Spin configurations for the leptonic decay of longitudinally polarized intermediate bosons. (a) $W^- \to e^- \bar{\nu}_e$; (b) $W^+ \to e^+ \nu_e$.

interactions, which is expected to play an important role in the search for intermediate bosons in high-energy $\bar{p}p$ collisions.

We have already mentioned that, whereas the introduction of the intermediate boson softens the divergence of the s-wave amplitude for the process $\nu_\mu e \to \mu \nu_e$, it gives rise to new divergences in other processes. The most celebrated example occurs in the reaction

$$\nu_e(q_1) + \bar{\nu}_e(q_2) \to W^+(k_+) + W^-(k_-). \qquad (6.2.25)$$

According to the picture we have developed until now, we need only evaluate the single diagram shown in Fig. 6-9. We define

$$P = q_1 - k_+ = k_- - q_2. \qquad (6.2.26)$$

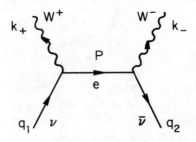

FIG. 6-9. Lowest-order contribution to the reaction $\nu \bar{\nu} \to W^+ W^-$, in the intermediate boson model.

The matrix element is

$$\mathcal{M} = -\frac{iG_F M_W^2}{\sqrt{2}} \bar{v}(v, q_2)\not{\epsilon}_-^*(1 - \gamma_5)\frac{(\not{P} + m)}{P^2 - m^2}\not{\epsilon}_+^*(1 - \gamma_5)u(v, q_1), \quad (6.2.27)$$

and in the c.m. frame the momentum four-vectors are

$$\left.\begin{aligned}
q_1 &\equiv (Q; 0, 0, Q) \\
q_2 &= (Q; 0, 0, -Q) \\
k_+ &= (Q; K \sin\theta, 0, K \cos\theta) \\
k_- &= (Q; -K \sin\theta, 0, -K \cos\theta)
\end{aligned}\right\}. \quad (6.2.28)$$

If $\varepsilon_\pm^\mu = (0; \hat{\boldsymbol{\varepsilon}}_\pm)$ represents the $W^\pm$ polarization in its rest frame, then in the c.m. frame we have, by virtue of a boost,

$$\varepsilon_\pm^\mu = \left[\frac{\mathbf{k}_\pm \cdot \hat{\boldsymbol{\varepsilon}}_\pm}{M_W}; \hat{\boldsymbol{\varepsilon}}_\pm + \frac{\mathbf{k}_\pm(\mathbf{k}_\pm \cdot \hat{\boldsymbol{\varepsilon}}_\pm)}{M_W(Q + M_W)}\right]. \quad (6.2.29)$$

For the case of longitudinal polarization, the c.m. polarization vector is thus

$$\varepsilon_\pm^\mu = \left(\frac{K}{M_W}; \frac{Q\hat{\mathbf{k}}_\pm}{M_W}\right), \quad (6.2.30)$$

which, in the limit of high energies $Q \simeq K \gg M_W$ approaches the limiting form

$$\lim_{K \to \infty} \varepsilon_\pm \to k_\pm/M_W. \quad (6.2.31)$$

The amplitude for the production of longitudinally polarized intermediate bosons at high energies thus becomes

$$\mathcal{M} = \frac{-iG_F}{\sqrt{2}} \bar{v}(v, q_2)\not{k}_-(1 - \gamma_5)\frac{(\not{P} + m)}{P^2 - m^2}\not{k}_+(1 - \gamma_5)u(v, q_1). \quad (6.2.32)$$

Using the Dirac equations

$$\left.\begin{aligned}
\not{q}_1 u(v, q_1) &= 0 \\
\bar{v}(v, q_2)\not{q}_2 &= 0
\end{aligned}\right\} \quad (6.2.33)$$

we replace

$$\not{k}_+ \to \not{k}_+ - \not{q}_1 = -\not{P} \quad (6.2.34)$$

and

$$\not{k}_- \to \not{k}_- - \not{q}_2 = \not{P} \quad (6.2.35)$$

to obtain

$$\mathcal{M} = \frac{iG_F}{\sqrt{2}} \bar{v}(v, q_2)\not{P}(1 - \gamma_5)\frac{\not{P}\not{P}}{P^2}(1 - \gamma_5)u(v, q_1), \quad (6.2.36)$$

where we have everywhere discarded the electron mass m. Since $\not{P}\not{P} = P^2$, we now have

$$\mathcal{M} = iG_F\sqrt{2}\bar{v}(v, q_2)\not{P}(1 + \gamma_5)u(v, q_1). \tag{6.2.37}$$

To display the high-energy behavior, it is simplest to evaluate

$$
\begin{aligned}
|\mathcal{M}|^2 &= 2G_F^2 \operatorname{tr}[\not{P}(1 - \gamma_5)\not{q}_1(1 + \gamma_5)\not{P}\not{q}_2] \\
&= 4G_F^2 \operatorname{tr}[(1 - \gamma_5)\not{q}_1\not{P}\not{q}_2\not{P}] \\
&= 16G_F^2(2q_1 \cdot P q_2 \cdot P - q_1 \cdot q_2 P^2) \\
&= 32G_F^2 Q^2 K^2 \sin^2\theta. \tag{6.2.38}
\end{aligned}
$$

The angular dependence is again readily understood in terms of angular momentum conservation. At high energies, the cross section becomes approximately

$$\sigma \simeq G_F^2 s / 3\pi, \tag{6.2.39}$$

in gross violation of unitarity.

Equivalently, we may remark that the second-order contribution to the neutrino–antineutrino scattering amplitude, given by the box diagram in Fig. 6-10, which is the square of the graph in Fig. 6-9, grows as s^2. Since the $k_\mu k_\nu$ piece of the W-boson propagator gives rise to more severe divergences in each order of perturbation theory, it seems imperative to find some means of eradicating the divergence order by order. One may accomplish this by inventing new physical particles, which will appear in new Feynman diagrams in which the couplings are astutely chosen to cause cancellations of the offending divergences.

In the case of the reaction $v\bar{v} \to W^+ W^-$, for which we have the diagram of Fig. 6-9, a possible solution[5] (though apparently not the one chosen by Nature) is to postulate an *antilepton* E^- with the same lepton number as

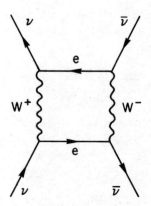

FIG. 6-10. Second-order contribution to neutrino–antineutrino elastic scattering, in the intermediate boson model.

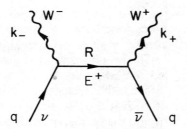

FIG. 6-11. Paraelectron-exchange contribution to the reaction $v\bar{v} \to W^+W^-$.

e^+ and $\bar{v}_e$, which gives rise to the diagram shown in Fig. 6-11. If the vEW coupling is chosen to be $-i[(G_E M_W^2)/\sqrt{2})]^{1/2}\gamma_\mu(1 - \gamma_5)$, a V − A form with arbitrary strength, the amplitude corresponding to Fig. 6-11 is simply

$$\mathcal{M} = \frac{-iG_E M_W^2}{\sqrt{2}}\bar{v}(v,q_2)\slashed{\epsilon}_+^*(1 - \gamma_5)\frac{(\slashed{R} + m_E)}{R^2 - m_E^2}\slashed{\epsilon}_-^*(1 - \gamma_5)u(v,q_1), \quad (6.2.40)$$

where

$$R \equiv q_1 - k_- = k_+ - q_2 = -P + q_1 - q_2, \quad (6.2.41)$$

and P was defined in (6.2.26). Neglecting fermion masses and taking both W^+ and W^- to be longitudinally polarized, we have, using (6.2.31),

$$\mathcal{M} = \frac{-iG_E}{\sqrt{2}}\bar{v}(v,q_2)\slashed{k}_+(1 - \gamma_5)\frac{\slashed{R}\slashed{k}_-}{R^2}(1 - \gamma_5)u(v,q_1). \quad (6.2.42)$$

Between spinors we may replace

$$k_- \to -\slashed{R} \quad (6.2.43)$$

and

$$k_+ \to -\slashed{P}, \quad (6.2.44)$$

so the amplitude becomes

$$\mathcal{M} = -iG_E\sqrt{2}\bar{v}(v,q_2)\slashed{P}(1 - \gamma_5)u(v,q_1), \quad (6.2.45)$$

which is precisely the negative of the original amplitude (6.2.37), provided that the new coupling constant is chosen equal to the old,

$$G_E = G_F. \quad (6.2.46)$$

This sort of order-by-order divergence cancellation is familiar in electro-dynamics (see Problems 3-5 and 6-7), where it is ensured by gauge invariance. Within the context of perturbation theory, such a cancellation is a pre-requisite to renormalizability, because cancellations between different orders would be unstable against small changes in the coupling constant.

It is remarkable that a full program of divergence cancellation can be carried out systematically for the weak and electromagnetic interactions.[6] More remarkable still is that systematic pursuit of the program leads to theories with precisely the interaction structure of spontaneously broken gauge theories. This perhaps reflects the deep connection between gauge invariance, "safe" high-energy behavior, and renormalizability. It should not be left unsaid that for many physicists (the author among them) the intricate cancellations that one witnesses in the course of a gauge theory calculation provide—together with the elegance of the gauge principle and the experimental support specific models have received, to be sure—an important reason for believing that gauge theories must have something to do with reality. On that subjective note, let us now regard from the gauge perspective a realistic model of the weak and electromagnetic interactions.

6.3   The Standard Model

The minimal model now regarded as "standard" is the result of a long development and many ingenious contributions, among which those of S. L. Glashow, S. Weinberg, and A. Salam are especially notable. We first consider the model in its purely leptonic form, which exhibits the motivation and the principal features.

We begin by designating the spectrum of fundamental fermions of the theory. It suffices for the moment to include only the electron and its neutrino, which form a left-handed "weak-isospin" doublet,

$$\mathsf{L} \equiv \begin{pmatrix} v \\ e \end{pmatrix}_{\mathsf{L}}, \qquad (6.3.1)$$

where the left-handed states are

$$v_{\mathsf{L}} = \tfrac{1}{2}(1 - \gamma_5)v$$
$$e_{\mathsf{L}} = \tfrac{1}{2}(1 - \gamma_5)e. \qquad (6.3.2)$$

The electron neutrino is known to be nearly massless. It is convenient (but by no means essential) to idealize it as exactly massless, in which case the right-handed state

$$v_{\mathsf{R}} = \tfrac{1}{2}(1 + \gamma_5)v = 0 \qquad (6.3.3)$$

does not exist. Thus we designate only one right-handed fermion,

$$\mathsf{R} \equiv e_{\mathsf{R}} = \tfrac{1}{2}(1 + \gamma_5)e, \qquad (6.3.4)$$

which is a weak-isospin singlet. This completes a specification of the weak charged currents. To incorporate electromagnetism, we define a "weak hypercharge," Y. Requiring that the Gell-Mann–Nishijima relation for electric charge,

$$Q = I_3 + \tfrac{1}{2}Y \qquad (6.3.5)$$

be satisfied leads to the assignments

$$Y_L = -1,$$
$$Y_R = -2. \tag{6.3.6}$$

By construction, the weak-isospin projection I_3 and the weak hypercharge Y are commuting observables,

$$[I_3, Y] = 0. \tag{6.3.7}$$

We now take the (product) group of transformations generated by I and Y to be the gauge group $SU(2)_L \otimes U(1)_Y$ of a gauge theory. To construct the theory, we introduce the gauge bosons

$$b_\mu^1, b_\mu^2, b_\mu^3 \qquad \text{for } SU(2)_L,$$
$$\mathscr{A}_\mu \qquad \text{for } U(1)_Y.$$

Evidently the Lagrangian for the theory may be written as

$$\mathscr{L} = \mathscr{L}_{\text{gauge}} + \mathscr{L}_{\text{leptons}}, \tag{6.3.8}$$

where the kinetic term for the gauge fields is

$$\mathscr{L}_{\text{gauge}} = -\tfrac{1}{4} F_{\mu\nu}^l F^{l\mu\nu} - \tfrac{1}{4} f_{\mu\nu} f^{\mu\nu}, \tag{6.3.9}$$

and the field-strength tensors are $\quad SU(2), \ p76.$

$$F_{\mu\nu}^l = \partial_\nu b_\mu^l - \partial_\mu b_\nu^l + g\varepsilon_{jkl} b_\mu^j b_\nu^k \tag{6.3.10}$$

for the $SU(2)_L$ gauge fields [cf. (4.2.37)] and

$$f_{\mu\nu} = \partial_\nu \mathscr{A}_\mu - \partial_\mu \mathscr{A}_\nu \tag{6.3.11}$$

for the $U(1)_Y$ gauge field [cf. (3.2.10)]. The matter term is

$$\mathscr{L}_{\text{leptons}} = \bar{R} i\gamma^\mu \left(\partial_\mu + \frac{ig'}{2} \mathscr{A}_\mu Y \right) R + \bar{L} i\gamma^\mu \left(\partial_\mu + \frac{ig'}{2} \mathscr{A}_\mu Y + \frac{ig}{2} \boldsymbol{\tau} \cdot \mathbf{b}_\mu \right) L. \tag{6.3.12}$$

The coupling constant for the weak-isospin group $SU(2)_L$ is called g, as in the Yang–Mills theory of Chapter 4, and the coupling constant for the weak-hypercharge group $U(1)_Y$ is denoted as $g'/2$, the factor $1/2$ being chosen to simplify later expressions.

The theory of weak and electromagnetic interactions described by the Lagrangian (6.3.8) is not a satisfactory one, for two immediately obvious reasons. It contains four massless gauge bosons $(b^1, b^2, b^3, \mathscr{A})$, whereas Nature has but one, the photon. In addition, the global $SU(2)$ invariance forbids a mass term for the electron. Our task is to modify the theory so that there will remain only a single conserved quantity (the electric charge) corresponding to one massless gauge boson (the photon), and the electron will acquire a mass.

To accomplish these things, we introduce a complex doublet of scalar fields

$$\phi \equiv \begin{pmatrix} \phi^+ \\ \phi^0 \end{pmatrix} \tag{6.3.13}$$

which transforms as an $SU(2)_L$ doublet and must therefore have weak hypercharge

$$Y_\phi = +1, \tag{6.3.14}$$

by virtue of the Gell-Mann–Nishijima relation (6.3.5). We add to the Lagrangian a term

$$\mathscr{L}_{\text{scalar}} = (\mathscr{D}^\mu \phi)^\dagger (\mathscr{D}_\mu \phi) - V(\phi^\dagger \phi), \tag{6.3.15}$$

where the covariant derivative is

$$\mathscr{D}_\mu = \partial_\mu + \frac{ig'}{2}\mathscr{A}_\mu Y + \frac{ig}{2}\boldsymbol{\tau} \cdot \mathbf{b}_\mu \tag{6.3.16}$$

and as usual the potential is

$$V(\phi^\dagger \phi) = \mu^2(\phi^\dagger \phi) + |\lambda|(\phi^\dagger \phi)^2. \tag{6.3.17}$$

We are also free to add an interaction term, which involves Yukawa couplings of the scalars to the fermions,

$$\mathscr{L}_{\text{Yukawa}} = -G_e[\overline{\mathsf{R}}(\phi^\dagger \mathsf{L}) + (\overline{\mathsf{L}}\phi)\mathsf{R}], \tag{6.3.18}$$

which is symmetric under local $SU(2)_L \otimes U(1)_Y$ transformations and is a Lorentz scalar.

Now let us imagine that $\mu^2 < 0$ and consider the consequences of spontaneous symmetry breaking. We choose as the vacuum expectation value of the scalar field

$$\langle\phi\rangle_0 = \begin{pmatrix} 0 \\ v/\sqrt{2} \end{pmatrix}, \tag{6.3.19}$$

where $v = \sqrt{-\mu^2/|\lambda|}$, which breaks both $SU(2)_L$ and $U(1)_Y$ symmetries, but preserves an invariance under the $U(1)_{\text{EM}}$ symmetry generated by the electric charge operator. Recall that a (would-be) Goldstone boson is associated with every generator of the gauge group that does not leave the vacuum invariant. The vacuum is left invariant by a generator $\mathscr{G}$ if

$$e^{i\alpha\mathscr{G}}\langle\phi\rangle_0 = \langle\phi\rangle_0. \tag{6.3.20}$$

For an infinitesimal transformation, the equation becomes

$$(1 + i\alpha\mathscr{G})\langle\phi\rangle_0 = \langle\phi\rangle_0, \tag{6.3.21}$$

so that the condition for $\mathscr{G}$ to leave the vacuum invariant is simply

$$\mathscr{G}\langle\phi\rangle_0 = 0. \tag{6.3.22}$$

For the generators of $SU(2)_L \otimes U(1)_Y$, we compute

$$\tau_1\langle\phi\rangle_0 = \begin{pmatrix} 0 & 1 \\ 1 & 0 \end{pmatrix}\begin{pmatrix} 0 \\ v/\sqrt{2} \end{pmatrix} = \begin{pmatrix} v/\sqrt{2} \\ 0 \end{pmatrix} \neq 0, \tag{6.3.23}$$

$$\tau_2\langle\phi\rangle_0 = \begin{pmatrix} 0 & -i \\ i & 0 \end{pmatrix}\begin{pmatrix} 0 \\ v/\sqrt{2} \end{pmatrix} = \begin{pmatrix} -iv/\sqrt{2} \\ 0 \end{pmatrix} \neq 0, \tag{6.3.24}$$

$$\tau_3\langle\phi\rangle_0 = \begin{pmatrix} 1 & 0 \\ 0 & -1 \end{pmatrix}\begin{pmatrix} 0 \\ v/\sqrt{2} \end{pmatrix} = \begin{pmatrix} 0 \\ -v/\sqrt{2} \end{pmatrix} \neq 0, \tag{6.3.25}$$

$$Y\langle\phi\rangle_0 = +1\langle\phi\rangle_0 \neq 0, \tag{6.3.26}$$

but

$$Q\langle\phi\rangle_0 = \tfrac{1}{2}(\tau_3 + Y)\langle\phi\rangle_0 = 0. \tag{6.3.27}$$

This is promising! All the original four generators are broken, but the linear combination corresponding to electric charge is not. The photon will therefore remain massless, while three other gauge bosons will acquire mass.

We next expand the Lagrangian about the minimum of the Higgs potential V by writing

$$\phi = \exp\left(\frac{i\boldsymbol{\zeta}\cdot\boldsymbol{\tau}}{2v}\right)\begin{pmatrix} 0 \\ (v+\eta)/\sqrt{2} \end{pmatrix} \tag{6.3.28}$$

and transforming at once to U-gauge:

$$\phi \to \phi' = \exp\left(\frac{-i\boldsymbol{\zeta}\cdot\boldsymbol{\tau}}{2v}\right)\phi = \begin{pmatrix} 0 \\ (v+\eta)/\sqrt{2} \end{pmatrix}, \tag{6.3.29}$$

$$\boldsymbol{\tau}\cdot\mathbf{b}_\mu \to \boldsymbol{\tau}\cdot\mathbf{b}'_\mu, \tag{6.3.30}$$

$$\mathscr{A}_\mu \to \mathscr{A}_\mu, \tag{6.3.31}$$

$$R \to R, \tag{6.3.32}$$

$$L \to L' = \exp\left(\frac{-i\boldsymbol{\zeta}\cdot\boldsymbol{\tau}}{2v}\right)L. \tag{6.3.33}$$

Had we followed literally the procedure of Section 5.4, we would have replaced the generator τ_3 in the expansion (6.3.28) of ϕ about $\langle\phi\rangle_0$ by the combination $K = (\tau_3 - Y)/2$ orthogonal to Q, which is strictly speaking the third broken generator. However, since Q leaves the vacuum invariant and $\tau_3 = K + Q$, the effect is the same.

We may now reëxpress the Lagrangian in terms of the U-gauge fields (6.3.29–33), omitting primes to avoid notational clutter, and investigate the consequences of spontaneous symmetry breaking. The Yukawa term in the

Lagrangian has become

$$\mathscr{L}_{\text{Yukawa}} = -G_e \frac{(v + \eta)}{\sqrt{2}} (\bar{e}_R e_L + \bar{e}_L e_R)$$

$$= \frac{-G_e v}{\sqrt{2}} \bar{e}e - \frac{G_e \eta}{\sqrt{2}} \bar{e}e, \tag{6.3.34}$$

so the electron has acquired a mass

$$m_e = G_e v/\sqrt{2}. \tag{6.3.35}$$

The scalar term in the Lagrangian now reads

$$\mathscr{L}_{\text{scalar}} = \frac{1}{2}(\partial^\mu \eta)(\partial_\mu \eta) - \mu^2 \eta^2$$

$$+ \frac{v^2}{8} [g^2 |b_\mu^1 - ib_\mu^2|^2 + (g' \mathscr{A}_\mu - g b_\mu^3)^2] + \cdots \tag{6.3.36}$$

plus interaction terms. We see at once that the η field has acquired a (mass)^2 $M_H^2 = -2\mu^2 > 0$; it is the physical Higgs boson. If we define the charged gauge fields

$$W_\mu^\pm \equiv \frac{b_\mu^1 \mp ib_\mu^2}{\sqrt{2}}, \tag{6.3.37}$$

the term proportional to $g^2 v^2$ is recognizable as a mass term for the charged vector bosons:

$$\frac{g^2 v^2}{8}(|W_\mu^+|^2 + |W_\mu^-|^2), \tag{6.3.38}$$

corresponding to charged intermediate boson masses

$$M_{W^\pm} = gv/2. \tag{6.3.39}$$

Finally, defining the orthogonal combinations

$$Z_\mu = \frac{-g' \mathscr{A}_\mu + g b_\mu^3}{\sqrt{g^2 + g'^2}} \tag{6.3.40}$$

and

$$A_\mu = \frac{g \mathscr{A}_\mu + g' b_\mu^3}{\sqrt{g^2 + g'^2}}, \tag{6.3.41}$$

we find that neutral intermediate boson has acquired a mass

$$M_{Z^0} = \sqrt{g^2 + g'^2}\, v/2$$

$$= M_W \sqrt{1 + g'^2/g^2}, \tag{6.3.42}$$

and that the field A_μ remains a massless gauge boson corresponding to the surviving $\exp[iQ\alpha(x)]$ symmetry. We have achieved, at least schematically, the desired particle content—plus a massive Higgs scalar we did not request.

Do the interactions also correspond to those in Nature? The interactions among the gauge bosons and leptons may be read off from $\mathscr{L}_{\text{leptons}}$ (6.3.12). For the charged gauge bosons we find

$$\mathscr{L}_{W-l} = -\frac{g}{\sqrt{2}}(\bar{\nu}_L\gamma^\mu e_L W^+_\mu + \bar{e}_L\gamma^\mu\nu_L W^-_\mu)$$

$$= -\frac{g}{2\sqrt{2}}[\bar{\nu}\gamma^\mu(1-\gamma_5)eW^+_\mu + \bar{e}\gamma^\mu(1-\gamma_5)\nu W^-_\mu], \quad (6.3.43)$$

which reproduces the low-energy phenomenology of the *ad hoc* intermediate boson model of Section 6.2, provided we identify the coupling constants as

$$g^2/8 = G_F M_W^2/\sqrt{2}. \quad (6.3.44)$$

With the aid of the expression (6.3.39) for the intermediate boson mass, we find that the vacuum expectation value parameter v is now determined as

$$v = (G_F\sqrt{2})^{-1/2} \simeq 246 \text{ GeV}, \quad (6.3.45)$$

so that the vacuum expectation value of the scalar field is

$$\langle\phi^0\rangle_0 = (G_F\sqrt{8})^{-1/2} \simeq 174 \text{ GeV}. \quad (6.3.46)$$

Similarly, the neutral gauge boson couplings to leptons are given by

$$\mathscr{L}_{0-l} = \frac{gg'}{\sqrt{g^2+g'^2}}\bar{e}\gamma^\mu e A_\mu$$

$$-\frac{\sqrt{g^2+g'^2}}{2}\bar{\nu}_L\gamma^\mu\nu_L Z_\mu$$

$$+\frac{Z_\mu}{\sqrt{g^2+g'^2}}\left[-g'^2\bar{e}_R\gamma^\mu e_R + \frac{(g^2-g'^2)}{2}\bar{e}_L\gamma^\mu e_L\right]. \quad (6.3.47)$$

Therefore we may indeed identify A_μ as the photon, provided that we set

$$gg'/\sqrt{g^2+g'^2} = e. \quad (6.3.48)$$

It is convenient to introduce a weak mixing angle θ_W to parametrize the mixing of the neutral gauge bosons. With the definition

$$g' = g\tan\theta_W, \quad (6.3.49)$$

whence

$$\sqrt{g^2+g'^2} = g/\cos\theta_W, \quad (6.3.50)$$

equations (6.3.40,41) may be rewritten as

$$Z_\mu = -\mathscr{A}_\mu\sin\theta_W + b^3_\mu\cos\theta_W, \quad (6.3.51)$$

$$A_\mu = \mathscr{A}_\mu\cos\theta_W + b^3_\mu\sin\theta_W, \quad (6.3.52)$$

which may be inverted to yield

$$\mathscr{A}_\mu = A_\mu \cos\theta_W - Z_\mu \sin\theta_W, \tag{6.3.53}$$

$$b_\mu^3 = A_\mu \sin\theta_W + Z_\mu \cos\theta_W. \tag{6.3.54}$$

In view of the identification (6.3.48), the coupling constants of the $SU(2)_L$ and $U(1)_Y$ gauge groups may be expressed as

$$g = e/\sin\theta_W \geq e, \tag{6.3.55}$$

$$g' = e/\cos\theta_W \geq e, \tag{6.3.56}$$

indicating that the previously disparate strengths of the weak and electromagnetic interactions are now related by a single parameter.

Taken together, the coupling constant identifications lead to

$$\begin{aligned} M_W^2 &= g^2/4G_F\sqrt{2} = e^2/4G_F\sqrt{2}\sin^2\theta_W \\ &= \pi\alpha/G_F\sqrt{2}\sin^2\theta_W \\ &= (37.3\ \text{GeV/c}^2)^2/\sin^2\theta_W, \end{aligned} \tag{6.3.57}$$

and

$$M_Z^2 = M_W^2/\cos^2\theta_W \geq M_W^2. \tag{6.3.58}$$

The feebleness of the weak interactions at low energies is thus laid to the large mass of the intermediate bosons, and not to an intrinsically small coupling constant. Note that the dimensionless coupling constant that endowed the electron with mass in (6.3.35) is both small,

$$\begin{aligned} G_e &= m_e\sqrt{2}/v = m_e 2^{3/4}G_F^{1/2} \\ &\simeq 3 \times 10^{-6}, \end{aligned} \tag{6.3.59}$$

and arbitrary.

It is convenient to rewrite the interaction Lagrangian (6.3.47) in terms of the weak mixing angle as

$$\begin{aligned} \mathscr{L}_{0-l} &= e\,\bar{e}\gamma^\mu e A_\mu - \frac{g}{2\cos\theta_W}\bar{v}_L\gamma^\mu v_L Z_\mu \\ &\quad - \frac{g}{2\cos\theta_W}[2\sin^2\theta_W \bar{e}_R\gamma^\mu e_R Z_\mu + (2\sin^2\theta_W - 1)\bar{e}_L\gamma^\mu e_L Z_\mu] \\ &= e\,\bar{e}\gamma^\mu e A_\mu - \frac{1}{\sqrt{2}}\left(\frac{G_F M_Z^2}{\sqrt{2}}\right)^{1/2}\bar{v}\gamma^\mu(1-\gamma_5)v Z_\mu \\ &\quad - \frac{1}{\sqrt{2}}\left(\frac{G_F M_Z^2}{\sqrt{2}}\right)^{1/2}[2\sin^2\theta_W\bar{e}\gamma^\mu(1+\gamma_5)e Z_\mu \\ &\quad + (2\sin^2\theta_W - 1)\bar{e}\gamma^\mu(1-\gamma_5)e Z_\mu] \end{aligned} \tag{6.3.60}$$

from which Feynman rules for the elementary vertices may readily be deduced. These are shown in Fig. 6-12.

The low-energy phenomenology of the model has been constructed to reproduce that of the effective Lagrangian description, which has been the subject of Sections 6.1 and 6.2. There are several novel elements as well: neutral weak current phenomena (which will be studied in Section 6.4), some definite predictions for the properties of gauge bosons, and good high-energy behavior—indeed, renormalizability. Let us look briefly at the last two of these.

In the $SU(2)_L \otimes U(1)_Y$ model the properties of the gauge bosons are correlated with those of the neutral current interactions by means of the weak mixing angle θ_W. The masses have already been expressed in terms

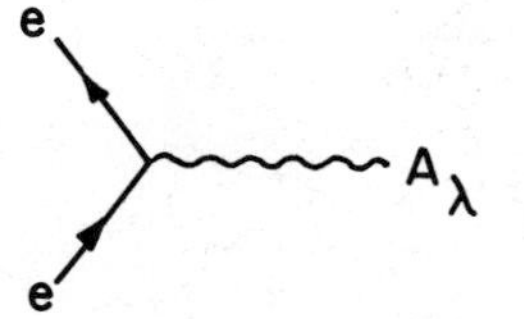

$$-iQ\,\bar{e}\gamma_\lambda\, e = ie\,\bar{e}\gamma_\lambda\, e$$

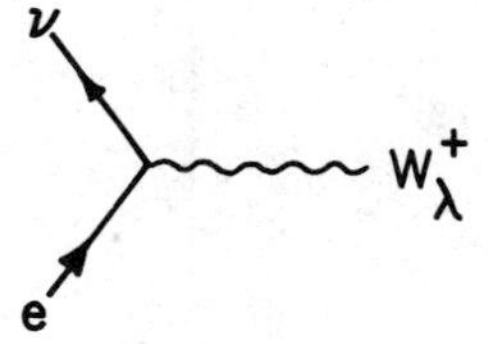

$$-i\left(\frac{G_F M_W^2}{\sqrt{2}}\right)^{1/2}\bar{\nu}\gamma_\lambda\,(1-\gamma_5)e$$

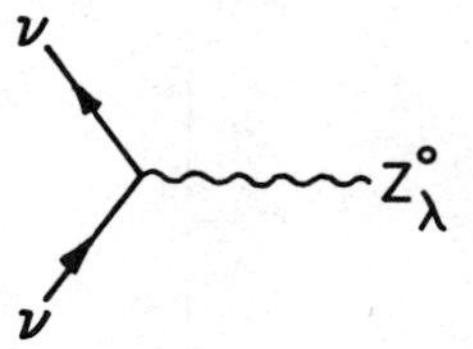

$$\frac{-i}{\sqrt{2}}\left(\frac{G_F M_Z^2}{\sqrt{2}}\right)^{1/2}\bar{\nu}\gamma_\lambda\,(1-\gamma_5)\nu$$

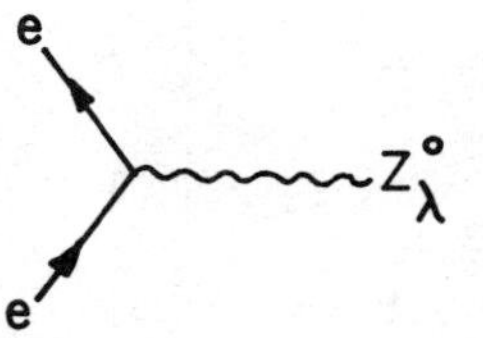

$$\frac{-i}{\sqrt{2}}\left(\frac{G_F M_Z^2}{\sqrt{2}}\right)^{1/2}\bar{e}\gamma_\lambda\left[R_e(1+\gamma_5)+L_e(1-\gamma_5)\right]e,$$

$$R_e = 2\sin^2\theta_W$$

$$L_e = 2\sin^2\theta_W - 1$$

FIG. 6-12. Feynman rules for fermion-gauge boson interactions in the Weinberg–Salam model.

of θ_W in equations (6.3.57) and (6.3.58). They are also plotted in Fig. 6-13(a). The expression for the partial decay rate of the $W^{\pm}$ (6.2.20) continues to hold because there has been no change in the charged-current phenomenology. By combining (6.3.57) and (6.2.20) we obtain the useful expression

$$\Gamma(W^- \to e^- \bar{\nu}_e) \simeq 23 \text{ MeV/sin}^3 \theta_W. \tag{6.3.61}$$

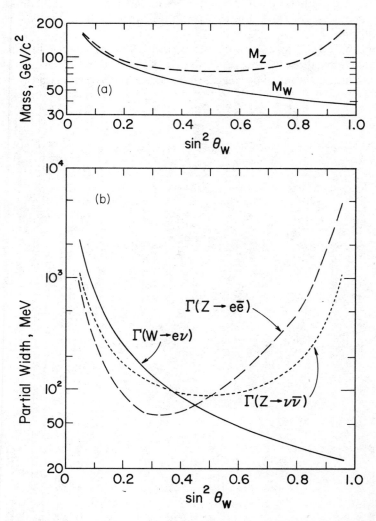

FIG. 6-13. Masses (a) and leptonic decay widths (b) of the intermediate bosons $W^{\pm}$ and Z^0 in the Weinberg–Salam model, as functions of the weak mixing angle.

The leptonic decay rates of the neutral gauge boson may be computed almost by transcription. Evidently

$$|\mathcal{M}(Z^0 \to v\bar{v})|^2 = \frac{1}{2}\frac{M_Z^2}{M_W^2}|\mathcal{M}(W \to ev)|^2, \tag{6.3.62}$$

neglecting the electron mass, so that

$$\Gamma(Z^0 \to v\bar{v}) = G_F M_Z^3/12\pi\sqrt{2}$$
$$\simeq 11.4 \text{ MeV}/(\sin\theta_W \cos\theta_W)^3. \tag{6.3.63}$$

Similarly we find for the decays into charged leptons

$$\Gamma(Z^0 \to e^+e^-) = \Gamma(Z^0 \to v\bar{v})[(2\sin^2\theta_W)^2 + (2\sin^2\theta_W - 1)^2], \tag{6.3.64}$$

where the two terms correspond to the contributions of right-handed and left-handed couplings. Our expectations for the partial decay widths are shown in Fig 6-13(b). For $\sin^2\theta_W = 0.2$ (a value currently indicated by neutral current experiments), the model yields

$$M_W \simeq 84 \text{ GeV}/c^2,$$
$$\Gamma(W \to ev) \simeq 250 \text{ MeV}, \tag{6.3.65}$$

and

$$M_Z \simeq 94 \text{ GeV}/c^2,$$
$$\Gamma(Z^0 \to v\bar{v}) \simeq 180 \text{ MeV}, \tag{6.3.66}$$
$$\Gamma(Z^0 \to e^+e^-) \simeq 90 \text{ MeV}.$$

1983 Cern.

These definite predictions are extremely inviting targets for experiments in the near future. We shall have more to say about experimental prospects in Section 6.4 and in Chapter 7.

As a first look at the high-energy properties of the theory let us reconsider the reaction

$$v_e\bar{v}_e \to W^+W^-, \tag{6.2.25}$$

the comportment of which was symptomatic of the diseases of the *ad hoc* intermediate boson theory. To do so we must determine the Feynman rules for interactions among the gauge bosons. The trilinear and quadrilinear terms in the kinetic term (6.3.9) of the Lagrangian, $\mathcal{L}_{\text{gauge}}$, give rise to the vertices shown in Fig. 6-14. The appearance of the $Z^0W^+W^-$ vertex, together with the $Z^0\bar{v}v$ vertex already discussed, indicates that there will be a new contribution to the amplitude for $v\bar{v} \to W^+W^-$: the s-channel Z^0-exchange

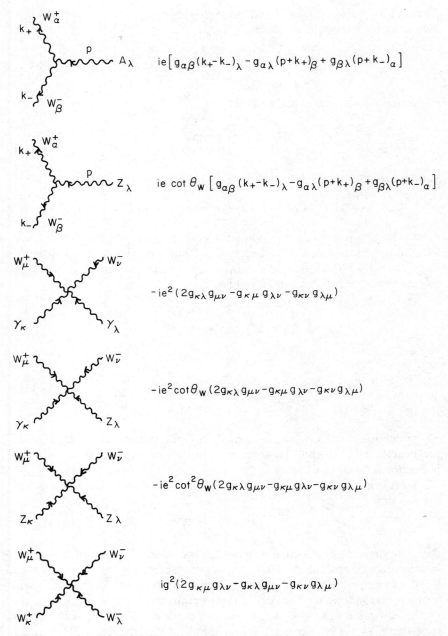

$$ie\left[g_{\alpha\beta}(k_+-k_-)_\lambda - g_{\alpha\lambda}(p+k_+)_\beta + g_{\beta\lambda}(p+k_-)_\alpha\right]$$

$$ie\cot\theta_W\left[g_{\alpha\beta}(k_+-k_-)_\lambda - g_{\alpha\lambda}(p+k_+)_\beta + g_{\beta\lambda}(p+k_-)_\alpha\right]$$

$$-ie^2(2g_{\kappa\lambda}g_{\mu\nu}-g_{\kappa\mu}g_{\lambda\nu}-g_{\kappa\nu}g_{\lambda\mu})$$

$$-ie^2\cot\theta_W(2g_{\kappa\lambda}g_{\mu\nu}-g_{\kappa\mu}g_{\lambda\nu}-g_{\kappa\nu}g_{\lambda\mu})$$

$$-ie^2\cot^2\theta_W(2g_{\kappa\lambda}g_{\mu\nu}-g_{\kappa\mu}g_{\lambda\nu}-g_{\kappa\nu}g_{\lambda\mu})$$

$$ig^2(2g_{\kappa\mu}g_{\lambda\nu}-g_{\kappa\lambda}g_{\mu\nu}-g_{\kappa\nu}g_{\lambda\mu})$$

FIG. 6-14. Feynman rules for interactions among gauge bosons in the Weinberg–Salam model.

graph shown in Fig. 6-15. Its contribution to the amplitude is

$$\mathcal{M}_Z(\nu\bar{\nu}\to W^+W^-) = \frac{-ig^2}{4(s-M_Z^2)}\bar{v}(v,q_2)\gamma_\mu(1-\gamma_5)u(v,q_1)\left(g^{\mu\nu}-\frac{S^\mu S^\nu}{M_Z^2}\right)$$

$$\times\, \varepsilon_+^{*\alpha}\varepsilon_-^{*\beta}[g_{\alpha\beta}(k_--k_+)_\nu + g_{\alpha\nu}(k_++S)_\beta - g_{\beta\nu}(k_-+S)_\alpha],$$

$$(6.3.67)$$

where

$$S \equiv q_1 + q_2 = k_+ + k_-. \qquad (6.3.68)$$

and the other kinematic quantities have been defined in (6.2.28–31). The $S^\mu S^\nu$ term is impotent between massless spinors. Thus we have only to evaluate

$$\mathcal{M}_Z = \frac{+ig^2}{4(s-M_Z^2)}\bar{v}(v,q_2)\gamma^\nu(1-\gamma_5)u(v,q_1)$$

$$\times\, [(\varepsilon_+^*\cdot\varepsilon_-^*)(k_--k_+) + (k_+\cdot\varepsilon_-^*)\varepsilon_+^* - (k\cdot\varepsilon_+^*)\varepsilon_-^* + (\varepsilon_-^*\cdot S)\varepsilon_+^* - (\varepsilon_+^*\cdot S)\varepsilon_-^*]_\nu$$

$$(6.3.69)$$

On substituting the asymptotic forms (6.2.31) of the polarization vectors for the longitudinally polarized intermediate bosons, and using

$$k_+\cdot S = k_-\cdot S = s/2, \qquad (6.3.70)$$

we find that

$$\mathcal{M}_Z = +\frac{ig^2}{8M_W^2}\bar{v}(v,q_2)(\not{k}_+ - \not{k}_-)(1-\gamma_5)u(v,q_1), \qquad (6.3.71)$$

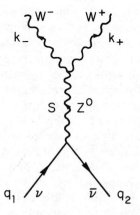

FIG. 6-15. Z^0-pole contribution to the reaction $\nu\bar{\nu}\to W^+W^-$ in the Weinberg–Salam model.

for $s \gg M_Z^2$. Then by virtue of (6.2.34) and (6.2.35) we have

$$
\begin{aligned}
\mathcal{M}_Z &= -\frac{ig^2}{4M_W^2} \bar{v}(v,q_2)\not\!P(1-\gamma_5)u(v,q_1) \\
&= -iG_F\sqrt{2}\bar{v}(v,q_2)\not\!P(1-\gamma_5)u(v,q_1),
\end{aligned}
\tag{6.3.72}
$$

which is to be added to the contribution of the electron-exchange graph of Fig. 6-9,

$$
\mathcal{M}_e = iG_F\sqrt{2}\bar{v}(v,q_2)\not\!P(1-\gamma_5)u(v,q_1).
\tag{6.3.73}
$$

The sum vanishes. Thus the p-wave s-channel resonance, the Z^0, has canceled the divergence in the p-wave scattering amplitude, and the amplitude is "asymptotically safe." Measurement of the cross section for the reaction

$$
e^+e^- \to W^+W^-,
\tag{6.3.74}
$$

for which similar divergence cancellations occur (as described in Section 6.5 and made explicit in Problem 6-8), has been advocated[7] as a probe of the gauge structure of the electroweak theory.

6.4 Neutral-Current Interactions

The prediction of neutral-current effects in gauge theories of the weak and electromagnetic interactions and the availability of high-energy neutrino beams spurred the search for experimental manifestations of the weak neutral current. In the summer of 1973, muonless events in deeply inelastic $\nu_\mu N$ and $\bar{\nu}_\mu N$ collisions were reported by the Gargamelle bubble chamber collaboration[8] working at the CERN Proton Synchrotron and by the Harvard–Pennsylvania-Wisconsin collaboration[9] working at Fermilab. Subsequently these observations have been confirmed[10] and extended, and the detailed study of neutral-current interactions has flowered so that an essentially complete determination of the neutral current couplings has been carried out. We shall return to deeply inelastic scattering in the next chapter, after having incorporated hadrons into the Weinberg–Salam model and reviewed the essential features of the parton model. For the moment, let us investigate the implications of the $SU(2)_L \otimes U(1)_Y$ gauge theory for purely leptonic interactions, for which the comparison of theory and experiment does not require the intervention of parton-model assumptions.

As a prelude to our review of the experimental consequences of the neutral current derived in the preceding section, let us incorporate additional leptons into the model. In view of the universality of the electromagnetic and weak charged-current interactions, this is evidently to be done merely by cloning the existing fermion structure. We add further left-handed weak-isospin doublets

$$L_\mu = \begin{pmatrix} v_\mu \\ \mu \end{pmatrix}_L, \qquad L_\tau = \begin{pmatrix} v_\tau \\ \tau \end{pmatrix}_L, \dots \qquad (6.4.1)$$

and right-handed singlets

$$R_\mu = \mu_R, \qquad R_\tau = \tau_R, \dots \qquad (6.4.2)$$

with the same weak hypercharge assignments as

$$L_e = \begin{pmatrix} v_e \\ e \end{pmatrix}_L, \qquad R_e = e_R. \qquad (6.4.3)$$

By omitting right-handed neutrinos, we are continuing to idealize the neutrinos as massless. The Yukawa interaction term (6.3.18) in the Lagrangian is generalized to

$$\mathscr{L}_{\text{Yukawa}} = -\sum_{i=e,\mu,\tau\dots} G_i[\bar{R}_i(\phi^\dagger L_i) + (\bar{L}_i\phi)R_i] \qquad (6.4.4)$$

This done, the Feynman rules for the interaction of the added leptons with gauge bosons are precisely those given for the electron family in Fig. 6-12, as universality is a direct consequence of the gauge symmetry. Mass terms and Higgs boson interactions for the charged leptons are generated just as those given in (6.3.34) for the electron. We are now prepared to calculate the consequences of the theory.

Before we do so, it will be useful for orientation to repeat the exercise carried out for charged-current interactions with arbitrary space-time structure in Section 6.1. This time we consider the reactions

$$\begin{aligned} v_\mu e &\to v_\mu e \\ \bar{v}_\mu e &\to \bar{v}_\mu e \end{aligned} \qquad (6.4.5)$$

which, though compatible with the general requirements of lepton number conservation, are not mediated by charged currents. Upon writing down the matrix elements for these processes, one notes immediately that the cross sections may be obtained from the earlier charged-current results by interchanging $p_2 \leftrightarrow q_2$ for $v_\mu e$ scattering or $p_1 \leftrightarrow q_2$ for $\bar{v}_\mu e$ scattering. Consequently there is no need to repeat the arithmetic in detail, and we may at once assemble the results in Table 6.2. Recall that the energy-loss parameter is $y = E'_e/E_v$.

As was the case for charged currents, the energy-loss distribution discriminates among the various forms for the interaction. Indeed, an interesting parameter of the neutral-current interaction is the mean value of the energy-loss parameter

$$\langle y \rangle \equiv \int_0^1 dy\, y\, d\sigma/dy \Big/ \int_0^1 dy\, d\sigma/dy, \qquad (6.4.6)$$

TABLE 6.2: DIFFERENTIAL CROSS SECTIONS FOR
NEUTRAL-CURRENT PROCESSES

Coupling	$\dfrac{\pi}{mE}\dfrac{d\sigma}{dy}(\nu_\mu e \to \nu_\mu e)$	$\dfrac{\pi}{mE}\dfrac{d\sigma}{dy}(\bar\nu_\mu e \to \bar\nu_\mu e)$
$|C_S|^2 + |C_P|^2$	$\frac{1}{2}y^2$	$\frac{1}{2}y^2$
$|C_T|^2$	$16(1 - y/2)^2$	$16(1 - y/2)^2$
$\mathrm{Re}[(C_S^* + C_P^*)C_T]$	$-4y(1 - y/2)$	$-4y(1 - y/2)$
$|C_{V-A}|^2$	4	$4(1 - y)^2$
$|C_{V+A}|^2$	$4(1 - y)^2$	4

the characteristic values of which are shown in Table 6.3. Just as for the charged-current interactions, however, a "confusion theorem" holds: for the spin-averaged cross section, it is always possible to find a combination of S, P, and T couplings that reproduces the distribution due to an arbitrary superposition of V ± A. The observation of $\gamma - Z^0$ interference effects in electron–positron annihilations and in electron–nucleon scattering argues in favor of the V ± A interpretation, and this is generally assumed in the "model-independent" analyses of neutral current interactions.

The leptonic neutral current interactions for which experiments may be contemplated are the neutrino reactions

$$\left.\begin{aligned} \nu_\mu e &\to \nu_\mu e \\ \bar\nu_\mu e &\to \bar\nu_\mu e \\ \nu_e e &\to \nu_e e \\ \bar\nu_e e &\to \bar\nu_e e \end{aligned}\right\} \tag{6.4.7}$$

TABLE 6.3: MEAN FRACTIONAL ENERGY
LOSS IN NEUTRAL-CURRENT PROCESSES

Coupling	$\langle y\rangle_{\nu_\mu e}$	$\langle y\rangle_{\bar\nu_\mu e}$
$|C_S|^2 + |C_P|^2$	0.75	0.75
$|C_T|^2$	0.393	0.393
$\mathrm{Re}[(C_S^* + C_P^*)C_T]$	0.625	0.625
$|C_{V-A}|^2$	0.5	0.25
$|C_{V+A}|^2$	0.25	0.5

and the electron–positron storage ring reactions

$$\left.\begin{array}{l} e^{+}e^{-} \rightarrow \mu^{+}\mu^{-} \\ e^{+}e^{-} \rightarrow \tau^{+}\tau^{-} \\ e^{+}e^{-} \rightarrow e^{+}e^{-} \end{array}\right\}.$$

(6.4.8)

We deal with these two classes in turn.

Feynman diagrams for the neutrino–electron scattering reactions are shown in Fig. 6-16. As has already been remarked, the reactions initiated

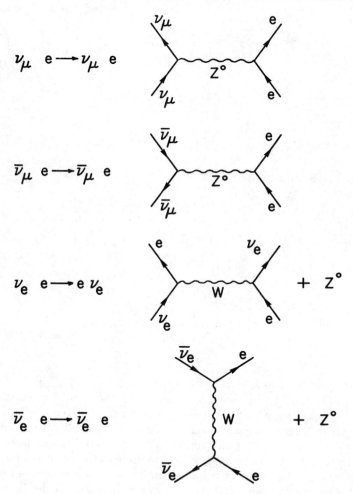

FIG. 6-16. Lowest-order contributions to neutrino–electron elastic scattering, in the Weinberg–Salam model.

by muon–neutrinos are pure neutral-current processes. At low energies (the condition $s \simeq 2mE \ll M_Z^2$ is met for any imaginable accelerator experiment), it is therefore possible to write down the cross sections at once, by referring to the Feynman rules of Fig. 6-12 and the general results of Table 6.2. They are

$$\frac{d\sigma}{dy}(\nu_\mu e \to \nu_\mu e) = \frac{4mE}{\pi}\frac{G_F^2}{8}[L_e^2 + R_e^2(1 - y)^2]$$

$$= \frac{G_F^2 mE}{2\pi}[(2x_W - 1)^2 + 4x_W^2(1 - y)^2], \qquad (6.4.9)$$

where we have introduced the useful notation

$$x_W = \sin^2 \theta_W, \qquad (6.4.10)$$

and

$$\frac{d\sigma}{dy}(\bar{\nu}_\mu e \to \bar{\nu}_\mu e) = \frac{G_F^2 mE}{2\pi}[L_e^2(1 - y)^2 + R_e^2]$$

$$= \frac{G_F^2 mE}{2\pi}[(2x_W - 1)^2(1 - y)^2 + 4x_W^2]. \qquad (6.4.11)$$

The total cross sections are therefore

$$\sigma(\nu_\mu e \to \nu_\mu e) = \frac{G_F^2 mE}{2\pi}\left[(2x_W - 1)^2 + \frac{4x_W^2}{3}\right]$$

$$= \frac{G_F^2 mE}{2\pi}\left(L_e^2 + \frac{R_e^2}{3}\right) \qquad (6.4.12)$$

and

$$\sigma(\bar{\nu}_\mu e \to \bar{\nu}_\mu e) = \frac{G_F^2 mE}{2\pi}\left[\frac{(2x_W - 1)^2}{3} + 4x_W^2\right]$$

$$= \frac{G_F^2 mE}{2\pi}\left(\frac{L_e^2}{3} + R_e^2\right). \qquad (6.4.13)$$

These are plotted as functions of $x_W = \sin^2 \theta_W$ in Fig. 6-17.

We now investigate the changes to the "charged-current" processes (6.4.7) that are brought about by the introduction of neutral currents. The charged-current cross sections were given in (6.1.65–68). The full calculation is instructive and will be left as an exercise (Problem 6-4), but the steps will be summarized here. At low energies corresponding to the point-coupling limit of s/M_Z^2, $s/M_W^2 \to 0$, and after a Fierz reordering of the W-boson exchange term, the Feynman amplitude for $\nu_e e$ scattering is seen to be identical to that for $\nu_\mu e$ scattering, with the replacement

$$L_e \to L_e + 2. \qquad (6.4.14)$$

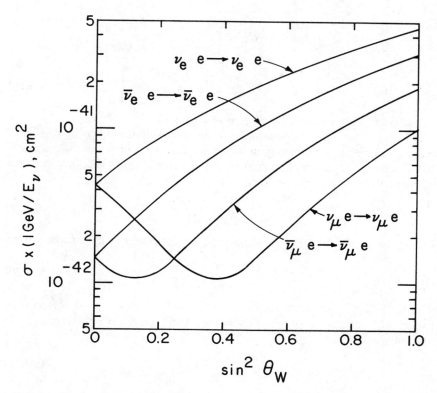

FIG. 6-17. Cross sections for neutrino–electron elastic scattering as functions of the weak mixing angle.

Without further calculation, we may therefore write

$$\frac{d\sigma}{dy}(\nu_e e \to \nu_e e) = \frac{G_F^2 mE}{2\pi}\left[(2x_W + 1)^2 + 4x_W^2(1 - y)^2\right] \qquad (6.4.15)$$

and

$$\sigma(\nu_e e \to \nu_e e) = \frac{G_F^2 mE}{2\pi}\left[(2x_W + 1)^2 + \frac{4x_W^2}{3}\right]$$

$$= \frac{G_F^2 mE}{2\pi}\left[(L_e + 2)^2 + \frac{R_e^2}{3}\right] \qquad (6.4.16)$$

In precisely the same way, the cross section for $\bar{\nu}_e e$ elastic scattering can be obtained from that for $\bar{\nu}_\mu e$ scattering, with the results

$$\frac{d\sigma}{dy}(\bar{\nu}_e e \to \bar{\nu}_e e) = \frac{G_F^2 mE}{2\pi}\left[(2x_W + 1)^2(1 - y)^2 + 4x_W^2\right] \qquad (6.4.17)$$

124   Electroweak Interactions of Leptons

and

$$\sigma(\overline{\nu}_e e \to \overline{\nu}_e e) = \frac{G_F^2 mE}{2\pi} \left[\frac{(2x_W + 1)^2}{3} + 4x_W^2 \right]$$

$$= \frac{G_F^2 mE}{2\pi} \left[\frac{(L_e + 2)^2}{3} + R_e^2 \right] \qquad (6.4.18)$$

These cross sections are also plotted in Fig. 6-17.

Before proceeding to a discussion of the available data, it is profitable to exploit further the general way in which we have arrived at the cross section. It has become traditional to express the measured cross sections in terms of the neutral current parameters

$$\left. \begin{array}{l} a_e = \frac{1}{2}(L_e - R_e) \\ v_e = \frac{1}{2}(L_e + R_e) \end{array} \right\} \qquad (6.4.19)$$

In the standard $SU(2) \otimes U(1)$ theory, these parameters take on the values

$$\left. \begin{array}{l} a_e = -\frac{1}{2} \\ v_e = -\frac{1}{2} + 2x_W \end{array} \right\} \qquad (6.4.20)$$

whereas in the V–A picture of Sections 6.1 and 6.2, $a_e = v_e = 0$. In terms of the newly defined parameters, the neutrino–electron scattering cross sections read

$$\sigma(\nu_\mu e \to \nu_\mu e) = \frac{2G_F^2 mE}{\pi} \left(\frac{a^2 + av + v^2}{3} \right) \qquad (6.4.21)$$

$$\sigma(\overline{\nu}_\mu e \to \overline{\nu}_\mu e) = \frac{2G_F^2 mE}{\pi} \left(\frac{a^2 - av + v^2}{3} \right) \qquad (6.4.22)$$

$$\sigma(\nu_e e \to \nu_e e) = \frac{2G_F^2 mE}{\pi} \left(\frac{a^2 + av + v^2}{3} + a + v + 1 \right) \qquad (6.4.23)$$

$$\sigma(\overline{\nu}_e e \to \overline{\nu}_e e) = \frac{2G_F^2 mE}{\pi} \left(\frac{a^2 - av + v^2 + a + v + 1}{3} \right), \qquad (6.4.24)$$

where subscripts on a and v have been suppressed.

Experimental results may be represented by ellipses on the a–v plane. The $\nu_\mu e$ and $\overline{\nu}_\mu e$ ellipses, which are centered at the origin, are perpendicular and intersect at four points. The four intersections reflect a $v \leftrightarrow a$ ambiguity and a sign ambiguity. The sign ambiguity can be resolved by considering $\nu_e e$ or $\overline{\nu}_e e$ results as well, since expressions (6.4.23,24) contain terms linear in $(v + a)$. The vector–axial vector ambiguity persists, and cannot be resolved with this class of data alone. Present experimental results are shown in Fig. 6-18. They are compatible with the predictions of the Weinberg–Salam model with

$$0.20 \le \sin^2 \theta_W \le 0.35. \qquad (6.4.25)$$

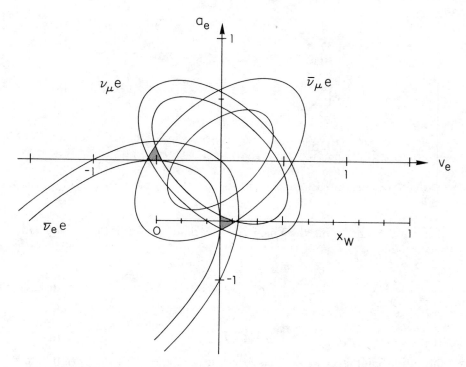

FIG. 6-18. Constraints on the neutral current parameters a_e and v_e from leptonic interactions (after F. Büsser, *Neutrino 81*, Proceedings of the 1981 International Conference on Neutrino Physics and Astrophysics, Maui, Hawaii, edited by R. J. Cence, E. Ma, and A. Roberts, High Energy Physics Group, University of Hawaii, Honolulu, 1981, Vol. II, p. 351).

We now turn briefly to the storage ring reactions (6.4.8). For electron–positron annihilations into μ-pairs or τ-pairs, the amplitude is given by the two Feynman graphs in Fig. 6-19, corresponding to *s*-channel γ and Z^0 poles. The scattering amplitude for the Bhabha process $e^+e^- \rightarrow e^+e^-$ also receives contributions from γ and Z^0 poles in the *t*-channel. The matrix element for $e^+e^- \rightarrow \mu^+\mu^-$, which serves as a useful prototype for all the inelastic processes, is given by

$$\mathcal{M}(e^+e^- \rightarrow \mu^+\mu^-) = -ie^2\bar{u}(\mu, q_-)\gamma_\lambda Q_\mu v(\mu, q_+)\frac{g^{\lambda\nu}}{s}\bar{v}(e, p_+)\gamma_\nu u(e, p_-)$$

$$+ \frac{i}{2}\left(\frac{G_F M_Z^2}{\sqrt{2}}\right)\bar{u}(\mu, q_-)\gamma_\lambda[R_\mu(1+\gamma_5) + L_\mu(1-\gamma_5)]v(\mu, q_+)$$

$$\times \frac{g^{\lambda\nu}}{s - M_Z^2}\bar{v}(e, p_+)\gamma_\nu[R_e(1+\gamma_5) + L_e(1-\gamma_5)]u(e, p_-),$$

$$(6.4.26)$$

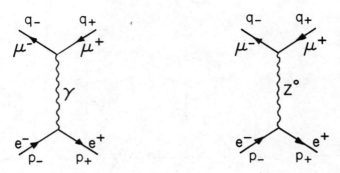

FIG. 6-19. Lowest-order contributions to the reaction $e^+e^- \to \mu^+\mu^-$, in the Weinberg–Salam model.

where in the c.m. frame the kinematic invariants (upon neglect of m^2 and μ^2 compared to s) are

$$
\left.
\begin{aligned}
p_+ \cdot p_- &= q_+ \cdot q_- = \tfrac{1}{2}s \\
p_- \cdot q_- &= p_+ \cdot q_+ = \tfrac{1}{4}s(1 - z) \\
p_+ \cdot q_- &= p_- \cdot q_+ = \tfrac{1}{4}s(1 + z)
\end{aligned}
\right\},
\tag{6.4.27}
$$

with

$$
z = \cos\theta_{\text{c.m.}}.
\tag{6.4.28}
$$

The muon charge $Q_\mu = -1$ has been inserted explicitly to permit later generalizations. A straightforward calculation leads to the spin-averaged cross section

$$
\frac{d\sigma}{dz}(e^+e^- \to \mu^+\mu^-) = \frac{\pi\alpha^2 Q_\mu^2}{2s}(1 + z^2)
$$

$$
- \frac{\alpha Q_\mu G_F M_Z^2(s - M_Z^2)}{8\sqrt{2}[(s - M_Z^2)^2 + M_Z^2\Gamma^2]}
$$

$$
\times \left[(R_e + L_e)(R_\mu + L_\mu)(1 + z^2) + 2(R_e - L_e)(R_\mu - L_\mu)z\right]
$$

$$
+ \frac{G_F^2 M_Z^4 s}{64\pi[(s - M_Z^2)^2 + M_Z^2\Gamma^2]}
$$

$$
\times \left[(R_e^2 + L_e^2)(R_\mu^2 + L_\mu^2)(1 + z^2) + 2(R_e^2 - L_e^2)(R_\mu^2 - L_\mu^2)z\right],
\tag{6.4.29}
$$

where the Z^0 propagator has been replaced with the form appropriate for an unstable particle of total width Γ. Written in this form, the spin-averaged cross section can easily be transformed into the cross section for definite initial- or final-state helicities. The cross section for transversely polarized colliding beams is also readily obtained using density matrix techniques. The first term in (6.4.29) is simply the electromagnetic contribution, which was calculated in Problem 1-5. The second is a weak-electromagnetic interference term, and the last represents the effect of the Z^0 diagram alone.

At energies far below the Z^0-boson mass, the final term is negligible. A quantity of particular experimental interest is the forward–backward asymmetry

$$A \equiv \frac{\int_0^1 dz\, d\sigma/dz - \int_{-1}^0 dz\, d\sigma/dz}{\int_{-1}^1 dz\, d\sigma/dz}. \qquad (6.4.30)$$

In the low-energy limit, the asymmetry is approximately given by

$$\lim_{s/M_Z^2 \to 0} A = \frac{3G_F s}{16\pi\alpha Q_\mu \sqrt{2}}(R_e - L_e)(R_\mu - L_\mu)$$

$$\simeq -6.7 \times 10^{-5}\left(\frac{s}{1\text{ GeV}^2}\right)(R_e - L_e)(R_\mu - L_\mu). \qquad (6.4.31)$$

For currently available storage rings, $s < 1400$ GeV^2, so that the asymmetry may be expected to reach

$$A \to -10\%(R_e - L_e)(R_\mu - L_\mu). \qquad (6.4.32)$$

In the Weinberg–Salam model,

$$R_e - L_e = R_\mu - L_\mu = +1, \qquad (6.4.33)$$

meaning that an asymmetry of approximately -10% is expected. More generally, if electron–muon universality is assumed, we may use the definition (6.4.19) to write

$$A(s \ll M_Z^2) = -3G_F s a^2/4\pi\alpha\sqrt{2}, \qquad (6.4.34)$$

whereupon a measurement of the asymmetry may resolve the $v \leftrightarrow a$ ambiguity in the "model-independent" determination of neutral-current parameters from neutrino-scattering experiments. The fact that the axial coupling enters quadratically in (6.4.34) serves as a reminder, should one be needed, that the forward–backward asymmetry is not parity-violating. At the present time, a number of experimental groups have reported asymmetries of about the expected size for both the $\mu^+\mu^-$ and $\tau^+\tau^-$ final states. Their results are collected in Table 6.4.

The e^+e^- annihilation channel is an extremely rich area for the study of electroweak effects, and an enormous amount of physics is contained in the "master formula" (6.4.29). We shall return to a fuller discussion of its implications in Section 7.2, following the incorporation of hadrons into the model.

TABLE 6.4:   FORWARD–BACKWARD ASYMMETRIES IN $e^+e^- \to l^+l^-$

Collaboration	Lepton pairs	$\langle\sqrt{s}\rangle$ (GeV)	Asymmetry (%)
CELLO[a]	$\mu^+\mu^-$	34.2	-6.4 ± 6.4
CELLO[b]	$\tau^+\tau^-$	34.2	-10.3 ± 5.2
JADE[c]	$\mu^+\mu^-$	33.5	-11.8 ± 3.8
MARK-J[d]	$\mu^+\mu^-$	34.6	-8.1 ± 2.1
TASSO[e]	$\mu^+\mu^-$	34.2	-16.1 ± 3.2
TASSO[e]	$\tau^+\tau^-$	34.2	-0.4 ± 6.6

[a] H. J. Behrend et al., Phys. Lett. **114B**, 287 (1982).
[b] H. J. Behrend, et al., Phys. Lett. **114B**, 282 (1982).
[c] W. Bartel et al., Phys. Lett. **108B**,140 (1982).
[d] B. Adeva et al., Phys. Rev. Lett. **48**, 1701 (1982).
[e] R. Brandelik et al., Phys. Lett. **110B**, 173 (1982).

6.5   The Higgs Boson

The introduction of a complex doublet of self-interacting auxiliary scalar fields has given rise to spontaneous breaking of the local $SU(2)_L \otimes U(1)_Y$ gauge symmetry. This has led, by means of the Higgs mechanism, to several agreeable consequences: the gauge bosons associated with the weak interactions have acquired masses, as have the charged fermions,[11] and a partial unification of the weak and electromagnetic interactions has been achieved. In place of three independent gauge coupling constants e, g, and g' associated with the $U(1)_{EM}$, $SU(2)_L$, and $U(1)_Y$ gauge groups, there are now only two— the electric charge e and the weak mixing angle θ_W. (Further reduction of the number of independent parameters will be discussed in Chapter 9.) The resulting theory has an improved high-energy behavior in that amplitudes corresponding to tree diagrams (Feynman diagrams without loops) are nondivergent, and indeed the theory can be made renormalizable. As we have begun to see, the low-energy phenomenology of the model is in excellent accord with experimental findings. An uninvited guest at these proceedings is the Higgs boson, the massive physical scalar particle that remains after spontaneous symmetry breaking. This section is devoted to elaborating the properties of this particle, which is an essential aspect of the theory as formulated. We shall discuss the interactions of the Higgs boson with fermions and gauge bosons, examine its role in divergence cancellation, consider prospects for the detection of the Higgs boson, and seek constraints on the Higgs boson mass. Finally, and very briefly, we shall consider dynamical symmetry-breaking alternatives to the standard Higgs scenario.

The interaction of the Higgs boson with fermions has already been given in the Yukawa Lagrangian (6.3.34), which leads to the Feynman rule shown

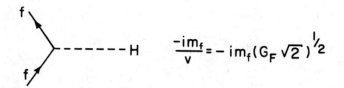

FIG. 6-20. Feynman rule for Higgs boson–fermion interactions.

in Fig. 6-20. We may calculate at once the rate for the decay

$$H \rightarrow f\bar{f}. \tag{6.5.1}$$

The invariant amplitude is given by

$$\mathcal{M} = -im(G_F\sqrt{2})^{1/2}\bar{u}(f,p_1)v(f,p_2), \tag{6.5.2}$$

where the momenta of the emitted fermions may be written as

$$\left.\begin{array}{l} p_1 = (M_H/2)(1;0,0,1) \\ p_2 = (M_H/2)(1;0,0,-1) \end{array}\right\} \tag{6.5.3}$$

for $m_f \ll M_H$. Consequently we compute

$$\begin{aligned} |\mathcal{M}|^2 &= G_F m_f^2 \sqrt{2}\,\mathrm{tr}(\not{p}_2\not{p}_1) \\ &= 4G_F m_f^2 \sqrt{2}\,p_1 \cdot p_2 \\ &= 2G_F M_H^2 m_f^2 \sqrt{2}. \end{aligned} \tag{6.5.4}$$

The differential decay rate is, according to (B.1.4),

$$\frac{d\Gamma}{d\Omega} = \frac{|\mathcal{M}|^2}{64\pi^2 M_H} = \frac{G_F M_H m_f^2}{16\pi^2 \sqrt{2}}, \tag{6.5.5}$$

which is isotropic as it must be for the decay of a spinless particle, and the total decay rate is

$$\Gamma(H \rightarrow f\bar{f}) = \frac{G_F M_H m_f^2}{4\pi\sqrt{2}}. \tag{6.5.6}$$

The dominant decay mode of a light ($M_H < 2M_W$) Higgs boson is therefore into pairs of the most massive fermion that is kinematically accessible.

A very similar calculation (posed as Problem 6-5) leads to the cross section for the reaction

$$e^+e^- \rightarrow H \rightarrow \text{all}. \tag{6.5.7}$$

It is simply

$$\sigma(e^+e^- \rightarrow H) = \frac{4\pi\Gamma(H \rightarrow e^+e^-)\Gamma(H \rightarrow \text{all})}{(s - M_H^2)^2 + M_H^2\Gamma(H \rightarrow \text{all})^2}, \tag{6.5.8}$$

for which the peak cross section at $s = M_H^2$ is

$$\sigma_{\text{peak}}(e^+e^- \to H) = \frac{4\pi}{M_H^2} \cdot \frac{\Gamma(H \to e^+e^-)}{\Gamma(H \to \text{all})}$$

$$\simeq \frac{5 \times 10^{-27} \text{ cm}^2}{[M_H/(1 \text{ GeV}/c^2)]^2} \cdot \frac{\Gamma(H \to e^+e^-)}{\Gamma(H \to \text{all})}. \quad (6.5.9)$$

This is an extremely discouraging prospect even for rather light Higgs bosons, because of the smallness of the $He\bar{e}$ coupling.

The Feynman rules for the interactions of Higgs bosons with gauge bosons and for their self-interactions may be obtained from the Higgs term $\mathscr{L}_{\text{scalar}}$ (6.3.15) in the Lagrangian. These are shown in Fig. 6-21. Because the Higgs boson–gauge boson couplings are not suppressed by small fermion masses, reactions involving gauge bosons may provide the most favorable means for Higgs boson production. The reaction

$$e^+e^- \to HZ^0, \quad (6.5.10)$$

which proceeds via the s-channel formation of a virtual Z^0, is a popular example. The cross section is

$$\sigma(e^+e^- \to HZ) = \frac{\pi\alpha^2}{24}\left(\frac{2K}{\sqrt{s}}\right)\frac{(K^2 + 3M_Z^2)}{(s - M_Z^2)^2 + M_Z^2\Gamma_Z^2}\frac{(1 - 4x_W + 8x_W^2)}{x_W^2(1 - x_W)^2} \quad (6.5.11)$$

where as usual $x_W = \sin^2\theta_W$ and K is the c.m. momentum of the emerging particles. At very high energies, for which $2K \to \sqrt{s}$, the ratio

$$\frac{\sigma(e^+e^- \to HZ)}{\sigma(e^+e^- \to \mu^+\mu^-)} \to \frac{(1 - 4x_W + 8x_W^2)}{128x_W^2(1 - x_W)^2} \quad (6.5.12)$$

approaches from below an asymptotic value of 16% for $x_W = 0.2$. Other possibilities for Higgs production will be found in the articles cited at the end of this chapter.

With the Feynman rules for Higgs boson interactions in hand we are in a position to investigate the role of the Higgs boson in the cancellation of high-energy, or ultraviolet, divergences. An illuminating example is provided by the reaction

$$e^+e^- \to W^+W^-, \quad (6.5.13)$$

which is closely related to the reaction $\nu\bar{\nu} \to W^+W^-$ that has frequently been the object of our interest. This reaction is described, in lowest order in the Weinberg–Salam theory, by the four Feynman graphs in Fig. 6-22. The leading divergence in the p-wave amplitude of the neutrino-exchange diagram Fig. 6-22(a) is canceled by the contributions of the direct-channel γ- and Z^0-exchange diagrams, in complete analogy with the two graphs (Figs. 6-9 and 6-15) for $\nu\bar{\nu} \to W^+W^-$. One may verify the cancellation

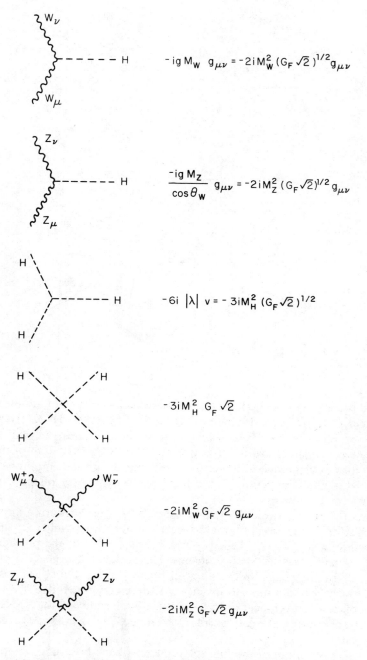

FIG. 6-21. Feynman rules for the interactions of Higgs bosons and gauge bosons.

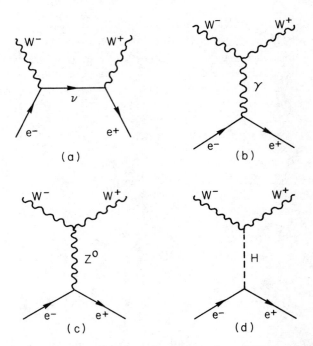

FIG. 6-22. Lowest-order contributions to the reaction $e^+e^- \to W^+W^-$, in the Weinberg–Salam model.

without carrying out the full calculation as follows. When fermion masses are neglected, the lepton-exchange diagrams of Figs. 6-9 and 6-22(a) represent equal amplitudes. In the very-high-energy approximation in which the γ and Z^0 propagators are equal, it is easy to show that the sum of the contributions of diagrams 6-22(b) and (c) is identical with the amplitude due to Fig. 6-15. The connection to the local gauge symmetry of the theory is easily traced.

This is not the whole story, however. The s-wave scattering amplitude, which exists in this case because the electrons are massive and may therefore be found in the "wrong" helicity state, grows as $s^{1/2}$ for the production of longitudinally polarized gauge bosons. This residual divergence is precisely canceled by the Higgs boson graph of Fig. 6-22(d). If the Higgs boson did not exist, we should have to invent something very much like it. From the point of view of divergence cancellations in S-matrix theory, the $Hf\bar{f}$ coupling must be proportional to m_f because "wrong helicity" amplitudes are always proportional to m_f.

Let us note some of the interrelations that have been uncovered in our discussions of high-energy behavior. Without spontaneous symmetry breaking and the Higgs boson, there would be no longitudinal gauge bosons and thus no extreme divergence difficulties. (Nor would there be a viable low-energy phenomenology of weak interactions.) The most severe divergences are eliminated by the gauge structure of the couplings among gauge bosons and leptons. A lesser, but still potentially fatal, divergence arises because the electron has acquired mass—thanks to the Higgs mechanism. Spontaneous symmetry breaking provides its own cure by supplying a Higgs boson to remove the last divergence. Here the common origin of the electron mass and the $He\bar{e}$ interaction in the Yukawa Lagrangian (6.3.34) is of crucial importance. One cannot but be impressed!

In spite of this tightly woven structure, some ambiguity persists. Nothing in the formulation of the Weinberg–Salam theory specifies the mass of the Higgs boson, and none of the applications to conventional processes that we have considered depends in any direct way upon the value of M_H. It may therefore appear that this is a completely free parameter of the theory, and this is indeed nearly the case. However, by imposing certain requirements of internal consistency upon the theory, one may narrow the range of possibilities somewhat.

It is relatively straightforward to obtain a sort of upper bound on M_H, by imposing the requirement that the theory make perturbative sense and thus that the amplitudes calculated in tree approximation satisfy partial-wave unitarity.[12] In gauge theories, the asymptotic growth of partial-wave amplitudes is regulated,[13] and all amplitudes are *at worst* in logarithmic violation of unitarity in lowest order. When the calculable higher-order corrections are applied, the amplitudes become properly finite. There is one possibly exceptional case: interactions in which the Higgs boson plays an important role.

For example, the quartic Higgs boson self-coupling is seen in Fig. 6-21 to be

$$-3iM_H^2 G_F/\sqrt{2}. \tag{6.5.14}$$

It may therefore happen that for large values of M_H certain partial-wave amplitudes, though "asymptotically safe," may exceed the numerical bounds imposed by unitarity. This is indeed the case for the s-wave scattering amplitudes for the scattering of Higgs bosons and longitudinally polarized gauge bosons. A systematic analysis of the four neutral channels $W_L^+ W_L^-$, HH, $Z_L Z_L$, and HZ_L (L for longitudinal) leads to the condition[14]

$$M_H^2 < 8\pi\sqrt{2}/3G_F \simeq (1 \text{ TeV}/c^2)^2 \tag{6.5.15}$$

for partial-wave unitarity to be respected by the tree diagrams. This condition provides a plausible upper bound on the Higgs boson mass. If it is respected (so that $M_H \ll 1 \text{ TeV}/c^2$), the weak interactions remain weak at all energies (except in the vicinity of gauge boson poles) in the sense that tree diagrams are reliable. If $M_H > 1$ TeV/c^2, partial-wave amplitudes become large, which is to say that weak interactions become *strong*. In the latter case, gauge boson interactions in the TeV regime may resemble the familiar hadronic interactions in the GeV range.

It was remarked in Section 5.5 that the analysis of spontaneous symmetry breaking presented there was entirely classical in that quantum corrections to the Higgs potential were neglected, and that one should verify in specific applications that these do not alter essential conclusions. Such considerations lead to an interesting *lower bound* on the mass of the Higgs boson in the Weinberg–Salam theory. When the one-loop corrections to the Higgs potential (6.3.17) are calculated (an infinite set of diagrams), the effective potential can be written in the form

$$V(\phi^\dagger\phi) = -\mu^2(\phi^\dagger\phi) + B(\phi^\dagger\phi)^2\log(\phi^\dagger\phi/M^2), \qquad (6.5.16)$$

where M has been chosen to absorb the erstwhile $|\lambda|(\phi^\dagger\phi)^2$ term. The coefficient B includes in principle the contributions of gauge boson, Higgs boson, and fermion loops, but in practice only the gauge boson loops are expected to contribute significantly. In this approximation,[15]

$$B = \frac{3(2M_W^4 + M_Z^4)}{64\pi^2\langle\phi^\dagger\phi\rangle_0}, \qquad (6.5.17)$$

where

$$\langle\phi^\dagger\phi\rangle_0 = v^2/2 = 1/2G_F\sqrt{2} \qquad (6.5.18)$$

is to be identified with the physical vacuum state.

The effective potential has a local minimum at a value of $\langle\phi^\dagger\phi\rangle_0$ given by

$$\langle\phi^\dagger\phi\rangle_0[\log(\langle\phi^\dagger\phi\rangle_0/M^2) + \tfrac{1}{2}] = \mu^2/2B, \qquad (6.5.19)$$

at which the value of the potential is

$$V(\langle\phi^\dagger\phi\rangle_0) = -B\langle\phi^\dagger\phi\rangle_0^2[\log(\langle\phi^\dagger\phi\rangle_0/M^2) + 1]. \qquad (6.5.20)$$

This local minimum will be an absolute minimum provided that

$$V(\langle\phi^\dagger\phi\rangle_0) < V(0) = 0, \qquad (6.5.21)$$

which is ensured if

$$\log(\langle\phi^\dagger\phi\rangle_0/M^2) > -1. \qquad (6.5.22)$$

Inserting this constraint in the expression for the mass of the physical Higgs boson,

$$M_H^2 \equiv V''|_{\langle\phi^\dagger\phi\rangle_0} = 4B\langle\phi^\dagger\phi\rangle_0[\ln(\langle\phi^\dagger\phi\rangle_0/M^2) + \tfrac{3}{2}], \qquad (6.5.23)$$

yields a lower bound[16] on the Higgs boson mass

$$M_H^2 > 2B\langle\phi^\dagger\phi\rangle_0 = \frac{3(2M_W^4 + M_Z^4)}{32\pi^2\langle\phi^\dagger\phi\rangle_0} = \frac{3G_F\sqrt{2}(2M_W^4 + M_Z^4)}{16\pi^2}. \qquad (6.5.24)$$

Expressing the intermediate boson masses in terms of the weak mixing angle by means of (6.3.55,56), we obtain

$$M_H^2 > \frac{3\alpha^2}{16 G_F \sqrt{2} x_W^2} \left[2 + \frac{1}{(1 - x_W)^2} \right]$$

$$= \frac{0.61 (\text{GeV}/c^2)^2}{x_W^2} \left[2 + \frac{1}{(1 - x_W)^2} \right]. \tag{6.5.25}$$

For $x_W = 0.2$, this expression yields the lower bound

$$M_H > 7.3 \text{ GeV}/c^2. \tag{6.5.26}$$

An interesting variation [17] of this argument consists in requiring that the renormalized mass parameter μ^2 in the effective potential (6.5.16) be set to zero. In this case all the curvature of the effective potential near its absolute minimum is due to radiative corrections. The condition $\mu^2 = 0$ implies through (6.5.19) that

$$\log(\langle \phi^\dagger \phi \rangle_0 / M^2) = -1/2 \tag{6.5.27}$$

which condition, when inserted in (6.5.23), gives the mass of the Higgs boson as

$$M_H^2 = \frac{3\alpha^2}{8 G_F \sqrt{2} x_W^2} \left[2 + \frac{1}{(1 - x_W)^2} \right], \tag{6.5.28}$$

so that M_H is $\sqrt{2}$ times the minimum allowed value. For $x_W = 0.2$ we therefore have the estimate

$$M_H \simeq 10.4 \text{ GeV}/c^2, \tag{6.5.29}$$

which apparently is within reach of the next generation of experiments.

The arbitrariness of the Higgs sector, with respect to both the mass of the Higgs boson and the original choice of the Higgs representation, misgivings[18] about field theories of elementary scalars, and the knowledge that the Ginzburg–Landau order parameter has a dynamical origin as a bound state of elementary fermions (Cooper pairs of electrons) all motivate the search for an alternative to the introduction of auxiliary scalar fields. The hope of the general approach known as dynamical symmetry breaking[19] is that scalar bound states will be generated by the dynamics of the situation and that these will play the role here assigned to the Higgs fields. Although many intriguing possibilities have been explored, no entirely satisfactory picture has yet emerged.

6.6   Renormalizability of the Theory

The quantization of non-Abelian gauge theories and the development of Feynman rules that are consistent when applied to diagrams containing closed loops is a highly nontrivial undertaking. So, too, is the demonstration, first given by 't Hooft,[20] that spontaneously broken gauge theories are renormalizable. A detailed discussion of quantization and renormalization lies outside the scope of this book, but two points have interesting repercussions and will therefore be developed briefly. These have to do with convenient choices of gauge and the role of anomalies.

To quantize a gauge theory and particularly to determine the vector boson propagators requires a choice of gauge, as we have already remarked for the case of QED in Section 3.2. The unitary gauge introduced for the first time in Section 5.3 is extremely convenient for examining the particle content of a spontaneously broken gauge theory. However, the form of the massive vector boson propagator that appears in this gauge,

$$\frac{-i(g_{\mu\nu} - q_\mu q_\nu/M^2)}{q^2 - M^2 + i\varepsilon}, \tag{6.6.1}$$

is ill-suited to the calculation of higher-order diagrams in perturbation theory. Already at the level of tree diagrams for gauge-boson scattering, complicated cancellations among diagrams that are individually ill-behaved are required for sensible results. The presence of the $q_\mu q_\nu/M^2$ term in the propagator means that finite amplitudes must result from the cancellation of contributions that are separately infinite, a procedure of some delicacy. Among 't Hooft's contributions was the insight that, because of the gauge invariance of the theory, one is entitled to choose one gauge to manifest the particle content and another to prove renormalizability, although each may be unsuited for the other's task. The general method is nicely illustrated in the Abelian Higgs model of Section 5.3. To the Lagrangian (5.3.11) of the Abelian Higgs model written in terms of shifted fields,

$$\mathcal{L} = \tfrac{1}{2}[(\partial_\mu\eta)(\partial^\mu\eta) + 2\mu^2\eta^2] + \tfrac{1}{2}(\partial_\mu\zeta)(\partial^\mu\zeta) \tag{6.6.2}$$
$$- \tfrac{1}{4}F_{\mu\nu}F^{\mu\nu} + MA_\mu(\partial^\mu\zeta) + \tfrac{1}{2}M^2A_\mu A^\mu$$

we add the gauge-fixing term

$$\mathcal{L}_{gf} = -(1/2\xi)(\partial^\mu A_\mu + \xi M\zeta)^2, \tag{6.6.3}$$

where the parameter ξ determines the specific gauge. $\mathcal{L}_{gf}$ is so constructed as to cancel the unwanted mixing term $MA_\mu\partial^\mu\zeta$ in the original Lagrangian. In general, the unphysical ζ field is now present in the Feynman rules of the theory, with a propagator given by

$$\frac{i}{q^2 - \xi M^2}. \tag{6.6.4}$$

The vector boson propagator may also be found by the usual procedure. It is given by

$$\frac{-i[g_{\mu\nu} - (1 - \xi)q_\mu q_\nu/(q^2 - \xi M^2)]}{q^2 - M^2 + i\varepsilon}. \tag{6.6.5}$$

Because the $q_\mu q_v$ term appears in combination with $1/q^2$ for any finite value of ξ, there is no reason to expect nonrenormalizable behavior in loop integrals. The apparent poles in (6.6.4,5) at $q^2 = \xi M^2$ must cancel in any S-matrix element. Let us note that $\xi = 1$ corresponds to the Feynman gauge familiar in electrodynamics, whereas $\xi = 0$ yields the Landau gauge in which the numerator of (6.6.5) assumes the form

$$g_{\mu v} - q_\mu q_v / q^2 \tag{6.6.6}$$

of a transverse projection operator. In the limit $\xi \to \infty$, the vector boson propagator approaches the familiar unitary gauge form (6.6.1) and the unphysical scalar decouples as its mass tends to infinity.

The demonstrations of renormalizability make extensive use of the intricate cancellations implied by gauge invariance, such as we have seen explicitly in operation at the tree level. It is essential to the proofs that these cancellations, which are related to local current conservation and are codified for QED in the Ward–Takahashi identities, also take place in diagrams containing closed loops. However, situations are known in field theory in which a (classical) local conservation law derived from gauge invariance with the aid of Noether's theorem holds at the tree level but is not respected by loop diagrams. Terms that violate the classical conservation laws are known as anomalies, and their properties have been extensively studied. The simplest example of a Feynman diagram leading to an anomaly is a fermion loop coupled to two vector currents and one axial current, as shown in Fig. 6-23. Because the weak interaction contains both vector and axial vector currents, there is a danger that such diagrams may arise in the Weinberg–Salam theory and destroy the renormalizability of the theory.

For a gauge theory, it is convenient to write the interaction Lagrangian in terms of left-handed and right-handed matter fields, as we have done in formulating the Weinberg–Salam model. The interaction terms may thus be written schematically as

$$\mathscr{L}_{\text{int}} = -gX_\mu^a(\bar{R}\gamma^\mu T_+^a R + \bar{L}\gamma^\mu T_-^a L), \tag{6.6.7}$$

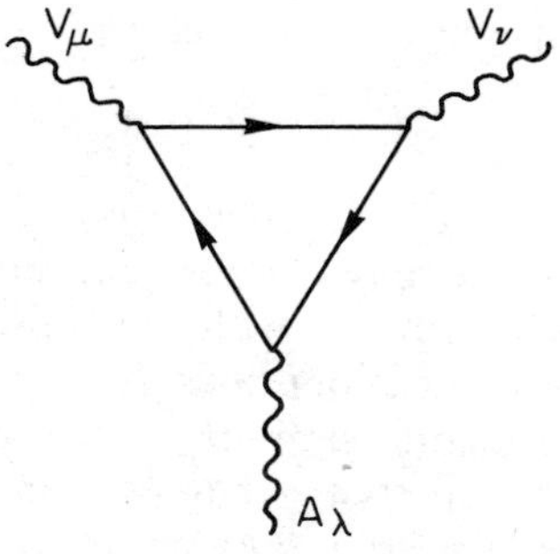

FIG. 6-23. Triangle anomaly for the axial-vector current.

where $T_\pm^a$ are generators of the appropriate representation of the gauge group and X_μ^a are the gauge fields. It may then be shown that the axial anomaly is proportional to

$$A^{abc} = A_+^{abc} - A_-^{abc}, \tag{6.6.8}$$

where the right-handed and left-handed contributions are given by

$$A_\pm^{abc} = \text{tr}(\{T_\pm^a, T_\pm^b\} T_\pm^c), \tag{6.6.9}$$

and do not depend on fermion masses. Evidently the theory will be anomaly-free either if there is a cancellation among the left-handed and right-handed sectors,

$$A_+^{abc} = A_-^{abc}, \tag{6.6.10}$$

in which case the theory is called "vectorlike," or if the two contributions vanish separately

$$A_+^{abc} = 0 = A_-^{abc}. \tag{6.6.11}$$

For the Weinberg–Salam theory, it is easy to verify that the only anomaly is proportional to

$$\text{tr}(\{\tau^a, \tau^b\} Y) = \text{tr}\{\tau^a, \tau^b\} \, \text{tr} \, Y \propto \sum_{\substack{\text{fermion} \\ \text{doublets}}} Y. \tag{6.6.12}$$

By virtue of the Gell-Mann–Nishijima relation, we may therefore express the condition for the absence of anomalies in terms of the electric charge as

$$\Delta Q = Q_R - Q_L = \left(\sum_{\substack{\text{right-handed} \\ \text{doublets}}} Q - \sum_{\substack{\text{left-handed} \\ \text{doublets}}} Q \right) = 0. \tag{6.6.13}$$

In the model of leptons that we have considered until now, built upon left-handed doublets such as

$$\mathsf{L}_e = \begin{pmatrix} \nu_e \\ e \end{pmatrix}_L, \tag{6.6.14}$$

we have

$$\Delta Q = -Q_L = 1, \tag{6.6.15}$$

for each lepton family. Thus the theory is bound to be nonrenormalizable.

To cancel the anomaly we may consider two possibilities. One is to add doublets of right-handed fermions with appropriate charge assignments. This course has the esthetic merit of restoring a measure of parity symmetry, but it lacks direct experimental support. The second is to add additional left-handed fermion doublets, with charges arranged to cancel the contribution of the leptons. This role falls naturally to the hadrons. Indeed, if we simply appended to the theory a left-handed doublet of nucleons,

$$\mathsf{L}_N \equiv \begin{pmatrix} p \\ n \end{pmatrix}_{\mathrm{L}}, \qquad (6.6.16)$$

the "left-handed charge" would become

$$Q_{\mathrm{L}} = 0 \qquad (6.6.17)$$

and the anomaly would be canceled. The nucleons are not elementary, however. One manifestation of the compositeness is that (even in the zero momentum transfer limit) the charged current interaction is characterized by $\gamma^\mu(1 - g_A\gamma_5)$, with $g_A \simeq 1.25 \neq 1$. We should not require any further encouragement to turn toward the quarks.

A single doublet of quarks

$$\mathsf{L}_q = \begin{pmatrix} u \\ d \end{pmatrix}_{\mathrm{L}} \qquad (6.6.18)$$

contributes to the left-handed charge an amount

$$Q_{\mathrm{L}}(q) = \tfrac{2}{3} - \tfrac{1}{3} = \tfrac{1}{3}, \qquad (6.6.19)$$

which is insufficient to cancel the lepton anomaly. If for each lepton doublet we include three quark doublets

$$\begin{pmatrix} u_{\mathrm{Red}} \\ d_{\mathrm{Red}} \end{pmatrix}_{\mathrm{L}}, \quad \begin{pmatrix} u_{\mathrm{Blue}} \\ d_{\mathrm{Blue}} \end{pmatrix}_{\mathrm{L}}, \quad \begin{pmatrix} u_{\mathrm{Green}} \\ d_{\mathrm{Green}} \end{pmatrix}_{\mathrm{L}} \qquad (6.6.20)$$

corresponding to the three quark colors suggested by numerous observations in Chapter 1, the anomaly will be canceled.[21] The cancellation generalizes at once to many lepton (and hence quark) doublets.

The requirement that the theory of electroweak interactions be renormalizable thus contributes new support to the idea of quark color, and makes the far-reaching suggestion that the spectra of leptons and quarks are related. We shall examine the consequences for the phenomenology of the weak interactions of quarks in the next chapter. Extended families of quarks and leptons will be central to a later attempt, in Chapter 9, at a (grand) unification of the strong, weak, and electromagnetic interactions.

Problems

6-1. Compute the differential and total cross sections for the reaction $\nu_\mu e \to \nu_\mu e$, retaining the electron mass and the effect of the Z^0 propagator. In what kinematic regimes does the result differ noticeably from the point-coupling limit (6.4.9,12)?

6-2. (a) Compute the decay rate for the disintegration of the tau lepton, $\tau^- \to e^- \bar{v}_e v_\tau$, neglecting the electron mass, and compare with the familiar result for muon decay.

(b) How would the rate change if, by virtue of an unconventional lepton number assignment, $v_\tau \equiv \bar{v}_e$?

6-3. Derive the Fierz reordering theorem

$$\bar{u}_3 \Gamma_i u_2 \bar{u}_1 \Gamma_i u_4 = \sum_{j=1}^{5} \lambda_{ij} \bar{u}_1 \Gamma_j u_2 \bar{u}_3 \Gamma_j u_4, \qquad (A.4.29)$$

where $\Gamma_i = (1, \gamma_\mu, \sigma_{\mu\nu}, \gamma_\mu \gamma_5, \gamma_5)$, and the 5×5 matrix λ_{ij} is given by (A.4.30). [Reference: V. B. Berestetskii, E. M. Lifshitz, and L. P. Pitaevski, *Relativistic Quantum Theory*, Part 1, translated by J. B. Sykes and J. S. Bell, Pergamon, Oxford, 1971, Section 22–28.]

6-4. Compute the differential and total cross sections for $v_e e$ and $\bar{v}_e e$ elastic scattering in the Weinberg–Salam model. Work in the limit of large M_W and M_Z, and neglect the electron mass with respect to large energies. The computation is done most gracefully by Fierz reordering one of the graphs, as indicated in Section 6.4. [Reference: G. 't Hooft, *Phys. Lett.* **37B**, 195 (1971).]

6-5. Calculate the cross section for the reaction $e^+ e^- \to H \to f \bar{f}$, where f is a massive lepton, and thus show that the cross section for the reaction $e^+ e^- \to H \to$ all can be written in the form (6.5.8), so long as $M_H < 2M_W$.

6-6. Show that for a heavy Higgs boson the rates for decay into a pair of intermediate vector bosons are given by

$$\frac{\Gamma(H \to W^+ W^-)}{M_H} = \frac{G_F M_W^2}{8\pi \sqrt{2}} \frac{(1-x)^{1/2}}{x} (3x^2 - 4x + 4)$$

and

$$\frac{\Gamma(H \to Z^0 Z^0)}{M_H} = \frac{G_F M_Z^2}{16\pi \sqrt{2}} \frac{(1-x')^{1/2}}{x'} (3x'^2 - 4x' + 4),$$

where $x = 4M_W^2 / M_H^2$ and $x' = 4M_Z^2 / M_H^2 = x/\cos^2 \theta_W$.

6-7. Consider the process $e^+ e^- \to \gamma\gamma$, which is described, in QED, by the two Feynman diagrams

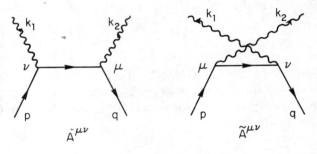

and for which the amplitude may be written $\mathcal{M} = \varepsilon_{1\nu}^{*}\varepsilon_{2\mu}^{*}(A^{\mu\nu} + \tilde{A}^{\mu\nu})$, where ε_i is the polarization vector of a photon and $A^{\mu\nu}(\tilde{A}^{\mu\nu})$ corresponds to the first (second) diagram.

(a) Calculate the tensors $A^{\mu\nu}$ and $\tilde{A}^{\mu\nu}$.

(b) Show that gauge invariance requires that

$$k_{1\nu}(A^{\mu\nu} + \tilde{A}^{\mu\nu}) = 0 = k_{2\mu}(A^{\mu\nu} + \tilde{A}^{\mu\nu}).$$

(c) Verify that these conditions are met, although the quantities $k_{1\nu}A^{\mu\nu}$, $k_{1\nu}\tilde{A}^{\mu\nu}$, $k_{2\mu}A^{\mu\nu}$, and $k_{2\mu}\tilde{A}^{\mu\nu}$ are all different from zero.

6-8.   Carry out the computation of the amplitudes for the reaction $e^{+}e^{-} \rightarrow W^{+}W^{-}$ described in Section 6.5, retaining the electron mass. Verify the role of the Higgs boson in the cancellation of divergences.

6-9.   Because the most serious high-energy divergences of a spontaneously broken gauge theory are associated with the longitudinal degrees of freedom of the gauge bosons, which arise from the auxiliary scalars, it is instructive to study the Higgs sector in isolation. Consider, therefore, the Lagrangian for the Higgs sector of the Weinberg–Salam model before the gauge couplings are turned on,

$$\mathcal{L}_{\text{scalar}} = (\partial^{\mu}\phi)^{\dagger}(\partial_{\mu}\phi) - \mu^{2}(\phi^{\dagger}\phi) - |\lambda|(\phi^{\dagger}\phi)^{2}.$$

(a) Choosing $\mu^{2} < 0$, investigate the effect of spontaneous symmetry breaking. Show that the theory describes three massless scalars (w^{+}, w^{-}, z^{0}) and one massive neutral scalar (h), which interact according to

$$\mathcal{L}_{\text{int}} = -|\lambda|vh(2w^{+}w^{-} + z^{2} + h^{2})$$
$$-(|\lambda|/4)(2w^{+}w^{-} + z^{2} + h^{2})^{2},$$

where $v^{2} = -\mu^{2}/|\lambda|$. In the language of the full Weinberg–Salam theory, $1/v^{2} = G_{F}\sqrt{2}$ and $\lambda = G_{F}M_{H}^{2}/\sqrt{2}$.

(b) Deduce the Feynman rules for interactions and compute the lowest-order (tree diagram) amplitude for the reaction $hz \rightarrow hz$.

(c) Compute the s-wave partial-wave amplitude in the high-energy limit and show that it respects partial-wave unitarity only if $M_{H}^{2} < 8\pi\sqrt{2}/G_{F}$. [Reference: B. W. Lee, C. Quigg, and H. B. Thacker, *Phys. Rev.* **D16**, 1519 (1977).]

6-10.   Glashow's $SU(2)_{L} \otimes U(1)_{Y}$ model of 1961 (see bibliography) has the same structure as the Weinberg model of 1967, but makes less definite predictions because the vector boson masses are not generated by a specific Higgs mechanism. To see how the choice of Higgs sector influences the predictions for vector meson masses, consider the following alternatives to the conventional weak-isospin doublet of complex scalars: (*i*) a weak isovector with $Y_{\phi} = 2$, which develops a vacuum expectation value $\langle\phi\rangle_{0} = (0, 0, v/\sqrt{2})^{T}$; (*ii*) a weak isovector with $Y_{\phi} = 0$, which develops a vacuum expectation value $\langle\phi\rangle_0 = (0, v/\sqrt{2}, 0)^T$; (iii) a weak isotensor with $Y_\phi = 4$, which develops a vacuum expectation value $\langle\phi\rangle_0 = (0,0,0,0, v/\sqrt{2})^T$. (The notation ^{T} denotes transpose.) For each case, consider a spontaneously broken gauge theory as constructed in Section 6.3.

(a) Examine the symmetry properties of the vacuum state under the generators of $SU(2)_L \otimes U(1)_Y$ to determine which gauge bosons will acquire mass.

(b) Expand $\mathscr{L}_{\text{scalar}}$ following the procedure of (6.3.36–42), and determine for each the ratio $\rho \equiv M_W^2/M_Z^2 \cos^2\theta_W$.

(c) Show that the low-energy effective Lagrangian for the neutral-current interactions of leptons can be written in the current–current form

$$\mathscr{L}_{NC} = -G_F \rho \sqrt{2} J_\lambda^0 J^{0\lambda},$$

where the neutral current is

$$J_\lambda^0 = \sum_i \bar{L}_i \gamma_\lambda \tau_3 L_i - 2\sin^2\theta_W J_\lambda^{(em)},$$

and the left-handed composite spinors have been defined in (6.3.1,2) and (6.4.1,3). This is to be compared with the effective Lagrangian (6.1.62) for charged-current interactions, which may be written as

$$\mathscr{L}_{CC} = -G_F \sqrt{2} J_{+\lambda} J_-^\lambda,$$

where

$$J_\pm^\lambda = \sum_i \bar{L}_i \gamma^\lambda \tau_\pm L_i.$$

6-11.   At very low momentum transfers, as in atomic physics applications, the nucleon appears elementary. If the effective Lagrangian for nucleon β-decay can be written in the limit of zero momentum transfer as

$$\mathscr{L}_\beta = -\frac{G_F}{2\sqrt{2}} \bar{e}\gamma_\lambda(1-\gamma_5)v\bar{p}\gamma^\lambda(1-g_A\gamma_5)n,$$

where $g_A = 1.254 \pm 0.007$ is the axial charge of the nucleon (renormalized from unity by the strong interactions), show that the eN neutral-current interactions may be represented by

$$\mathscr{L}_{ep} = \frac{G_F}{2\sqrt{2}} \bar{e}\gamma_\lambda(1-4x_W-\gamma_5)e\bar{p}\gamma^\lambda(1-4x_W-g_A\gamma_5)p$$

and

$$\mathscr{L}_{en} = -\frac{G_F}{2\sqrt{2}} \bar{e}\gamma_\lambda(1-4x_W-\gamma_5)e\bar{n}\gamma^\lambda(1-g_A\gamma_5)n.$$

Now perform, for a nucleus regarded as a noninteracting collection of Z protons and N neutrons, the nonrelativistic reduction of the implied nuclear matrix elements. Show that, for a heavy nucleus, the dominant parity-violating contribution to the electron–nucleus amplitude will be of the form

$$\mathcal{M}_{\text{p.v.}} = -\frac{iG_{\text{F}}}{2\sqrt{2}} Q^W \bar{e}\rho_N(\mathbf{r})\gamma_5 e,$$

where ρ_N is the nucleon density as a function of the electron coordinate $\mathbf{r}$, and the weak charge is

$$Q^W = Z(1 - 4x_{\text{W}}) - N.$$

For Further Reading

WEAK INTERACTIONS BEFORE GAUGE THEORIES. Among introductions to the V − A theory and its consequences, exemplary treatments appear in

E. D. Commins, *Weak Interactions*, McGraw-Hill, New York, 1973.

J. D. Jackson, *Elementary Particle Physics and Field Theory*, Volume 1 of the 1962 Brandeis Summer Institute Lectures, edited by K. W. Ford, Benjamin, New York, 1963, p. 263.

R. Marshak, Riazuddin, and C. P. Ryan, *Theory of Weak Interactions in Particle Physics*, Wiley-Interscience, New York, 1969.

L. B. Okun, *Weak Interactions of Elementary Particles*, Pergamon, Oxford, 1965.

Useful instruction in computational techniques is to be found in

R. P. Feynman, *Theory of Fundamental Processes*, Benjamin, New York, 1961.

The universal V − A interaction was proposed by

R. P. Feynman and M. Gell-Mann, *Phys. Rev.* **109**, 193 (1958).

S. S. Gerstein and Y. B. Zeldovich, *Zh. Eksp. Teor. Fiz.* **29**, 698 (1955) [English translation: *Sov. Phys.—JETP* **2**, 576 (1956)].

E. C. G. Sudarshan and R. E. Marshak, *Phys. Rev.* **109**, 1860 (1958).

J. J. Sakurai, *Nuovo Cim.* **7**, 649 (1958).

Establishment of the V − A interaction was a major achievement of the 1950s. Several of the key experiments are reported in

W. B. Hermannsfelt et al., *Phys. Rev. Lett.* **1**, 61 (1958).

T. Fazzini et al., *Phys. Rev. Lett.* **1**, 247 (1958).

G. Impeduglia et al., *Phys. Rev. Lett.* **1**, 249 (1958).

H. L. Anderson et al., *Phys. Rev. Lett.* **2**, 53 (1959).

M. Goldhaber, L. Grodzins, and A. W. Sunyar, *Phys. Rev.* **109**, 1015 (1958).

The history is reviewed in

P. K. Kabir (editor), *The Development of Weak Interaction Theory*, Gordon and Breach, New York, 1963.

E. J. Konopinski, *The Theory of Beta Radioactivity*, Oxford University Press, London, 1966.

A recent survey of the limits on right-handed charged current interactions has been given by

M. Strovink, *Weak Interactions as Probes of Unification*, Blacksburg, 1980, edited by G. B. Collins, L. N. Chang, and J. R. Ficenec, American Institute of Physics, New York, 1981, p. 46.

GAUGE THEORIES OF WEAK AND ELECTROMAGNETIC INTERACTIONS. Very thorough intellectual histories have been presented by

M. Veltman, *Proceedings of the 6th International Symposium on Electron and Photon Interactions at High Energies*, edited by H. Rollnik and W. Pfeil, North-Holland, Amsterdam, 1974, p. 429.

S. Coleman, *Science* **206**, 1290 (1979).

Also of interest are the Nobel lectures of S. L. Glashow, A. Salam, and S. Weinberg cited in the bibliography to Chapter 1, and the retrospective essay by
  B. W. Lee, in *Gauge Theories and Neutrino Physics*, Physics Reports Reprint Book Series Vol. 2, edited by M. Jacob, North-Holland, Amsterdam, 1978, p. 147.
The evolution of the intermediate boson hypothesis has been reviewed by
  P. Q. Hung and C. Quigg, *Science* **210**, 1205 (1980).
Among notable contributions in the development of the "standard model" are
  O. Klein, *Les Nouvelles Théories de la Physique*, Proceedings of a symposium held in Warsaw, 30 May–3 June 1938, Institut International de la Coöpération Intellectuelle, Paris, 1938, p. 6.
  J. Schwinger, *Ann. Phys. (NY)* **2**, 407 (1957).
  S. A. Bludman, *Nuovo Cim.* **9**, 443 (1958).
  S. L. Glashow, *Nucl. Phys.* **22**, 579 (1961).
  A. Salam and J. C. Ward, *Phys. Lett.* **13**, 168 (1964).
A number of these works appear in the well-chosen reprint collections,
  C. H. Lai (editor), *Gauge Theories of the Weak and Electromagnetic Interactions*, World Scientific, Singapore, 1981.
  C. H. Lai and R. N. Mohapatra (editors), *Gauge Theories of the Fundamental Interactions*, World Scientific, Singapore, 1981.
In addition to the textbooks and monographs cited in earlier chapters, the articles by
  M. A. B. Bég and A. Sirlin, *Ann. Rev. Nucl. Sci.* **24**, 379 (1974).
  E. S. Fradkin and I. V. Tyutin, *Riv. Nuovo Cim.* **4**, 1 (1974).
  R. Gatto, *Proceedings of the International School of Physics ⟨⟨Enrico Fermi⟩⟩*, Course LXXI, edited by M. Baldo-Ceolin, North-Holland, Amsterdam, 1979, p. 1.
provide self-contained discussions of the Weinberg–Salam model. Many technical issues are superbly treated in
  R. Balian and J. Zinn-Justin (editors), *Methods in Field Theory*, 1975 Les Houches Lectures, North-Holland, Amsterdam, 1976; World Scientific, Singapore, 1981.
NEUTRAL CURRENTS. An interesting account of the discovery of neutral currents from the perspective of a historian of science is given by
  P. Galison, *Rev. Mod. Phys.* **55**, 477 (1983).
The status of "model-independent" determinations of the neutral current couplings is assessed by
  J. E. Kim, P. Langacker, M. Levine, and H. H. Williams, *Rev. Mod. Phys.* **53**, 211 (1981).
  P. Q. Hung and J. J. Sakurai, *Ann. Rev. Nucl. Part. Sci.* **31**, 375 (1981).
THE HIGGS BOSON. The expected properties of light Higgs bosons have been detailed by
  J. Ellis, M. K. Gaillard, and D. Nanopoulos, *Nucl. Phys.* **B106**, 292 (1976).
  A. I. Vainshtein, V. I. Zakharov, and M. A. Shifman, *Usp. Nauk Fiz.* **131**, 537 (1980) [English translation: *Sov. Phys.—Uspekhi* **23**, 429 (1980)].
A strategy for bounding the mass of the Higgs boson from above based on the calculability of radiative corrections, as opposed to the partial-wave unitarity methods described in the text, has been developed by
  M. Veltman, *Acta Phys. Polon.* **B8**, 475 (1977); *Nucl. Phys.* **B123**, 89 (1977); *Phys. Lett.* **91B**, 95 (1980); *Quarks and Leptons*, Cargèse 1979, edited by M. Lévy et al., Plenum, New York, 1980, p. 1.
Related bounds on heavy fermion masses are also implied by tree-unitarity considerations. For these, see
  M. S. Chanowitz, M. A. Furman, and I. Hinchliffe, *Nucl. Phys.* **B153**, 402 (1979).
DYNAMICAL SYMMETRY BREAKING. A convenient summary and reprint collection is
  E. Farhi and R. Jackiw (editors), *Dynamical Gauge Symmetry Breaking*, World Scientific, Singapore, 1982.

Detailed reviews of the approach to dynamical symmetry breaking known as technicolor have been given by

E. Farhi and L. Susskind, *Phys. Rep.* **74**, 277 (1981).

R. Kaul, *Rev. Mod. Phys.* **55**, 449 (1983).

S-MATRIX "DERIVATION" OF SPONTANEOUSLY BROKEN GAUGE THEORIES. The program of divergence cancellation for the tree diagrams is thoroughly explained by

C. H. Llewellyn Smith, *Phenomenology of Particles at High Energies*, 14th Scottish Universities Summer School in Physics, 1973, edited by R. L. Crawford and R. Jennings, Academic, New York and London, 1974, p. 459.

RENORMALIZATION OF SPONTANEOUSLY BROKEN GAUGE THEORIES. Among the classic papers on the subject are

G. 't Hooft, *Nucl. Phys.* **B33**, 173 (1971); **B35**, 167 (1971).

B. W. Lee and J. Zinn-Justin, *Phys. Rev.* **D5**, 3121, 3137, 3155 (1972); **D7**, 1049 (1973).

G. 't Hooft and M. Veltman, *Nucl. Phys.* **B44**, 189 (1972); **B50**, 318 (1972).

D. A. Ross and J. C. Taylor, *Nucl. Phys.* **B51**, 125 (1973).

C. Becchi, A. Rouet, and R. Stora, *Ann. Phys. (NY)* **98**, 287 (1976).

Important for the renormalization program are the generalized Ward–Takahashi identities presented in

J. C. Taylor, *Nucl. Phys.* **B33**, 436 (1971).

A. A. Slavnov, *Teor. Mat. Fiz.* **10**, 153 (1972) [English translation: *Theor. Math. Phys.* **10**, 99 (1972)].

ANOMALIES. Surveys of the axial anomaly problem are given in

S. L. Adler, *Lectures on Elementary Particles and Quantum Field Theory*, Brandeis Summer Institute 1970, edited by S. Deser, M. Grisaru, and H. Pendleton, MIT Press, Cambridge, Massachusetts, 1970, Vol. 1, p. 1.

R. Jackiw, *Lectures on Current Algebra and Its Applications*, by S. B. Treiman, R. Jackiw, and D. J. Gross, Princeton University Press, Princeton, New Jersey, 1972, p. 97.

THE NEUTRAL CURRENT IN ATOMIC PHYSICS. Experiments to probe the neutral current at small momentum transfers are reviewed by

E. D. Commins and P. H. Bucksbaum, *Ann. Rev. Nucl. Part. Sci.* **30**, 1 (1980).

FEYNMAN RULES. The rules given in this chapter are complete only for diagrams without loops. A convenient source for the rules in a general gauge of the type described in Section 6.6 is

K. Fujikawa, B. W. Lee, and A. I. Sanda, *Phys. Rev.* **D6**, 2923 (1972).

References

[1] S. Weinberg, *Phys. Rev. Lett.* **19**, 1264 (1967); A. Salam, *Elementary Particle Theory: Relativistic Groups and Analyticity* (Nobel Symposium No. 8), edited by N. Svartholm, Almqvist and Wiksell, Stockholm, 1968, p. 367.

[2] E. Fermi, *Ric. Sci.* **4**, 491 (1933); reprinted in E. Fermi, *Collected Papers*, edited by E. Segrè *et al.*, University of Chicago Press, Chicago, 1962, Vol. 1, p. 538; and *Z. Phys.* **88**, 161 (1934) [English translation: F. L. Wilson, *Am. J. Phys.* **36**, 1150 (1968)].

[3] R. E. Shrock and L.-L. Wang, *Phys. Rev. Lett.* **41**, 1692 (1978).

[4] N. Armenise *et al.* (Gargamelle Collaboration), *Phys. Lett.* **84B**, 137 (1979); M. Jonker *et al.* (CHARM Collaboration), *ibid.* **93B**, 203 (1980).

[5] A model of this kind was proposed in a gauge theory framework by H. Georgi and S. L. Glashow, *Phys. Rev. Lett.* **28**, 1494 (1972).

[6] C. H. Llewellyn Smith, *Phys. Lett.* **46B**, 233 (1973); S. Joglekar, *Ann. Phys. (NY)* **83**, 427 (1974).

⁷ O. P. Sushkov, V. V. Flambaum, and I. B. Khriplovich, *Yad. Fiz.* **20**, 1016 (1974) [English translation: *Sov. J. Nucl. Phys.* **20**, 537 (1975)]; W. Alles, C. Boyer, and A. J. Buras, *Nucl. Phys.* **B119**, 125 (1977); K. Gaemers and G. Gounaris, *Z. Phys.* **C1**, 259 (1979).

⁸ F. J. Hasert *et al.*, *Phys. Lett.* **46B**, 121, 138 (1973); *Nucl. Phys.* **B73**, 1 (1974).

⁹ A. Benvenuti *et al.*, *Phys. Rev. Lett.* **32**, 800 (1974).

¹⁰ B. Aubert *et al.*, *Phys. Rev. Lett.* **34**, 1454, 1457 (1974); S. J. Barish *et al.*, *ibid.* **33**, 468 (1974); B. C. Barish *et al.*, *ibid.* **34**, 538 (1975).

¹¹ The masslessness of the neutrinos is a consequence of the absence of right-handed neutrinos in the original enumeration of fermions. It can easily be avoided if experiment so dictates.

¹² Logarithmic violations of unitarity that occur at exponentially high energies $\sim M_W e^{1/\alpha}$ will be of no concern to us here.

¹³ J. M. Cornwall, D. N. Levin, and G. Tiktopoulos, *Phys. Rev. Lett.* **30**, 1268 (1973); *Phys. Rev.* **D10**, 1145 (1974).

¹⁴ B. W. Lee, C. Quigg, and H. B. Thacker, *Phys. Rev.* **D16**, 1519 (1977).

¹⁵ S. Coleman and E. Weinberg, *Phys. Rev.* **D7**, 1888 (1973); S. Weinberg, *ibid.* **D7**, 2887 (1973); E. Gildener and S. Weinberg, *ibid.* **D13**, 3333 (1976).

¹⁶ A. D. Linde, *Pis'ma Zh. Eksp. Teor. Fiz.* **23**, 73 (1976) [English translation: *Sov. Phys.— JETP Letters* **23**, 64 (1976)]; S. Weinberg, *Phys. Rev. Lett.* **36**, 294 (1973).

¹⁷ See the first and third papers of ref. 15.

¹⁸ K. G. Wilson, *Phys. Rev.* **D3**, 1818 (1971), particularly Section V.

¹⁹ Discussions directly relevant to gauge theories include R. Jackiw, "Dynamical Symmetry Breaking," in *Laws of Hadronic Matter*, 1973 Erice School, edited by A. Zichichi, Academic Press, New York and London, 1975, p. 225; S. Weinberg, *Phys. Rev.* **D13**, 974 (1976), **D19**, 1277 (1978); L. Susskind, *ibid.* **D20**, 2619 (1979).

²⁰ G. 't Hooft, *Nucl. Phys.* **B33**, 173 (1971); **B35**, 167 (1971).

²¹ C. Bouchiat, J. Iliopoulos, and P. Meyer, *Phys. Lett.* **38B**, 519 (1972). A thorough analysis of the anomaly problem in the Weinberg–Salam model was made by D. J. Gross and R. Jackiw, *Phys. Rev.* **D6**, 477 (1972). General conditions for anomaly cancellation were investigated by H. Georgi and S. L. Glashow *ibid.* **D6**, 429 (1972).

CHAPTER 7

ELECTROWEAK INTERACTIONS OF QUARKS

In this chapter we implement the extension to the hadronic sector of the standard model of weak and electromagnetic interactions. This will be accomplished through the medium of the quark model, as is natural in view of the experimental suggestions that quarks and leptons are comparably elementary. Because of the similarity to leptons, construction of a theory at the quark level is relatively straightforward and yet nontrivial. It was remarked in Section 6.6 that an anomaly-free, and hence renormalizable, theory could be formulated if each weak-isospin doublet of leptons were accompanied by a color triplet of weak-isospin doublets of quarks. The necessity of enlarging the quark spectrum beyond u, d, and s was already noted in Section 1.3. We are now in a position to perceive more clearly the need for the Glashow–Iliopoulos–Maiani mechanism and to describe its generalization to theories with many quark and lepton doublets.

The difficulty in describing completely the weak interactions of quarks is that the quarks are bound within hadrons. We lack a comprehensive theory of hadron structure and cannot yet give a full account of the influence of the strong interactions upon weak transition amplitudes. For semileptonic processes, the parton model has been found to give a reliable approximate description. Its development here serves two purposes: the first, to provide a tool for analyzing hadronic weak interactions and testing weak-interaction models; and the second, to recognize the challenge to understanding that the success of the parton model poses for a theory of the strong interactions. The role of the parton model in the development of quantum chromodynamics will be taken up further in Chapter 8.

In the parton-model approximation we shall complete our description of the properties of the intermediate vector bosons and then turn to the study

147

of several high-energy processes. The subject of electron–positron annihilation is treated at some length, both for the Z^0-formation reaction and for the study of $\gamma - Z^0$ interference already introduced in Section 6.4. We then develop both the general kinematics and the parton-model description of deeply inelastic lepton–nucleon scattering. This leads to an analysis of weak neutral-current effects in neutrino–nucleon and electron–nucleon scattering, and to a brief assessment of the possible interest in high-energy electron–proton colliders. Finally, we discuss expectations for the production of intermediate vector bosons in hadron–hadron collisions.

7.1   The Weinberg–Salam–Glashow–Iliopoulos–Maiani Model

We shall now arrive at the minimal internally consistent theory of the weak and electromagnetic interactions that is (but for the omission of heavy fermions) consistent with experiment. Consider first the direct generalization of the theory of electrons developed in Chapter 6. We noted in Section 6.6 that an $SU(2)_L \otimes U(1)_Y$ theory based upon the electron doublet

$$\mathsf{L}_e = \begin{pmatrix} \nu_e \\ e \end{pmatrix}_L \tag{7.1.1}$$

and quark doublets

$$\begin{pmatrix} u_{\text{Red}} \\ d_{\text{Red}} \end{pmatrix}_L, \qquad \begin{pmatrix} u_{\text{Blue}} \\ d_{\text{Blue}} \end{pmatrix}_L, \qquad \begin{pmatrix} u_{\text{Green}} \\ d_{\text{Green}} \end{pmatrix}_L \tag{7.1.2}$$

would be free of anomalies and therefore have the benign ultraviolet behavior required for renormalizability. For many purposes in weak interaction theory, color merely serves as a device for counting multiplets. Whenever it is unnecessary to retain the color indices we shall suppress them with the understanding that the quarks are color triplets. The hadronic sector of the theory is thus built upon a single (color triplet of) left-handed weak-isospin doublet(s)

$$\mathsf{L}_q = \begin{pmatrix} u \\ d \end{pmatrix}_L \tag{7.1.3}$$

with weak hypercharge

$$Y(q_L) = \tfrac{1}{3}, \tag{7.1.4}$$

and two (color-triplet) right-handed weak-isospin singlets

$$\begin{aligned} \mathsf{R}_u &= u_R = \tfrac{1}{2}(1 + \gamma_5)u \\ \mathsf{R}_d &= d_R = \tfrac{1}{2}(1 + \gamma_5)d \end{aligned} \Bigg\}, \tag{7.1.5}$$

with weak hypercharge

$$\begin{aligned} Y(u_R) &= \tfrac{4}{3} \\ Y(d_R) &= -\tfrac{2}{3} \end{aligned} \Bigg\}. \tag{7.1.6}$$

The complex doublet of Higgs bosons is as usual

$$\phi = \begin{pmatrix} \phi^+ \\ \phi^0 \end{pmatrix},$$
(7.1.7)

with

$$Y_\phi = +1,$$
(7.1.8)

and its complex conjugate is the $SU(2)_L$ doublet

$$\bar{\phi} = i\sigma_2 \phi^* = \begin{pmatrix} \bar{\phi}^0 \\ -\phi^- \end{pmatrix},$$
(7.1.9)

with

$$Y_{\bar{\phi}} = -1.$$
(7.1.10)

The only change in form from the theory of leptons is in the interaction term

$$\mathscr{L}_{\text{Yukawa}} = -G_u[(\bar{L}_q \bar{\phi})u_R + \bar{u}_R(\bar{\phi}^\dagger L_q)] \\ - G_d[(\bar{L}_q \phi)d_R + \bar{d}_R(\phi^\dagger L_q)]$$
(7.1.11)

which will generate, upon spontaneous symmetry breaking, masses for both up and down quarks.

We need not repeat all the arithmetic of Section 6.3 to complete the construction of the theory, but may simply refer to our earlier results for the consequences. Evidently the charged-current interaction is, by construction,

$$\mathscr{L}_{W-q} = -\frac{g}{2\sqrt{2}}[\bar{u}\gamma^\mu(1 - \gamma_5)dW_\mu^+ + \bar{d}\gamma^\mu(1 - \gamma_5)uW_\mu^-],$$
(7.1.12)

and the weak neutral-current interaction is of the now familiar form [compare (6.3.60)]

$$\mathscr{L}_{Z-q} = -\frac{g}{4\cos\theta_{\text{W}}}\{\bar{u}\gamma^\mu[(1 - \gamma_5)\tau_3 - 2x_{\text{W}}Q]uZ_\mu \\ + \bar{d}\gamma^\mu[(1 - \gamma_5)\tau_3 - 2x_{\text{W}}Q]dZ_\mu\},$$
(7.1.13)

where as usual $x_{\text{W}} = \sin^2\theta_{\text{W}}$. The gauge boson and Higgs boson sectors are unchanged from the theory of leptons, and the Yukawa couplings G_u and G_d may be chosen to reproduce the quark masses.

The theory of the weak and electromagnetic interactions of v_e, e, u, and d that we have described is formally neat and internally self-consistent. However, the low-energy phenomenology is incorrect in a small but significant respect: the hadronic charged current is described not by the quark doublet (7.1.3), but rather by the quark-model transcription of the Cabibbo current, namely

$$L_u = \begin{pmatrix} u \\ d_\theta \end{pmatrix}_L,$$
(7.1.14)

with

$$d_\theta = d\cos\theta_{\text{C}} + s\sin\theta_{\text{C}},$$
(7.1.15)

where θ_C is the Cabibbo angle, which has been determined[1] as

$$\cos\theta_C = 0.9737 \pm 0.0025. \tag{7.1.16}$$

The implied $SU(2)_L \otimes U(1)_Y$ model is defective in several ways. It is esthetically unsatisfying that the combination of charge $-1/3$ quarks orthogonal to (7.1.15),

$$s_\theta = s\cos\theta_C - d\sin\theta_C \tag{7.1.17}$$

is apparently superfluous. More urgently, the weak neutral-current interaction now contains terms proportional to

$$\sin\theta_C\cos\theta_C[\bar{d}\gamma^\mu(1-\gamma_5)s + \bar{s}\gamma^\mu(1-\gamma_5)d]Z_\mu, \tag{7.1.18}$$

which correspond to flavor-changing (specifically, strangeness-changing) neutral currents. This is phenomenologically unacceptable because stringent experimental limits have been placed on the rates of decays mediated by strangeness-changing neutral currents such as the decay

$$K^+ \to \pi^+\nu\bar{\nu}, \tag{7.1.19}$$

which may be interpreted in terms of the elementary transition

$$\bar{s} \to \bar{d}\nu\bar{\nu}, \tag{7.1.20}$$

and for which the branching ratio[2] is

$$\frac{\Gamma(K^+ \to \pi^+\nu\bar{\nu})}{\Gamma(K^+ \to \text{all})} < 1.4 \times 10^{-7}. \tag{7.1.21}$$

Similarly, the observed rate for the decay $K_L \to \mu^+\mu^-$,

$$\Gamma(K_L \to \mu^+\mu^-) = (3.4 \pm 0.7) \times 10^{-9}\,\Gamma(K^+ \to \mu^+\nu_\mu), \tag{7.1.22}$$

can be understood in terms of QED and the known $K_L \to \gamma\gamma$ transition rate, and leaves little room for an elementary $\bar{s}d \to \mu^+\mu^-$ transition. A similar conclusion may be drawn from the smallness of observables linked to $|\Delta S| = 2$ transition amplitudes, such as the $K_L - K_S$ mass difference

$$M(K_L) - M(K_S) = (0.5349 \pm 0.0022) \times 10^{10}\,\hbar\,\text{sec}^{-1}$$
$$\simeq 3.5 \times 10^{-6}\,\text{eV}$$
$$\simeq 7 \times 10^{-15}\,M(K^0) \tag{7.1.23}$$

Thus, in the Weinberg–Salam model and, more generally, in models that entail neutral-current interactions proportional to the third component of weak isospin, it becomes important to guard against the appearance of strangeness-changing neutral currents.

An elegant solution to the problem of flavor-changing neutral currents was devised by Glashow, Iliopoulos, and Maiani.[3] The key observation, as we have already seen in Section 1.3, is that by introducing a new "charmed" quark c, the weak-isospin partner of s_θ, one cancels the offending terms (7.1.18) in the neutral-current interaction with quarks. We therefore consider an $SU(2)_L \otimes U(1)_Y$ theory in which the fundamental fermions are the leptons

$$\mathsf{L}_e = \begin{pmatrix} \nu_e \\ e \end{pmatrix}_L, \qquad \mathsf{L}_\mu = \begin{pmatrix} \nu_\mu \\ \mu \end{pmatrix}_L, \qquad \mathsf{R}_e = e_R, \qquad \mathsf{R}_\mu = \mu_R \tag{7.1.24}$$

and the quarks

$$\mathsf{L}_u = \begin{pmatrix} u \\ d_\theta \end{pmatrix}_L, \qquad \mathsf{L}_c = \begin{pmatrix} c \\ s_\theta \end{pmatrix}_L,$$

$$\mathsf{R}_u = u_R, \qquad \mathsf{R}_d = d_R, \qquad \mathsf{R}_c = c_R, \qquad \mathsf{R}_s = s_R. \tag{7.1.25}$$

The usual construction then yields an anomaly-free theory with flavor-preserving neutral currents, Cabibbo universality for the charged currents, and an agreeable lepton–hadron symmetry. The success of the remarkable inference that a not-very-massive charmed quark must exist is well known, and thoroughly documented elsewhere. The resulting Feynman rules for gauge boson–quark interactions are presented in Fig. 7-1.

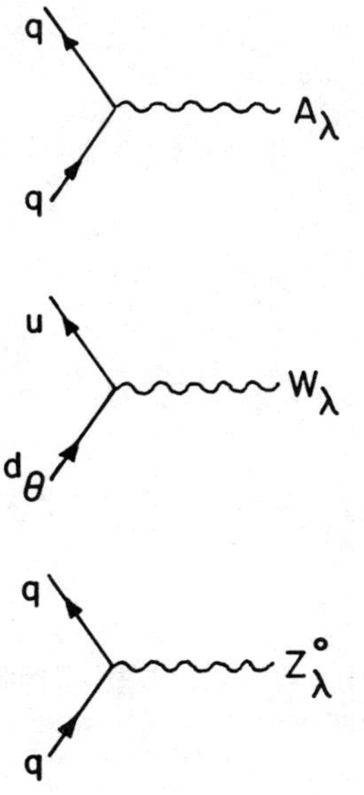

$$-iQ_q \bar{q}\gamma_\lambda q$$

$$-i\left(\frac{G_F M_W^2}{\sqrt{2}}\right)^{1/2} \bar{u}\gamma_\lambda(1-\gamma_5)d_\theta$$

$$\frac{-i}{\sqrt{2}}\left(\frac{G_F M_Z^2}{\sqrt{2}}\right)^{1/2} \bar{q}\gamma_\lambda\left[R_q(1+\gamma_5)+L_q(1-\gamma_5)\right]q$$

$$R_q = -2Q_q \sin^2\theta_W$$

$$L_q = \tau_3 - 2Q_q \sin^2\theta_W$$

FIG. 7-1. Feynman rules for interactions of quarks and gauge bosons in the Weinberg–Salam model.

The flavor-conserving property of the neutral-current interactions is easily generalized to the case of many quark generations. This is of more than academic interest because of the observation of a fifth quark (the b-quark) in the Υ family of meson resonances as well as the existence of the third charged lepton $\tau(1782)$ (cf. Tables 1.1 and 1.3). Suppose that there are n left-handed quark doublets

$$\begin{pmatrix} u \\ d' \end{pmatrix}_L, \qquad \begin{pmatrix} c \\ s' \end{pmatrix}_L, \qquad \begin{pmatrix} t \\ b' \end{pmatrix}_L, \ldots, \tag{7.1.26}$$

where the primes denote generalized Cabibbo mixing among the quarks of charge $-1/3$. We write all the quarks in terms of a composite $2n$-component spinor

$$\Psi = \begin{pmatrix} \vdots \\ t \\ c \\ u \\ \text{---} \\ d \\ s \\ b \\ \vdots \end{pmatrix} \tag{7.1.27}$$

and express the structure of the charged current as

$$J_+^\lambda = \frac{1}{\sqrt{2}} \bar\Psi \gamma^\lambda (1 - \gamma_5) \mathcal{O} \Psi, \tag{7.1.28}$$

where the $2n \times 2n$ matrix $\mathcal{O}$ is of the form

$$\mathcal{O} = \begin{pmatrix} 0 & U \\ 0 & 0 \end{pmatrix} \tag{7.1.29}$$

and U is the unitary $n \times n$ matrix that describes quark mixing. In the four-quark theory, the mixing matrix assumes the familiar form

$$U = \begin{pmatrix} \cos\theta_C & \sin\theta_C \\ -\sin\theta_C & \cos\theta_C \end{pmatrix}. \tag{7.1.30}$$

The weak-isospin contribution to the neutral current is then

$$J_3^\lambda = \tfrac{1}{2} \bar\Psi \gamma^\lambda (1 - \gamma_5) [\mathcal{O}, \mathcal{O}^\dagger] \Psi, \tag{7.1.31}$$

but since

$$[\mathcal{O}, \mathcal{O}^\dagger] = \begin{pmatrix} I & 0 \\ 0 & -I \end{pmatrix}, \tag{7.1.32}$$

the neutral current will be flavor-diagonal with a weak isospin contribution simply proportional to τ_3, as expected:

$$J_3^\lambda = \tfrac{1}{2}\overline{\Psi}\gamma^\lambda(1 - \gamma_5)\tau_3\Psi. \tag{7.1.33}$$

For most of our explicit discussion, we shall address the case of four quarks, leaving the obvious generalizations to more families of quarks to be made as needed.

In this spirit, let us take up once more the question of fermion masses. As was the case for the leptons, the generation of fermion masses by the Higgs mechanism is both possible (which is a virtue) and entirely *ad hoc* (which is not). The scalar–fermion interaction Lagrangian may be written in gauge-invariant form as

$$\begin{aligned}
\mathscr{L}_{\text{Yukawa}} = &-G_1\big[(\overline{L}_u\overline{\phi})u_R + \overline{u}_R(\overline{\phi}^\dagger L_u)\big] - G_2\big[(\overline{L}_u\phi)d_R + \overline{d}_R(\phi^\dagger L_u)\big] \\
&- G_3\big[(\overline{L}_u\phi)s_R + \overline{s}_R(\phi^\dagger L_u)\big] - G_4\big[(\overline{L}_c\overline{\phi})c_R + \overline{c}_R(\overline{\phi}^\dagger L_c)\big] \\
&- G_5\big[(\overline{L}_c\phi)d_R + \overline{d}_R(\phi^\dagger L_c)\big] - G_6\big[(\overline{L}_c\phi)s_R + \overline{s}_R(\phi^\dagger L_c)\big].
\end{aligned} \tag{7.1.34}$$

Because we have exercised our freedom to express all the mixing in terms of the charge $-1/3$ quarks, the other conceivable terms, $\overline{L}_u\phi c_R$ and $\overline{L}_c\overline{\phi}u_R$, may be omitted. Replacing the scalar field ϕ by its expectation value

$$\langle\phi\rangle_0 = \begin{pmatrix} 0 \\ v/\sqrt{2} \end{pmatrix}, \tag{7.1.35}$$

we obtain a series of mass terms as in (6.3.34). The Yukawa couplings $G_1, G_2, \dots, G_6$ must be chosen so that u, d, s, and c are mass eigenstates with the correct quark masses:

$$\left.\begin{aligned}
G_1 &= m_u\sqrt{2}/v \\
G_2 &= m_d\cos\theta_{\text{C}}\sqrt{2}/v \\
G_3 &= m_s\sin\theta_{\text{C}}\sqrt{2}/v \\
G_4 &= m_c\sqrt{2}/v \\
G_3 &= -G_2\tan\theta_{\text{C}} \\
G_6 &= G_3\cot\theta_{\text{C}}
\end{aligned}\right\} \tag{7.1.36}$$

We are evidently at liberty to do this. Clearly any symmetry principle relating the Yukawa couplings G_i may imply connections between the Cabibbo mixing angle and the quark masses.

Having seen how to construct an internally consistent theory of the weak and electromagnetic interactions of four quarks and four leptons that is not in gross disagreement with experimental systematics, let us now proceed to enumerate the important consequences of the theory. We begin with the properties of the gauge bosons themselves. The results

$$M_W^2 = \pi\alpha/G_F\sqrt{2}\sin^2\theta_W \qquad (6.3.57)$$

and

$$M_Z^2 = M_W^2/\cos^2\theta_W \qquad (6.3.58)$$

are unaffected, to leading order, by the introduction of quarks into the theory. To compute the nonleptonic decay rates of the intermediate bosons, we appeal to the parton model in analogy with the successful description of the reaction

$$e^+e^- \to \text{hadrons} \qquad (7.1.37)$$

in terms of the elementary process

$$e^+e^- \to \gamma_{\text{virtual}} \to q\bar{q} \qquad (7.1.38)$$

in which, as discussed in Section 1.2, the quark and antiquark in the semifinal state are assumed to materialize with probability one into the observed jets of hadrons. The partial widths for nonleptonic $W^\pm$ and Z^0 decays may then be related at once to those for the leptonic decays. For the charged intermediate boson we have

$$\begin{aligned}\Gamma(W^+ \to u\bar{d}) &= 3\cos^2\theta_C\Gamma(W \to l\nu)\\\Gamma(W^+ \to u\bar{s}) &= 3\sin^2\theta_C\Gamma(W \to l\nu)\end{aligned} \Bigg\} , \qquad (7.1.39)$$

where the factor of 3 accounts for the quark colors and the Cabibbo angle factors are a consequence of the Feynman rules. As in the discussion of leptonic decays, we have idealized the products as massless, an approximation that may readily be undone. The leptonic decay rate continues to be given by

$$\Gamma(W \to l\nu) = G_F M_W^3/6\pi\sqrt{2}. \qquad (6.2.20)$$

More generally, if D_q is the number of color-triplet $SU(2)_L$ doublets of quarks into which the intermediate boson can decay, the rate for the inclusive decay of $W \to$ hadrons is given by

$$\Gamma(W \to \text{hadrons}) = 3D_q\Gamma(W \to l\nu). \qquad (7.1.40)$$

Similarly, the total width is given by

$$\Gamma(W \to \text{all}) = (D_l + 3D_q)\Gamma(W \to l\nu), \qquad (7.1.41)$$

where D_l is the number of energetically accessible lepton doublets. Thus for the four-quark, four-lepton theory developed in this section, we expect a muonic branching ratio of

$$\Gamma(W \to \mu\nu)/\Gamma(W \to \text{all}) = \tfrac{1}{8}, \qquad (7.1.42)$$

while with three doublets each of quarks and leptons, we expect

$$\Gamma(W \to \mu\nu)/\Gamma(W \to \text{all}) = \tfrac{1}{12}. \qquad (7.1.43)$$

These estimates imply, together with the current estimates of the intermediate boson mass and leptonic width,

$$M_W \simeq 84 \text{ GeV/c}^2,$$

$$\Gamma(W \to l\nu) \simeq 250 \text{ MeV},$$

(6.3.65)

a total width of 2–3 GeV. Hadronic decays are expected to occur as back-to-back jets in the W-boson rest frame. The normalized angular distribution of the product quark, the direction of which defines a jet axis, is the same as that of the electron described in (6.2.19) and (6.2.23,24).

Apart from the predictions (6.3.57) and (6.3.65) for the intermediate boson mass and partial width, none of the results for decays of the charged intermediate boson depends sensitively upon the specific model of the weak and electromagnetic interactions. Rather, they all follow from the intermediate boson picture developed in Section 6.2, augmented by the idea of quark–lepton universality and by the parton model for nonleptonic decays. It is worthwhile to remark that the latter has been tested for virtual $W^\pm$ decays by the successful estimate of the branching ratios of the τ-lepton noted in Section 1.2. The properties of the neutral intermediate boson, in contrast, are evidently quite specific to the theory, although no less easy to compute. Comparing the Feynman rules for $Z^0 \to \nu\bar{\nu}$ and $Z^0 \to e^+ e^-$ given in Fig. 6-12 with those for $Z^0 \to q\bar{q}$ given in Fig. 7-1, we have at once

$$\Gamma(Z^0 \to q\bar{q}) = 3(R_q^2 + L_q^2)\Gamma(Z^0 \to \nu\bar{\nu}),$$

(7.1.44)

so that

$$\Gamma(Z^0 \to \text{hadrons})$$

$$= D_q\Gamma(Z^0 \to u\bar{u} + d\bar{d})$$

$$= 3D_q\left[\left(1 - \frac{8x_W}{3} + \frac{32x_W^2}{9}\right) + \left(1 - \frac{4x_W}{3} + \frac{8x_W^2}{9}\right)\right]\Gamma(Z^0 \to \nu\bar{\nu})$$

$$= 3D_q\left(2 - 4x_W + \frac{40x_W^2}{9}\right)\Gamma(Z^0 \to \nu\bar{\nu}),$$

(7.1.45)

whereas

$$\Gamma(Z^0 \to e^+ e^-) = (1 - 4x_W + 8x_W^2)\Gamma(Z^0 \to \nu\bar{\nu}).$$

(6.3.64)

In this case the scale is set by the neutrinic decay rate

$$\Gamma(Z^0 \to \nu\bar{\nu}) = G_F M_Z^3/12\pi\sqrt{2}.$$

(6.3.63)

For the four-quark, four-lepton theory we may therefore estimate the branching ratio

$$\Gamma(Z^0 \to e^+ e^-)/\Gamma(Z^0 \to \text{all}) \simeq 0.046$$

(7.1.46)

for $x_W = 0.2$. The corresponding estimate for the six-quark, six-lepton theory is

$$\Gamma(Z^0 \to e^+ e^-)/\Gamma(Z^0 \to \text{all}) \simeq 0.031,$$

(7.1.47)

still for $x_W = 0.2$. The angular distribution of the decay products is also easily read off the results for W-decay. For the current estimate of the Z^0 mass and leptonic decay width,

$$M_{Z^0} \simeq 94 \text{ GeV}/c^2,$$
$$\Gamma(Z^0 \to e^+e^-) \simeq \tfrac{1}{2}\Gamma(Z^0 \to \nu\bar{\nu}) \simeq 90 \text{ MeV}, \tag{6.3.66}$$

we anticipate a total width $\Gamma(Z^0 \to \text{all})$ of 2–3 GeV, comparable to that of the W-boson.

We have obtained a reasonably complete description of the intermediate bosons in the standard model of weak and electromagnetic interactions. In the succeeding sections, as we develop other consequences of the theory, we shall be alert for experimental situations in which the production of intermediate bosons is particularly favorable.

7.2   Electron–Positron Annihilations

Because of the simplicity and definiteness of the parton model in this situation, we begin our discussion of electroweak processes involving hadrons by considering electron–positron annihilations. The contribution of direct-channel γ- and Z^0-exchange diagrams (Fig. 6-19) to the reaction

$$e^+e^- \to \mu^+\mu^- \tag{7.2.1}$$

has already been presented in Equation (6.4.29). The differential cross section for the parton-model reaction

$$e^+e^- \to q\bar{q} \to \text{hadrons} \tag{7.2.2}$$

differs only in an overall factor of 3 for quark color and in the replacement of the electromagnetic and weak couplings of the muon by those of the quark in question. It is thus

$$\frac{d\sigma}{dz}(e^+e^- \to q\bar{q}) = \frac{3\pi\alpha^2 Q_q^2}{2s}(1 + z^2)$$

$$- \frac{3\alpha Q_q G_F M_Z^2(s - M_Z^2)}{8\sqrt{2}[(s - M_Z^2)^2 + M_Z^2\Gamma_Z^2]}$$
$$\times [(R_e + L_e)(R_q + L_q)(1 + z^2) + 2(R_e - L_e)(R_q - L_q)z]$$
$$+ \frac{3G_F^2 M_Z^4 s}{64\pi[(s - M_Z^2)^2 + M_Z^2\Gamma_Z^2]}$$
$$\times [(R_e^2 + L_e^2)(R_q^2 + L_q^2)(1 + z^2) + 2(R_e^2 - L_e^2)(R_q^2 - L_q^2)z], \tag{7.2.3}$$

where $z = \cos\theta_{c.m.}$ measures the angle between incoming and outgoing fermions. Although in displaying quantitative consequences we shall utilize the couplings prescribed by the standard model, it is worth emphasizing that the forms (6.4.29) and (7.2.3) are valid for any theory containing a single neutral weak boson with vector and axial-vector couplings. Therefore we shall revert from time to time to the notation of the "model-independent" analysis introduced in Section 6.4.

The cross section expected for the production of various species in the standard model with $x_W = \sin^2\theta_W = 0.2$ is plotted in Fig. 7-2 in terms of the quantity

$$R \equiv \sigma(e^+e^- \rightarrow f\bar{f})/\sigma_{QED}(e^+e^- \rightarrow \mu^+\mu^-). \tag{7.2.4}$$

The Z^0 has been assumed to decay without any kinematic suppression into three families of quarks and leptons, and no radiative corrections have been made to the cross section. The Z^0 peak is quite unmistakable. At the peak, the cross section for each fermion species is

$$\begin{aligned}\sigma_{peak}(e^+e^- \rightarrow f\bar{f}) &= G_F^2 M_Z^4 (L_e^2 + R_e^2)(L_f^2 + R_f^2)N_c/24\pi\Gamma_Z^2 \\ &= G_F^2 M_Z^4 (v_e^2 + a_e^2)(v_f^2 + a_f^2)N_c/6\pi\Gamma_Z^2,\end{aligned} \tag{7.2.5}$$

where N_c is the number of colors of the produced fermion and the result is also written in the notation

$$\left.\begin{aligned}a_i &= \tfrac{1}{2}(L_i - R_i) \\ v_i &= \tfrac{1}{2}(L_i + R_i)\end{aligned}\right\} \tag{7.2.6}$$

used for model-independent analyses. Thus we may write the ratio (7.2.4) at the intermediate boson peak as

$$R_{peak}(e^+e^- \rightarrow f\bar{f}) = \frac{G_F^2 M_Z^6 (L_e^2 + R_e^2)(L_f^2 + R_f^2)N_c}{32\pi^2\alpha^2\Gamma_Z^2}. \tag{7.2.7}$$

In the standard model, with

$$\Gamma_Z = \frac{G_F M_Z^3 \sqrt{2}}{3\pi}\left(1 - 2x_W + \frac{8x_W^2}{3}\right) \cdot D, \tag{7.2.8}$$

where D represents the number of quark and lepton doublets, we therefore have

$$R_{peak}(e^+e^- \rightarrow \nu\bar{\nu}) = \frac{9(1 - 4x_W + 8x_W^2)}{64\alpha^2 D^2\left(1 - 2x_W + \dfrac{8x_W^2}{3}\right)} \tag{7.2.9}$$

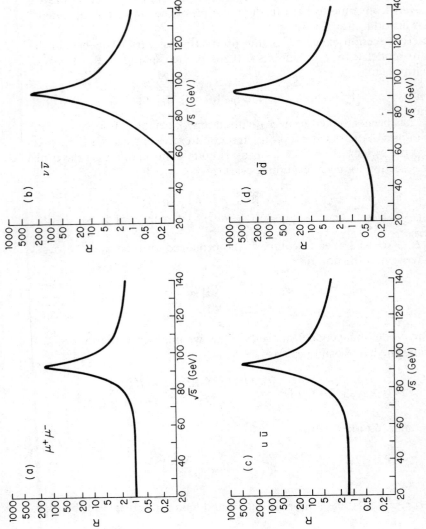

FIG. 7-2. Energy dependence of the ratio $R \equiv \sigma(e^+e^- \to f\bar{f})/\sigma_{\mathrm{QED}}(e^+e^- \to \mu^+\mu^-)$ in the standard model with $x_W = 0.2$ and three generations of quarks and leptons: (a) l^+l^- final state, (b) $\nu\bar{\nu}$ final state, (c) $u\bar{u}$ (charge $2/3$ quarks) final state, (d) $d\bar{d}$ (charge $-1/3$ quarks) final state.

and

$$R_{\text{peak}}(e^+e^- \rightarrow \mu^+\mu^-) = (1 - 4x_W + 8x_W^2)R_{\text{peak}}(e^+e^- \rightarrow \nu\bar{\nu}), \quad (7.2.10)$$

$$R_{\text{peak}}(e^+e^- \rightarrow u\bar{u}) = 3\left(1 - \frac{8x_W}{3} + \frac{32x_W^2}{9}\right)R_{\text{peak}}(e^+e^- \rightarrow \nu\bar{\nu}), \quad (7.2.11)$$

$$R_{\text{peak}}(e^+e^- \rightarrow d\bar{d}) = 3\left(1 - \frac{4x_W}{3} + \frac{8x_W^2}{9}\right)R_{\text{peak}}(e^+e^- \rightarrow \nu\bar{\nu}). \quad (7.2.12)$$

The resulting peak cross sections and ratios are given in Table 7.1, for $x_W = 0.2$ and three generations of quarks and leptons. The visible cross section is quite enormous with respect to the QED reference cross section of about 10 picobarns:

$$R_{\text{peak}}(e^+e^- \rightarrow \text{charged leptons or hadrons}) \simeq 4270. \quad (7.2.13)$$

Although this will be reduced somewhat in practice by radiative corrections and uncertainty in beam energy, one may still anticipate a copious yield of intermediate bosons in the "Z^0 factories" now being planned. It is amusing to note that a careful comparison of the total width of the Z^0 as inferred from its line shape with the visible cross section at the peak may serve to determine the number of universally coupled neutrino species. The prospect of high yields encourages the hope that rare decays of the Z^0 may be amenable to experimental study. Of these, channels such as $Z \rightarrow H\mu^+\mu^-$ and $Z \rightarrow H\gamma$ may provide relatively clean tags for the production of Higgs bosons.[4]

An observable that may be accessible to present-day experimentation is the forward–backward asymmetry

$$A = \frac{\int_0^1 dz(d\sigma/dz) - \int_{-1}^0 dz(d\sigma/dz)}{\int_{-1}^1 dz(d\sigma/dz)} \quad (7.2.14)$$

TABLE 7.1: Cross Sections for Electron–Positron Annihilation at the Z^0 Peak[a]

Channel	σ_{peak} (nb)	R_{peak}
$\nu\bar{\nu}$	3.05	305
$\mu^+\mu^-$	1.59	159
$u\bar{u}$	5.57	557
$d\bar{d}$	7.04	704

[a] $x_W = 0.2$, three quark and lepton generations.

already discussed for the charged leptons in Section 6.4. In the low-energy limit, the asymmetry is approximately given by

$$A(q\bar{q}) = \frac{3G_F s}{16\pi\alpha Q_q\sqrt{2}}(R_e - L_e)(R_q - L_q). \qquad (7.2.15)$$

In the standard model, $(R_i - L_i) = \tau_3^{(i)}$, so that

$$R_e - L_e = R_d - L_d = -(R_u - L_u) = 1 \qquad (7.2.16)$$

and therefore, comparing with (6.4.31), we find

$$\left.\begin{array}{l} A(u\bar{u}) = \frac{3}{2}A(\mu^+\mu^-) \\ A(d\bar{d}) = 3A(\mu^+\mu^-) \end{array}\right\}, \qquad (7.2.17)$$

where u and d are to be understood as generic symbols for the quarks of charge $+2/3$ and $-1/3$, respectively. Thus, in experiments at $s \simeq 1400$ GeV^2 one may anticipate asymmetries of

$$\left.\begin{array}{l} A(u\bar{u}) \simeq -15\% \\ A(d\bar{d}) \simeq -30\% \end{array}\right\} \qquad (7.2.18)$$

in the standard model. Whether this can be observed indirectly as a hadronic charge asymmetry or by tagging the decays of short-lived particles bearing the quantum numbers charm and beauty is an experimental issue that remains for the moment to be understood. If the latter can be achieved, it will open the way to an independent determination of the neutral-current couplings of the heavy quarks. In model-independent notation (7.2.6), the charge asymmetry at low energies may be written as

$$A(q\bar{q}) = 3G_F s a_e a_q/4\pi\alpha Q_q\sqrt{2}. \qquad (7.2.19)$$

Although the low-energy charge asymmetry is (at least in principle) a sensitive probe of neutral-current couplings, it is not directly sensitive to the *position* of the Z^0-boson pole, as the absence of M_Z from equations (6.4.31,34) and (7.2.15,19) makes plain. The evolution of the asymmetry with increasing energy is, however, influenced by the Z^0 propagator, as the linear rise with s of the magnitude of the asymmetry is damped. The behavior anticipated in the standard model with $x_W = 0.2$ is illustrated in Fig. 7-3, where the Z^0 has again been assumed to decay into three families of quarks and leptons.

On the Z^0 peak, the asymmetry may be expressed as

$$\begin{aligned} A_{\text{peak}}(f\bar{f}) &= \frac{3(L_e^2 - R_e^2)(L_f^2 - R_f^2)}{4(L_e^2 + R_e^2)(L_f^2 + R_f^2)} \\ &= \frac{3a_e v_e a_f v_f}{(v_e^2 + a_e^2)(v_f^2 + a_f^2)}. \end{aligned} \qquad (7.2.20)$$

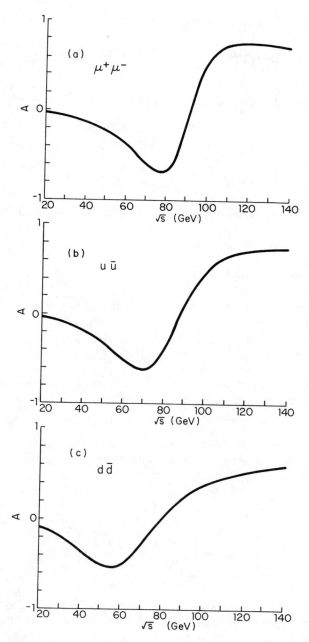

FIG. 7-3. Energy dependence of the forward–backward asymmetry in electron–positron annihilations into charged fermions, in the standard model with three generations of quarks and leptons and $x_W = 0.2$: (a) l^+l^-, (b) $u\bar{u}$, (c) $d\bar{d}$ final states.

In the standard model we therefore anticipate

$$A_{\text{peak}}(\mu^+\mu^-) = \frac{3}{4}\left(\frac{1-4x_{\text{W}}}{1-4x_{\text{W}}+8x_{\text{W}}^2}\right)^2, \tag{7.2.21a}$$

$$A_{\text{peak}}(u\bar{u}) = \frac{3}{4}\left(\frac{1-4x_{\text{W}}}{1-4x_{\text{W}}+8x_{\text{W}}^2}\right)\left(\frac{1-8x_{\text{W}}/3}{1-8x_{\text{W}}/3+32x_{\text{W}}^2/9}\right), \tag{7.2.21b}$$

$$A_{\text{peak}}(d\bar{d}) = \frac{3}{4}\left(\frac{1-4x_{\text{W}}}{1-4x_{\text{W}}+8x_{\text{W}}^2}\right)\left(\frac{1-4x_{\text{W}}/3}{1-4x_{\text{W}}/3+8x_{\text{W}}^2/9}\right). \tag{7.2.21c}$$

If the weak mixing parameter x_{W} is indeed close to 0.2, the peak asymmetries will be quite small (0.11, 0.22, 0.28, respectively), because the vector coupling of the electron vanishes at $x_{\text{W}} = 0.25$.

Parity violation in the neutral current interaction is manifested as a net polarization

$$P = (\sigma_{\text{R}} - \sigma_{\text{L}})/\sigma \tag{7.2.22}$$

of the produced fermions. In the standard model with $x_{\text{W}} = 0.2$, the effect is small for the cross section integrated over all angles, except in the neighborhood of the Z^0. This is shown in Fig. 7-4. On the Z^0 peak, the net polarization is given by

$$\begin{aligned}P_{\text{peak}}(f\bar{f}) &= (R_f^2 - L_f^2)/(R_f^2 + L_f^2) \\ &= -2a_f v_f/(v_f^2 + a_f^2).\end{aligned} \tag{7.2.23}$$

This means that the electron parameters are in principle accessible in the ratio

$$A_{\text{peak}}(f\bar{f})/P_{\text{peak}}(f\bar{f}) = -3a_e v_e/2(a_e^2 + v_e^2) \tag{7.2.24}$$

without the intervention of universality assumptions. The peak polarizations predicted by the standard model are appreciable for $x_{\text{W}} = 0.2$,

$$P_{\text{peak}}(\mu^+\mu^-) = \frac{4x_{\text{W}}-1}{1-4x_{\text{W}}+8x_{\text{W}}^2} \to -0.385, \tag{7.2.25a}$$

$$P_{\text{peak}}(u\bar{u}) = \frac{8x_{\text{W}}/3-1}{1-8x_{\text{W}}/3+32x_{\text{W}}^2/9} \to -0.766, \tag{7.2.25b}$$

$$P_{\text{peak}}(d\bar{d}) = \frac{4x_{\text{W}}/3-1}{1-4x_{\text{W}}/3+8x_{\text{W}}^2/9} \to -0.954, \tag{7.2.25c}$$

but are likely to be difficult to measure in any case. The availability of longitudinally polarized electron or positron beams would greatly expand the range of possibilities, as one may learn from the basic formula (7.2.3).

Because the manifestations of the weak interactions become significant only at high energies, electron–positron collisions have not yet contributed

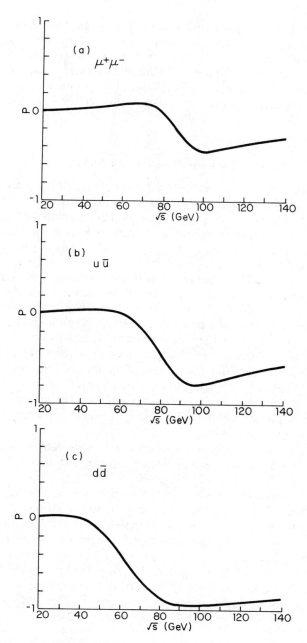

FIG. 7-4. Energy dependence of the net polarization of charged fermions produced in electron–positron annihilations, in the standard model with $x_W = 0.2$ and three generations of quarks and leptons: (a) l^+l^-, (b) $u\bar{u}$, (c) $d\bar{d}$ final states.

decisively to our knowledge of the structure of the weak neutral current interaction. (The importance of studies of the charm-changing charged current, of not finding the Z^0 or other large effects at low energies, etc., is not to be minimized, of course.) As energies nearer to the expected mass of the Z^0 become accessible to experiment, the situation may be expected to change dramatically, as the examples of this section have shown.

7.3 Deeply Inelastic Lepton–Hadron Scattering

When hadron structure is examined on a very short time or distance scale, as is possible in large-momentum-transfer collisions, hadrons are found to behave as if they were composed of essentially noninteracting pointlike structures. The pointlike constituents of hadrons have been named partons by Feynman. The basic idea of the parton model is analogous to the well-traveled notion that for many purposes a nucleus may be regarded as a collection of structureless and noninteracting nucleons, but with a critical difference. Nucleons are rather easily liberated from nuclei, but the separation of a hadron into its constituent partons has never been observed. The parton model has thus had a paradoxical aspect, which may be summarized in the following question: How could partons be quasifree within hadrons if they interact so strongly that they cannot be separated? The question becomes more urgent with each success of the model. Only with the development of the theory of strong interactions known as quantum chromodynamics has emerged the outline of a resolution of this paradox, as we shall see in Chapter 8. The general problem of hadron structure still appears far from a complete solution.

Our principal concern in this section is not with hadron structure, but with the interactions of weak and electromagnetic currents with the fundamental constituents. Thus, we shall develop and apply the parton model in an intuitive fashion that is close to the original spirit in which the model was formulated. Loftier formulations may be found among the suggested readings at the end of the chapter. The heuristic approach has the advantage of leading immediately to a transparent physical picture of violent collisions, which is not misleading. The development of this section will have three general goals: to formulate in a general language the kinematics and observables of inclusive lepton–hadron scattering, to specialize to the parton model and verify its approximate validity, and to investigate what can be learned about the electroweak interactions in deeply inelastic lepton–hadron scattering.

We wish to consider inclusive reactions of the general form

$$l + N \rightarrow l' + \text{anything}, \tag{7.3.1}$$

for which the kinematic notation is indicated in Fig. 7-5. From the four-momenta designated there we may form the useful invariants

$$s = (l + P)^2 \tag{7.3.2}$$

$$Q^2 \equiv -q^2 = -(l - l')^2, \tag{7.3.3}$$

$$v = q \cdot P/M, \tag{7.3.4}$$

where M is the target mass and

$$W^2 = 2Mv + M^2 - Q^2 \tag{7.3.5}$$

is the square of the invariant mass of the produced hadronic system "anything." In the laboratory frame, in which (neglecting the lepton mass)

$$\left.\begin{aligned} l^\mu &= (E; 0, 0, E)\\ l'^\mu &= (E'; E' \sin\theta, 0, E' \cos\theta)\\ P^\mu &= (M; 0, 0, 0) \end{aligned}\right\} \tag{7.3.6}$$

we may write

$$\begin{aligned} Q^2 &= 2EE'(1 - \cos\theta)\\ &= 4EE' \sin^2(\theta/2), \end{aligned} \tag{7.3.7}$$

and recognize as an energy-loss variable

$$v = E - E'. \tag{7.3.8}$$

From the connection (7.3.5) it is seen that the line

$$Q^2 = 2Mv \tag{7.3.9}$$

in the Q^2–$2Mv$ plane corresponds to elastic scattering, for which

$$W^2 = M^2. \tag{7.3.10}$$

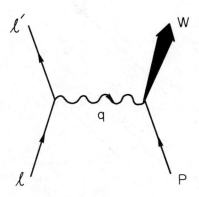

FIG. 7-5. Kinematics of deeply inelastic scattering.

Similarly, the excitation of a system with mass M^*, such as a discrete resonance, occurs along a line

$$Q^2 = 2M\nu + M^2 - M^{*2}. \qquad (7.3.11)$$

These kinematic regimes are indicated in Fig. 7-6.

As a first specific example, consider the case of inelastic electron–proton scattering with an unpolarized beam and target. The square of the invariant amplitude, averaged over initial spins and summed over final spins, may be written as

$$\overline{|\mathcal{M}|^2} = \frac{e^4}{(q^2)^2} L^{\mu\nu} W_{\mu\nu}, \qquad (7.3.12)$$

where the anticipated factors for the spin average have been absorbed into the definitions of the tensors $L^{\mu\nu}$ and $W_{\mu\nu}$, which describe the structures of the leptonic and hadronic vertices, respectively. The lepton tensor is prescribed by QED as

$$
\begin{aligned}
L^{\mu\nu} &= \tfrac{1}{2}\operatorname{tr}[\bar{u}(l')\gamma^\mu u(l)\bar{u}(l)\gamma^\nu u(l')] \\
&= \tfrac{1}{2}\operatorname{tr}[\gamma^\mu(m + \not{l})\gamma^\nu(m + \not{l}')] \\
&= 2[l^\mu l'^\nu + l'^\nu l'^\mu - g^{\mu\nu}(l \cdot l' - m^2)].
\end{aligned} \qquad (7.3.13)
$$

In the absence of any information about the hadronic vertex, we may write the general form

$$
\begin{aligned}
W_{\mu\nu} = &\; V_1 g_{\mu\nu} + V_2 P_\mu P_\nu + V_3 (P_\mu q_\nu + P_\nu q_\mu) \\
&+ V_4 (P_\mu q_\nu - P_\nu q_\mu) + V_5 q_\mu q_\nu + V_6 \varepsilon_{\mu\nu\alpha\beta} P^\alpha q^\beta.
\end{aligned} \qquad (7.3.14)
$$

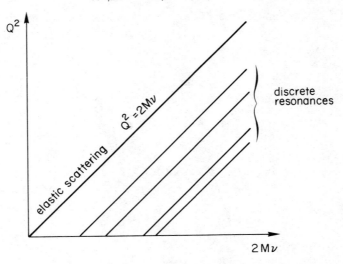

FIG. 7-6. The Q^2–$2M\nu$ plane in deeply inelastic scattering.

Because $L^{\mu\nu}$ is symmetric under interchange of the indices μ and ν, the antisymmetric terms proportional to V_4 and V_6 cannot contribute to the cross section. Current conservation requires that

$$q^\mu W_{\mu\nu} = 0 = q^\nu W_{\mu\nu}. \qquad (7.3.15)$$

This implies that

$$V_1 q_\nu + V_2(q \cdot P)P_\nu + V_3(q \cdot P q_\nu + q^2 P_\nu) + V_5 q^2 q_\nu = 0, \qquad (7.3.16)$$

from which the coefficients of P_ν and q_ν must vanish separately, so that

$$\left.\begin{array}{r} V_1 + (q \cdot P)V_3 + q^2 V_5 = 0 \\ (q \cdot P)V_2 + q^2 V_3 = 0 \end{array}\right\}. \qquad (7.3.17)$$

Eliminating

$$V_3 = -(q \cdot P)V_2/q^2 \qquad (7.3.18)$$

and

$$V_5 = \frac{(q \cdot p)^2 V_2}{q^4} - \frac{V_1}{q^2} \qquad (7.3.19)$$

we find that

$$W_{\mu\nu} = V_1\left(g_{\mu\nu} - \frac{q_\mu q_\nu}{q^2}\right) + V_2\left[P_\mu - \frac{(q \cdot P)q_\mu}{q^2}\right]\left[P_\nu - \frac{(q \cdot P)q_\nu}{q^2}\right], \qquad (7.3.20)$$

which is conventionally written as

$$W_{\mu\nu} = -W_1\left(g_{\mu\nu} - \frac{q_\mu q_\nu}{q^2}\right) + \frac{W_2}{M^2}\left[P_\mu - \frac{(q \cdot P)q_\mu}{q^2}\right]\left[P_\nu - \frac{(q \cdot P)q_\nu}{q^2}\right]. \qquad (7.3.21)$$

The objects $W_{1,2}$ are known as the structure functions for (unpolarized) inelastic electron–proton scattering. All that may be learned about hadron structure from such collisions is contained in these two structure functions, which depend, in principle, upon P and q or, in Lorentz-invariant language, upon the invariants Q^2 and ν.

To make contact with observables, we neglect lepton masses, so that

$$l \cdot l' = q \cdot l' = -q \cdot l = Q^2/2 \qquad (7.3.22)$$

and form

$$L^{\mu\nu}W_{\mu\nu} = 2W_1(Q^2, \nu)Q^2 + W_2(Q^2, \nu)\left[\frac{4(l \cdot P)(l' \cdot P)}{M^2} - Q^2\right]. \qquad (7.3.23)$$

In the laboratory frame, where

$$\begin{array}{l} l \cdot P = ME, \\ l' \cdot P = ME', \end{array} \qquad (7.3.24)$$

we therefore obtain

$$L^{\mu\nu}W_{\mu\nu} = 4EE'[2W_1(Q^2, \nu)\sin^2(\theta/2) + W_2(Q^2, \nu)\cos^2(\theta/2)]. \qquad (7.3.25)$$

The differential cross section in the laboratory frame is given by

$$\frac{d^2\sigma}{dE'\,d\Omega'} = \frac{1}{16\pi^2}\frac{E'}{E}|\mathcal{M}|^2 = \frac{(4\pi\alpha)^2}{16\pi^2 Q^4}\frac{E'}{E}L^{\mu\nu}W_{\mu\nu}$$

$$= \frac{4\alpha^2 E'^2}{Q^4}\left[2W_1(Q^2,v)\sin^2\left(\frac{\theta}{2}\right) + W_2(Q^2,v)\cos^2\left(\frac{\theta}{2}\right)\right]. \quad (7.3.26)$$

It is often more convenient to express the differential cross section with respect to the invariants v and Q^2 as

$$\frac{d^2\sigma}{dQ^2\,dv} = \frac{\pi}{EE'}\frac{d^2\sigma}{dE'\,d\Omega'} = \frac{4\pi\alpha^2}{Q^4}\frac{E'}{E}\left[2W_1(Q^2,v)\sin^2\left(\frac{\theta}{2}\right) + W_2(Q^2,v)\cos^2\left(\frac{\theta}{2}\right)\right].$$

$$(7.3.27)$$

In experimental settings one may maintain the same values of Q^2 and v upon changing E, E', and θ, and may thus in principle separate the two structure functions W_1 and W_2.

It is interesting to compare the cross section (7.3.26) with more familiar results for elastic scattering. The Rosenbluth formula for elastic electron-proton scattering, which was derived in Problem 3-3, may be written in the form

$$\frac{d^2\sigma}{d\Omega'\,dE'} = \frac{4\alpha^2 E'^2}{Q^4}\left\{2Q^2\left[\frac{\Gamma_1(Q^2)}{2M} + \Gamma_2(Q^2)\right]^2\sin^2\left(\frac{\theta}{2}\right)\right.$$

$$\left. + \left[\Gamma_1^2(Q^2) + Q^2\Gamma_2^2(Q^2)\right]\cos^2\left(\frac{\theta}{2}\right)\right\}$$

$$\times \delta\left(v - \frac{Q^2}{2M}\right), \quad (7.3.28)$$

which has precisely the same structure. For a structureless "Dirac proton," also considered in Problem 3-3, the form factors are simply

$$\left.\begin{array}{l}\Gamma_1(Q^2) = 1\\ \Gamma_2(Q^2) = 0\end{array}\right\}, \quad (7.3.29)$$

so we may identify the structure functions for scattering from a pointlike spin-1/2 particle as

$$\left.\begin{array}{l}W_1^{pt}(Q^2,v) = (Q^2/4M^2)\delta(v - Q^2/2M)\\ W_2^{pt}(Q^2,v) = \delta(v - Q^2/2M)\end{array}\right\}. \quad (7.3.30)$$

The dimensionless combinations

$$\left.\begin{array}{l}2MW_1^{pt}(Q^2,v) = (Q^2/2Mv)\delta(Q^2/2Mv - 1)\\ vW_2^{pt}(Q^2,v) = \delta(Q^2/2Mv - 1)\end{array}\right\} \quad (7.3.31)$$

are seen to depend upon the kinematic invariants only through the dimensionless ratio

$$x = Q^2/2Mv. \tag{7.3.32}$$

Similarly, by examining the form of the differential cross section for electron–elementary scalar scattering derived in Problem 3-1, or by considering directly the structure of the photon–scalar vertex, one sees at once that

$$\left. \begin{array}{l} W_2^{(0)}(Q^2, v) = \delta(v - Q^2/2M) \\ W_1^{(0)}(Q^2, v) = 0 \end{array} \right\}. \tag{7.3.33}$$

To conclude these essentially kinematic developments, let us relate the structure functions to the cross sections for the absorption of virtual photons. Writing the flux of virtual photons as Φ, we may define the cross section for absorption of photons of helicity λ as

$$\sigma_\lambda = \frac{4\pi^2\alpha}{\Phi} \varepsilon_\lambda^{\mu*} W_{\mu\nu} \varepsilon_\lambda^\nu, \tag{7.3.34}$$

following the conventions appropriate to real photons. For photons of helicity ± 1, for which

$$\varepsilon_{\pm 1}^\mu = (0; -1, \mp i, 0)/\sqrt{2}, \tag{7.3.35}$$

we compute

$$\sigma_T \equiv \frac{1}{2}(\sigma_{+1} + \sigma_{-1}) = \frac{4\pi^2\alpha}{\Phi} W_1, \tag{7.3.36}$$

whereas for scalar photons, with

$$\varepsilon_S^\mu = (\sqrt{Q^2 + v^2}; 0, 0, v)/\sqrt{Q^2}, \tag{7.3.37}$$

$$\sigma_S = \frac{4\pi^2\alpha}{\Phi} \left[-W_1 + \left(\frac{1 + v^2}{Q^2}\right) W_2 \right]. \tag{7.3.38}$$

The requirement that these cross sections be nonnegative leads to the restrictions

$$\left. \begin{array}{l} W_1 \geq 0 \\ W_2(1 + v^2/Q^2) \geq W_1 \end{array} \right\}. \tag{7.3.39}$$

Finally we note that the ratio

$$\frac{\sigma_S}{\sigma_T} = \frac{W_2}{W_1}\left(1 + \frac{v^2}{Q^2}\right) - 1 \tag{7.3.40}$$

is sensitive to the spin of the target. For a structureless spin-1/2 target, with structure functions given by (7.3.30), the ratio vanishes as $v \to \infty$

$$\sigma_S/\sigma_T = 2M/v \xrightarrow[v \to \infty]{} 0, \qquad \text{spin-}\tfrac{1}{2}. \tag{7.3.41}$$

In contrast, for a spinless target with structure functions given by (7.3.33), the transverse cross section vanishes so that

$$\sigma_S/\sigma_T \to \infty, \qquad \text{spin zero.} \qquad (7.3.42)$$

This completes our discussion of the kinematics of electron–proton scattering. We now turn to the parton model itself.

The transition to the parton model is most easily made in the "infinite momentum frame," in which the longitudinal momentum of the target (proton, for example) is extremely large. The proton is regarded as a collection of N free partons, each carrying a fraction x_i ($i = 1, 2, \ldots, N$) of the longitudinal momentum of the proton, as shown in Fig. 7-7. Assuming the mass of a parton to be insignificant both before and after a collision, and that the transverse momentum of an incident parton is negligible, we may write the four-momentum of an individual parton as

$$p_i^\mu = x_i P^\mu. \qquad (7.3.43)$$

Then if interactions between the partons can be neglected, so that the individual current–parton interactions may be treated incoherently, we may write the contribution to W_2 due to scattering from a single parton of charge e_i as[5]

$$W_2^{(i)}(Q^2, \nu; x_i) = x_i e_i^2 \delta\left(\frac{q \cdot p_i}{M} - \frac{Q^2}{2M}\right)$$

$$= x_i e_i^2 \delta\left(\frac{x_i q \cdot P}{M} - \frac{Q^2}{2M}\right)$$

$$= e_i^2 \delta\left(\nu - \frac{Q^2}{2M x_i}\right), \qquad (7.3.44)$$

which reproduces, as it must, the Rutherford cross section

$$d\sigma/dQ^2 = 4\pi\alpha^2 e_i^2/Q^4 \qquad (7.3.45)$$

at high energies.

The incoherence assumption, or impulse approximation, means that the structure function for electron–proton scattering is simply the sum over

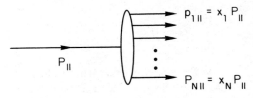

FIG. 7-7. Parton-model depiction of the proton in the infinite momentum frame.

the contributions of individual partons:

$$W_2(Q^2, v) = \sum_i \int_0^1 dx_i f_i(x_i) W_2^{(i)}(Q^2, v; x_i), \tag{7.3.46}$$

where $f_i(x_i)$ gives the probability of finding the ith parton with momentum fraction x_i. The integration over dx_i is readily carried out using the rule

$$\int dx\, \delta(f(x)) = 1/(\partial f/\partial x)\Big|_{f(x)=0}. \tag{7.3.47}$$

We find at once

$$W_2(Q^2, v) = \sum_i e_i^2 f_i(x) x/v, \tag{7.3.48}$$

where the scaling variable $x = Q^2/2Mv$ has already been defined in (7.3.32). Consequently the dimensionless quantity

$$v W_2(Q^2, v) = \sum_i e_i^2 f_i(x) x, \tag{7.3.49}$$

obtained as an incoherent sum over the contributions of individual partons, is identified as a function of a single dimensionless variable x. It is convenient to define the scaling form as the combination

$$F_2(x) \equiv v W_2(Q^2, v). \tag{7.3.50}$$

Knowing the relation between W_1 and W_2 for structureless particles of spin zero or one-half, we may construct from this result the general form of the cross section. It is again convenient to define a dimensionless form based on W_1, which will evidently depend only upon the scaling variable x. It is conventionally written as

$$F_1(x) \equiv M W_1(Q^2, v), \tag{7.3.51}$$

which is given by

$$F_1(x) = \begin{cases} 0, & \text{spin-0 partons} \\ (1/2x)F_2(x), & \text{spin-}\tfrac{1}{2}\text{ partons.} \end{cases} \tag{7.3.52}$$

That $v W_2(Q^2, v)$ should become independent of Q^2 for fixed values of x as $v, Q^2 \to \infty$ was anticipated by Bjorken[6] and demonstrated in classic experiments by the SLAC–MIT Collaboration.[7] This behavior, which we have seen to be quite natural within the parton model, is in sharp contrast to the comportment of elastic scattering or resonance excitation, which are characterized by rapidly falling form factors. Experiments have also shown the ratio σ_S/σ_T to be small,[8] suggesting that the charged partons carry spin-1/2. The fact that deeply inelastic electron scattering is readily interpreted as the scattering from structureless, spin-1/2 constituents suggests that the charged partons be identified as quarks. What are the implications of this identification?

To fully exploit the scaling behavior of the cross section, it is convenient to introduce the inelasticity parameter

$$y \equiv v/E, \tag{7.3.53}$$

which evidently satisfies

$$0 \le y \le 1. \tag{7.3.54}$$

The cross section may then be expressed as

$$\frac{d^2\sigma}{dx\,dy} = \frac{2Mv^2}{y}\frac{d^2\sigma}{dQ^2\,dv} = \frac{4\pi\alpha^2 s}{Q^4}[F_2(x)(1-y) + F_1(x)xy^2], \tag{7.3.55}$$

plus a term of order M/E, which may be safely neglected at high energies. According to (7.3.49), the structure functions are directly related to the quark–parton distribution functions. For ep scattering, we have that

$$F_2^{ep}(x)/x = \tfrac{4}{9}(u(x) + \bar{u}(x)) + \tfrac{1}{9}(d(x) + \bar{d}(x))$$
$$+ \tfrac{1}{9}(s(x) + \bar{s}(x)) + \cdots, \tag{7.3.56}$$

with $u(x) = f_u(x)$, etc. The neutron structure functions may be obtained by applying an isospin rotation, which amounts to $u \to d$, whereupon

$$F_2^{en}(x)/x = \tfrac{4}{9}(d(x) + \bar{d}(x)) + \tfrac{1}{9}(u(x) + \bar{u}(x)) + \cdots, \tag{7.3.57}$$

where $u(x), d(x), \ldots$, refer always to the quark content of the proton. The expressions (7.3.56, 57) lead at once to the bounds.

$$\tfrac{1}{4} \le F_2^{en}(x)/F_2^{ep}(x) \le 4. \tag{7.3.58}$$

Recent data on this ratio are shown in Fig. 7-8. The parton-model bounds are respected, and the data have the following simple interpretation. For small values of x, where the sea of quark–antiquark pairs may be expected to dominate, so that

$$u(x) \simeq \bar{u}(x) \simeq d(x) \simeq \bar{d}(x), \qquad x \simeq 0, \tag{7.3.59}$$

then one anticipates a ratio close to unity, as is observed. At large x, it is the so-called valence quarks that must dominate, which is to say that

$$u(x), d(x) \gg \bar{u}(x), \bar{d}(x), \qquad x \to 1. \tag{7.3.60}$$

If isospin symmetry were exact for the valence quark distributions as $x \to 1$ (a simple but not compelling hypothesis), which would imply

$$u(x) = 2d(x), \qquad x \to 1, \tag{7.3.61}$$

then the ratio $F_2^{en}(x)/F_2^{ep}(x)$ would be expected to approach $2/3$. The fact that the data appear to approach the parton-model lower bound of $1/4$ may be taken as a suggestion that

$$d(x)/u(x) \to 0, \qquad x \to 1. \tag{7.3.62}$$

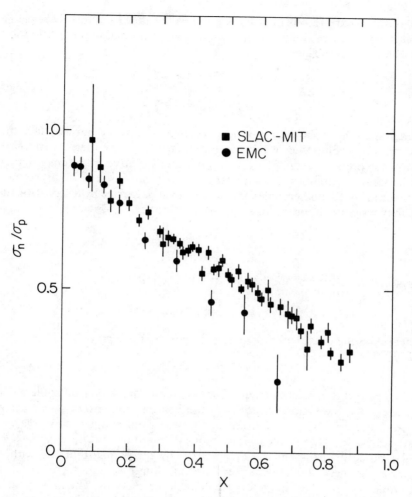

FIG. 7-8. x-dependence of the ratio $\sigma(en)/\sigma(ep)$ of cross sections for deeply inelastic scattering on nucleons. [From A. Bodek *et al.*, SLAC-MIT Collaboration, *Phys. Rev.* **D20**, 1471 (1979), and J. J. Aubert, *et al.*, European Muon Collaboration, *Phys. Lett.* **123B**, 123 (1983).]

The parton-model expressions imply a number of useful sum rules, which must be satisfied by the quark distributions. The electric charge sum rule for the proton reads

$$\int_0^1 dx\left[\tfrac{2}{3}(u(x) - \overline{u}(x)) - \tfrac{1}{3}(d(x) - \overline{d}(x))\right] = 1, \qquad (7.3.63)$$

and that for the neutron is

$$\int_0^1 dx\left[\tfrac{2}{3}(d(x) - \overline{d}(x)) - \tfrac{1}{3}(u(x) - \overline{u}(x))\right] = 0, \qquad (7.3.64)$$

where the contributions of strange and heavier quarks vanish because the nucleon carries no net strangeness, charm, etc. Combining these definitions we obtain the sum rules

$$\left.\begin{array}{c} \int_0^1 dx(u(x) - \bar{u}(x)) = 2 \\ \int_0^1 dx(d(x) - \bar{d}(x)) = 1 \end{array}\right\},\tag{7.3.65}$$

which are simply the parton-model restatements of the familiar fact that the proton is composed of two (net) up quarks and one (net) down quark. Similar expressions can be derived for other additive quantum numbers such as baryon number and strangeness.

Since the total momentum of the proton must be carried by its constituents, we may write the momentum sum rule

$$\sum_{\substack{i = \text{parton} \\ \text{species}}} \int_0^1 dx \, x f_i(x) = 1.\tag{7.3.66}$$

Neglecting the strange and heavier quarks, we may write

$$F_2^{ep}(x) + F_2^{en}(x) = \frac{5x}{9}(u(x) + \bar{u}(x) + d(x) + \bar{d}(x)),\tag{7.3.67}$$

whereupon the fractional momentum carried by the quarks is[9]

$$\tfrac{9}{5} \int_0^1 dx(F_2^{ep}(x) + F_2^{en}(x)) \simeq 0.45.\tag{7.3.68}$$

Unless our neglect of strange quarks was grossly in error (see Problem 7-7), we are led to conclude that 55% of the proton momentum is carried by neutral partons. As we shall see in the following chapter, this role falls naturally to the gluons, the gauge bosons of quantum chromodynamics.

To proceed further in the analysis of the quark distributions without making strongly model-dependent assumptions, we must make use of information from the charged-current weak interactions. The most general form of the cross section for the inclusive reaction

$$v + N \to \mu + \text{anything}\tag{7.3.69}$$

may be derived (Problem 7-8) by the same methods used to derive the form (7.3.27) for deeply inelastic electron scattering. There is the important difference that the lepton tensor in this case has a V − A structure, and the cross section is slightly complicated by the violation of parity. The general result is

$$\frac{d^2\sigma^v}{dQ^2 \, dv} = \frac{G_F^2}{2\pi} \frac{E'}{E} \left[2W_1^v \sin^2\left(\frac{\theta}{2}\right) + W_2^v \cos^2\left(\frac{\theta}{2}\right) + W_3^v \frac{(E + E')}{M} \sin^2\left(\frac{\theta}{2}\right) \right],\tag{7.3.70}$$

where the final term arises from the parity-violating $\varepsilon_{\mu\nu\alpha\beta}P^\alpha q^\beta$ term in the general expansion (7.3.14) of the hadronic vertex. The cross section is conveniently recast in terms of the scaling variables x and y as

$$\frac{d^2\sigma^{\nu,\bar\nu}}{dx\,dy} = \frac{G_F^2 ME}{\pi}[\mathscr{F}_1(x)xy^2 + \mathscr{F}_2(x)(1-y) \pm \mathscr{F}_3(x)xy(1-y/2)], \quad (7.3.71)$$

where the dimensionless structure functions $\mathscr{F}_{1,2}(x)$ are defined in analogy with $F_{1,2}(x)$ in the electromagnetic case and

$$\mathscr{F}_3(x) \equiv \nu W_3^\nu(x). \tag{7.3.72}$$

Without further analysis we obtain the parton-model prediction that

$$\sigma_{\text{total}}(\nu N \to l + \text{anything}) \propto E. \tag{7.3.73}$$

The experimental results compiled in Fig. 7-9 show this to be the case to excellent approximation for neutrino and antineutrino scattering at energies from 10 GeV to more than 200 GeV, with

$$\left.\begin{array}{l}\sigma_{\text{total}}(\nu N)/E \simeq 6 \times 10^{-39} \text{ cm}^2/\text{GeV}\\ \sigma_{\text{total}}(\bar\nu N)/E \simeq 3 \times 10^{-39} \text{ cm}^2/\text{GeV}\end{array}\right\}. \tag{7.3.74}$$

A direct calculation (also part of Problem 7-8) shows that

$$x\mathscr{F}_3(x)/\mathscr{F}_2(x) = \begin{cases}+1, & \text{fermion target}\\ -1, & \text{antifermion target.}\end{cases} \tag{7.3.75}$$

When combined with the earlier finding [compare (7.3.52)] that

$$2x\mathscr{F}_1(x) = \mathscr{F}_2(x), \tag{7.3.76}$$

this leads to the conclusion that

$$\frac{d^2\sigma}{dx\,dy}(\nu q) = \frac{G_F^2 ME}{\pi}\mathscr{F}_2(x)\{1\}, \tag{7.3.77}$$

while

$$\frac{d^2\sigma}{dx\,dy}(\bar\nu q) = \frac{G_F^2 ME}{\pi}\mathscr{F}_2(x)\{(1-y)^2\}, \tag{7.3.78}$$

which reproduces the angular distributions computed in Section 6.1.

As we did for the electromagnetic case, we may compute the contributions of various elementary processes to the structure functions. One finds directly that, in the approximation in which $\cos\theta_C = 1$,

$$\mathscr{F}_2^\nu(x) = 2x(d(x) + \bar u(x))$$
$$\mathscr{F}_3^\nu(x) = 2(d(x) - \bar u(x)) \tag{7.3.79}$$

and

$$\mathscr{F}_2^{\bar\nu}(x) = 2x(u(x) + \bar d(x))$$
$$\mathscr{F}_3^{\bar\nu}(x) = 2(u(x) - \bar d(x)). \tag{7.3.80}$$

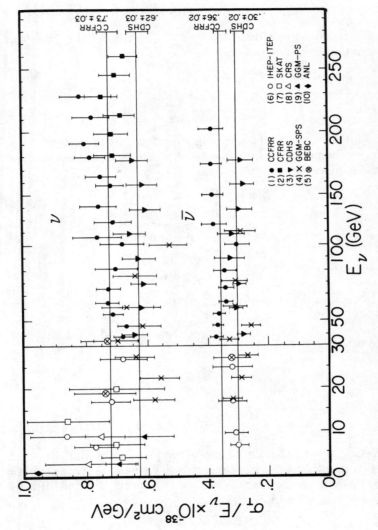

FIG. 7-9. Slopes σ_{total}/E_ν for the muon neutrino and antineutrino charged-current total cross section as a function of incident neutrino energy. The error bars include both statistical and systematic errors. The straight lines are averages for the Caltech–Columbia–Fermilab–Rochester–Rockefeller and CERN–Dortmund–Heidelberg–Saclay measurements. (Compiled by M. Shaevitz.)

In lieu of an extended analysis, we may content ourselves with a few elementary remarks. The combination

$$\mathscr{F}_2^{\nu p} + \mathscr{F}_2^{\nu n} = 2x(u(x) + \bar{u}(x) + d(x) + \bar{d}(x)) \tag{7.3.81}$$

is simply proportional to the analogous quantity for electromagnetic scattering:

$$\frac{F_2^{ep} + F_2^{en}}{\mathscr{F}_2^{\nu p} + \mathscr{F}_2^{\nu n}} = \frac{5}{18}, \tag{7.3.82}$$

up to neglect of strange quarks. This is nicely supported by the data,[10] and reinforces the conclusion that approximately half the momentum of the proton is carried by gluons that are inert with respect to the weak and electromagnetic interactions.

Let us notice finally that the structure function $\mathscr{F}_3$ measures the difference between quark and antiquark contributions, whereas $\mathscr{F}_2$ measures the sum. Thus, a comparison of these two structure functions leads to an assessment of the importance of the quark–antiquark sea with respect to the valence component. More specifically, we have, for example, that

$$\mathscr{F}_3^{\nu p} + \mathscr{F}_3^{\nu n} = 2(u - \bar{u} + d - \bar{d}), \tag{7.3.83}$$

whence the baryon number sum rule

$$\int_0^1 dx(\mathscr{F}_3^{\nu p}(x) + \mathscr{F}_3^{\nu n}(x)) = 6. \tag{7.3.84}$$

Typical parton distributions extracted from the data on deeply inelastic lepton–hadron scattering are shown in Fig. 7-10.

All this discussion has been prologue to our central interest in this section, which is a consideration of the neutral current interactions of hadrons. For

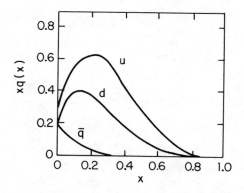

FIG. 7-10. Representative momentum distributions of partons within the proton [after R. D. Field and R. P. Feynman, *Phys. Rev.* **D15**, 2590 (1977)].

an isoscalar nucleon,

$$N \equiv \tfrac{1}{2}(p + n), \tag{7.3.85}$$

we may write the charged-current cross sections as

$$\frac{d^2\sigma}{dx\,dy}(vN \to \mu^- X) = \frac{G_F^2 ME}{\pi} x[(u(x) + d(x)) + (\overline{u}(x) + \overline{d}(x))(1 - y)^2], \tag{7.3.86}$$

and

$$\frac{d^2\sigma}{dx\,dy}(\overline{v}N \to \mu^+ X) = \frac{G_F^2 ME}{\pi} x[(u(x) + d(x))(1 - y)^2 + (\overline{u}(x) + \overline{d}(x))]. \tag{7.3.87}$$

On comparing the Feynman rules for charged-current and neutral-current interactions with quarks, we may discern at once that

$$\frac{d^2\sigma}{dx\,dy}(vN \to vX) = \frac{G_F^2 ME}{4\pi} x\{(L_u^2 + L_d^2)[(u(x) + d(x)) + (\overline{u}(x) + \overline{d}(x))(1 - y)^2]$$
$$+ (R_u^2 + R_d^2)[(u(x) + d(x))(1 - y)^2 + (\overline{u}(x) + \overline{d}(x))]\}. \tag{7.3.88}$$

and

$$\frac{d^2\sigma}{dx\,dy}(\overline{v}N \to \overline{v}X) = \frac{G_F^2 ME}{4\pi} x\{(L_u^2 + L_d^2)[(u(x) + d(x))(1 - y)^2 + (\overline{u}(x) + \overline{d}(x))]$$
$$+ (R_u^2 + R_d^2)[(u(x) + d(x)) + (\overline{u}(x) + \overline{d}(x))(1 - y)^2]\}. \tag{7.3.89}$$

It is useful for orientation to make the idealization that the antiquark distributions may be neglected, whereupon compact expressions follow for the neutral-current to charged-current ratios

$$R_v \equiv \frac{\sigma(vN \to vX)}{\sigma(vN \to \mu^- X)} = \frac{1}{2} - x_W + \frac{20x_W^2}{27}, \tag{7.3.90}$$

$$R_{\overline{v}} \equiv \frac{\sigma(\overline{v}N \to \overline{v}X)}{\sigma(\overline{v}N \to \mu^+ X)} = \frac{1}{2} - x_W + \frac{20x_W^2}{9}. \tag{7.3.91}$$

Thus, for $x_W = 0.2$, the ratios are

$$R_v = 0.33, \tag{7.3.92}$$

$$R_{\overline{v}} = 0.39. \tag{7.3.93}$$

These simplified predictions are shown together with a compilation of data in Fig. 7-11.

The detailed comparison of the model with experiment requires attention to many subtleties: the isotopic composition of heavy targets, the con-

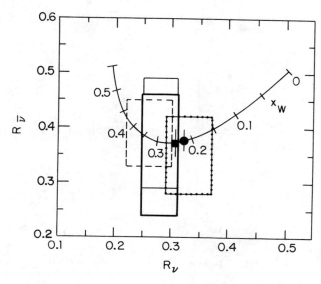

FIG. 7-11. Comparison of the ratios R_v and $R_{\bar{v}}$ of neutral current to charged-current cross sections in the standard model (after F. Büsser, *Neutrino 81*, Proceedings of the 1981 International Conference on Neutrino Physics and Astrophysics, Maui, Hawaii, edited by R. J. Cence, E. Ma, and A. Roberts, High Energy Physics Group, University of Hawaii, Honolulu, 1982, Vol. II, p. 351). Data are from the CITF (thin line), BEBC (thick line), GGM (broken line), HPW (dotted line), CHARM (circle), and CDHS (square) experiments.

tributions of the quark–antiquark sea including strange quarks, threshold effects associated with the onset of charm production via the elementary reactions

$$\left.\begin{array}{l} v + d_\theta \rightarrow \mu^- + c \\ \bar{v} + \bar{d}_\theta \rightarrow \mu^+ + \bar{c} \end{array}\right\}, \qquad (7.3.94)$$

and so on. Because of the extensive body of data that has been accumulated in deeply inelastic scattering, these complications can now be rather well controlled. One has as well the opportunity to include measurements made in hydrogen and deuterium, which separate the neutral-current cross sections on protons and neutrons, as well as data on exclusive final states. All results in neutrino scattering are compatible with the standard model, with a weak mixing parameter x_W close to 0.2.

Until now, we have discussed separately reactions mediated by γ, $W^\pm$, or Z^0 exchange. The deeply inelastic scattering of charged leptons provides another opportunity (as in $e^+ e^- \rightarrow f\bar{f}$) to observe $\gamma - Z^0$ interference effects and thus to investigate the neutral-current couplings of leptons and quarks.

Within the parton model, we may write

$$\frac{1}{s}\frac{d^2\sigma}{dx\,dy}(e^\mp p \to e^\mp X)$$

$$= \frac{4\pi\alpha^2}{Q^4} e_q^2(q+\bar{q})x\left(1-y+\frac{y^2}{2}\right)$$

$$- \frac{2\alpha G_F M_Z^2 e_q}{Q^2 \sqrt{2}(Q^2+M_Z^2)}\left[(L_e+R_e)(L_q+R_q)(q+\bar{q})x\left(1-y+\frac{y^2}{2}\right)\right.$$

$$\left. \pm (L_e-R_e)(L_q-R_q)(q-\bar{q})xy\left(1-\frac{y}{2}\right)\right]$$

$$+ \frac{1}{2\pi}\left(\frac{G_F M_Z^2}{\sqrt{2}}\right)^2 \frac{1}{(Q^2+M_Z^2)^2}\left[(L_e^2+R_e^2)(L_q^2+R_q^2)(q+\bar{q})\right.$$

$$\left. \times x\left(1-y+\frac{y^2}{2}\right)\pm(L_e^2-R_e^2)(L_q^2-R_q^2)(q-\bar{q})xy\left(1-\frac{y}{2}\right)\right], \quad (7.3.95)$$

where $q(x)$ and $\bar{q}(x)$ are the distributions of quarks and antiquarks within the proton, and a summation over quark species is implied. As for the analogous expression (6.4.29) for electron–positron annihilations, the first term is of electromagnetic origin, the third represents the effect of the weak neutral current, and the second is a weak-electromagnetic interference term. The weak neutral current gives rise to both parity violations and charge asymmetries, which therefore suggest particularly sensitive experimental approaches to the problem of neutral-current structure. To estimate the magnitude of these effects at low energies, it is permissible to neglect the final term because $Q^2 \ll M_Z^2$. Furthermore, if we choose $x_W = 1/4$, not far from the best value suggested by other experiments, the electron's neutral-current coupling is purely axial, so only the first piece of the interference term need be considered. Finally, by selecting an isoscalar target, we ensure that for the target as a whole

$$\langle u(x)+\bar{u}(x)\rangle_{\text{target}} = \langle d(x)+\bar{d}(x)\rangle_{\text{target}}. \quad (7.3.96)$$

If the contributions of heavy quarks are neglected, we find easily that the asymmetry

$$A = \frac{\sigma_R-\sigma_L}{\sigma_R+\sigma_L} \simeq \frac{-G_F Q^2(R_e-L_e)}{4\pi\alpha\sqrt{2}}\frac{\displaystyle\sum_{q=u,d}(L_q+R_q)e_q}{\displaystyle\sum_{q=u,d}e_q^2}$$

$$= \frac{-G_F Q^2}{4\pi\alpha\sqrt{2}}\cdot\frac{9}{5}\left(1-\frac{20x_W}{9}\right). \quad (7.3.97)$$

In this approximation, we expect the asymmetry to be independent of y, and to be extremely small and negative. For $x_w = 1/4$,

$$A/Q^2 \simeq -7.2 \times 10^{-5}. \qquad (7.3.98)$$

Precisely such an effect was measured in a meticulously executed experiment carried out by the SLAC–Yale Collaboration,[11] from which the data are reproduced in Fig. 7-12. A less idealized analysis than the one we have just given yields, from this experiment alone, a best value of the weak mixing parameter of

$$x_W = 0.224 \pm 0.020, \qquad (7.3.99)$$

consistent with other determinations. This experiment was of considerable importance in gaining acceptance for the standard model because, in addition to being the first to demonstrate directly parity violation in the neutral current, it came at a time when conflicting results from atomic physics experiments had suggested the necessity of a more involved structure for the neutral current.

At higher energies the parity violations and charge asymmetries implied by the standard model become more dramatic. Typical expectations for a very-high-energy electron–proton colliding beams device are shown in

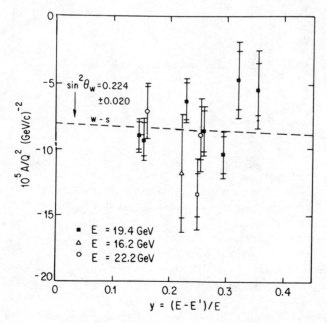

FIG. 7-12. The y-dependence of the parity-violating asymmetry A in deeply inelastic electron–deuteron scattering (from ref. 11).

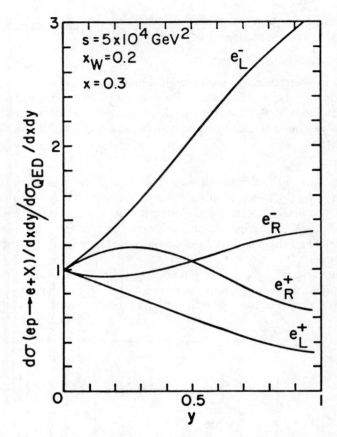

FIG. 7-13. Effects of γ-Z interference in longitudinally polarized $e^{\pm}p$ scattering in the standard model (from ref. 12).

Fig. 7-13. Effects of one part in 10^4 at $Q^2 \simeq 1$ GeV^2 have now become of order unity. The benefits of having lepton beams with well-controlled longitudinal polarization are evident. Obtaining such beams is an open challenge in accelerator design.

Most of our discussion of the consequences of the standard model for lepton–hadron interactions has taken place in the context of the point-coupling limit $Q^2/M^2_{W,Z} \ll 1$. The prospect of studying weak interactions at extremely high energies raises the possibility of observing the damping effect of the intermediate boson propagator upon the total cross section and other observables. It is straightforward to calculate in the parton model the total charged-current cross sections for high-energy electron–proton scattering shown in Fig. 7-14. If there were no intermediate bosons and if Bjorken

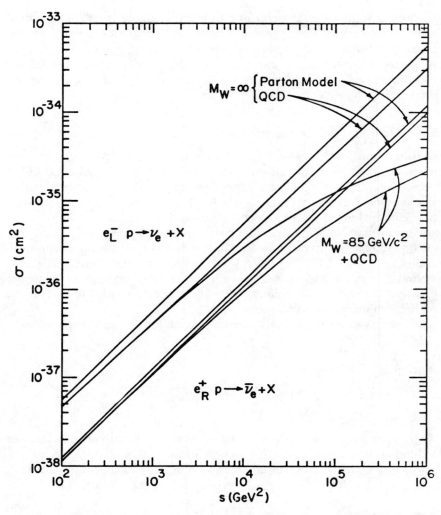

FIG. 7-14. Charged-current total cross sections for $e^{\pm}p$ scattering with "correct" helicity leptons. Predictions are shown for the four-fermion theory in the parton model with and without QCD scaling violations, and for the standard weak-interaction model including scaling violations (from ref. 12).

scaling were perfect, the cross section would rise linearly with s. Scaling violations expected on the basis of quantum chromodynamics (the origin of which will be discussed in the following chapter) modify the cross sections somewhat, but the effects are not dramatic on this scale. An intermediate boson with a mass of 85 GeV/c^2 causes a pronounced damping of the total cross section—by a factor of 2 or more at $s = 10^5$ GeV^2. Differential cross sections show a still greater sensitivity. The challenge is to develop instruments with which to observe these characteristic effects. The absence of perceptible propagator effects in the νN charged-current total cross sections shown in Fig. 7-9 places a lower bound $M_W \geq 30$ GeV/c^2 on the intermediate boson mass.

7.4   Hadron–Hadron Interactions

We shall now look very briefly at manifestations of the weak and electro-magnetic interactions in hadron–hadron collisions. The conciseness of the presentation does not reflect a limited interest in the experimental possi-bilities but merely the fact that—at the level of the parton model at least—no new concepts are involved, beyond those we have already encountered in situations that are somewhat simpler to analyze. Indeed, the highest c.m. energy attained in accelerator experiments is 540-GeV proton–antiproton collisions in the CERN Super Proton Synchrotron, and a similar device at Fermilab is planned to reach 2000 GeV. These very high energies engender a special interest in hadronic experiments, particularly with respect to the search for intermediate vector bosons.

In the quark–parton model, the cross section for the hadronic reaction

$$a + b \rightarrow c + \text{anything} \tag{7.4.1}$$

is given schematically by

$$\sigma(a + b \rightarrow c + X) = \sum_{\substack{\text{parton} \\ \text{species } i,j}} f_i^{(a)}(x_a) f_j^{(b)}(x_b) \hat{\sigma}(i + j \rightarrow c + X'), \tag{7.4.2}$$

where $f_i^{(a)}(x_a)$ is the probability of finding parton i with momentum fraction x_a of hadron a, and $\hat{\sigma}$ is the cross section for the elementary process leading to the desired final state. The summation runs over all contributing parton configurations. If we denote the invariant mass of the i–j system as

$$\mathcal{M} = \sqrt{s\tau} \tag{7.4.3}$$

and its longitudinal momentum in the hadron–hadron c.m. frame by

$$p = x\sqrt{s}\,/\,2, \tag{7.4.4}$$

then it is clear that the kinematic variables $x_{a,b}$ of the elementary process are related to those of the hadronic process by

$$x_{a,b} = \tfrac{1}{2}[(x^2 + 4\tau)^{1/2} \pm x]. \tag{7.4.5}$$

Drell and Yan[13] treated in this spirit the reaction

$$a + b \rightarrow l^+ l^- + \text{anything}, \tag{7.4.6}$$

in which a lepton pair of invariant mass $\mathscr{M}$ is produced with c.m. momentum fraction x by the elementary reaction

$$q + \bar{q} \to \gamma \to l^+ l^-. \tag{7.4.7}$$

The differential cross section is given by

$$\frac{d\sigma}{d\mathscr{M}^2 \, dx} = \left(\frac{4\pi\alpha^2}{3\mathscr{M}^4}\right) F(\tau, x), \tag{7.4.8}$$

where the first factor is familiar as the cross section for the reaction $e^+ e^- \to \mu^+ \mu^-$ and the second is particular to the parton-model subprocess. It takes the form

$$F(\tau, x) = \frac{x_a x_b}{(x^2 + 4\tau)^{1/2}} \, g(x_a, x_b), \tag{7.4.9}$$

where the first factor is a Jacobian determinant connecting the variables (x_a, x_b) with $(x, \mathscr{M}^2)$, and information about the quark distributions within the hadrons is contained in the function

$$g(x_a, x_b) = \tfrac{1}{3} \sum_{\text{flavors } i} e_i^2 [q_i^{(a)}(x_a) \bar{q}_i^{(b)}(x_b) + \bar{q}_i^{(a)}(x_a) q_i^{(b)}(x_b)], \tag{7.4.10}$$

where e_i is the quark charge. The factor $1/3$ is a consequence of color: the quark and antiquark that annihilate into a virtual photon must have the same color as well as flavor, but we have not distinguished colors in measuring the quark distributions in deeply inelastic scattering. The parton model thus provides a link between lepton–hadron and hadron–hadron processes.

The Drell–Yan picture of dilepton production carries a number of significant implications. The most general of these is the scaling prediction that the combination

$$\mathscr{M}^4 \frac{d\sigma}{d\mathscr{M}^2} = \frac{4\pi\alpha^2 \tau}{3} \int_\tau^1 dx \, \frac{g(x, \tau/x)}{x} \tag{7.4.11}$$

should be a function of the dimensionless variable τ alone. Although there are important strong-interaction (i.e., QCD) corrections to the parton model for this process, the scaling behavior has been established experimentally to good approximation. A recent compilation is shown in Fig. 7-15. The peaking of the cross section at small values of τ reflects the largeness of the quark distributions at small x, which is apparent in Fig. 7-10.

The cross section for the reaction

$$a + b \to W^\pm + \text{anything} \tag{7.4.12}$$

can also be computed directly in the Drell–Yan picture. In this case the elementary reactions are

$$u\bar{d}_\theta \to W^+,$$
$$\bar{u} d_\theta \to W^-. \tag{7.4.13}$$

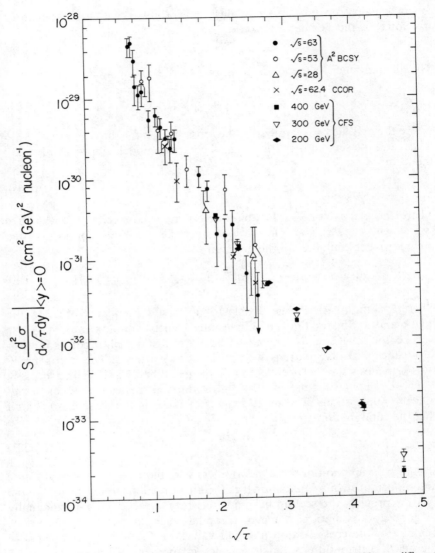

FIG. 7-15. Test of Drell–Yan scaling in the reaction $p + N \to \mu^+\mu^-$ + anything at different incident beam momenta. [Data are from the Athens–Athens–Brookhaven–CERN–Syracuse–Yale Collaboration: $\sqrt{s} = 28$ GeV, J. H. Cobb *et al.*, *Phys. Lett.* **72B**, 273 (1977); $\sqrt{s} = 53$ and 63 GeV, C. Kourkoumelis *et al.*, *ibid.* **91B**, 475 (1980); from the CERN–Columbia–Oxford–Rockefeller Collaboration, A. L. S. Angelis *et al.*, *ibid.* **87B**, 398 (1979); and from the Columbia–Fermilab–Stony Brook Collaboration, A. Ito *et al.*, *Phys. Rev.* **D23**, 604 (1981).]

The integrated cross section is

$$\sigma(a + b \to W^{\pm} + \cdots) = G_F \pi \sqrt{2} \tau \int_{\tau}^{1} dx \, W_{ab}^{(\pm)}(x, \tau/x)/x, \qquad (7.4.14)$$

where

$$W_{ab}^{(+)} = \tfrac{1}{3}\{[u^{(a)}(x_a)\bar{d}^{(b)}(x_b) + \bar{d}^{(a)}(x_a)u^{(b)}(x_b)]\cos^2\theta_C$$
$$+ [u^{(a)}(x_a)\bar{s}^{(b)}(x_b) + \bar{s}^{(a)}(x_a)u^{(b)}(x_b)]\sin^2\theta_C\}. \qquad (7.4.15)$$

A cross section approaching 10^{-32} cm^2 is anticipated in high-energy proton–antiproton collisions, for an intermediate boson of mass 85 GeV/c^2 as predicted in the standard model. Such a cross section appears within experimental reach. The production and leptonic decay of the charged intermediate boson owes an interesting experimental signature to the $V - A$ structure of the charged current. If we idealize the proton as composed entirely of quarks and the antiproton as composed entirely of antiquarks, the intermediate bosons will be produced with spins aligned along the direction of the incident antiproton because only left-handed fermions and right-handed antifermions participate in the formation reactions (7.4.13). Then, for the same reason (as we have already discussed in Section 6.2), the decay products that are fermions will be emitted opposite to the direction of W-polarization. As a consequence, the negative muons from the decay

$$W^- \to \mu^- \bar{\nu}_\mu \qquad (7.4.16)$$

will be preferentially emitted in the direction defined by the incident protons, whereas positive muons (from W^+-decay) will tend to follow the antiproton direction. The resulting charge asymmetry, which is opposite to what might be expected for the background of strong-interaction processes, should provide important evidence that a weak-interaction process has been observed.

Evidently the cross section for the production of neutral intermediate bosons may be estimated in precisely the same manner from the calculable rates for the elementary reactions

$$u\bar{u} \to Z^0,$$
$$d\bar{d} \to Z^0, \qquad (7.4.17)$$
$$s\bar{s} \to Z^0.$$

The cross section may be written as

$$\sigma(a + b \to Z^0 + \cdots) = \frac{G_F \pi}{\sqrt{2}} \tau \int_{\tau}^{1} dx \, Z_{ab}(x, \tau/x)/x, \qquad (7.4.18)$$

where

$$Z_{ab}(x_a, x_b) = \tfrac{1}{3} \sum_q [q^{(a)}(x_a)\bar{q}^{(b)}(x_b) + \bar{q}^{(a)}(x_a)q^{(b)}(x_b)](L_q^2 + R_q^2), \qquad (7.4.19)$$

where L_q and R_q are the familiar neutral-current couplings. Again cross sections approaching 10^{-32} cm^2 are expected in high-energy $\bar{p}p$ collisions. In spite of the small (few percent) branching ratio, the charged leptonic decays

$$Z^0 \to l^+l^- \tag{7.4.20}$$

should provide an excellent experimental signature.

Although many other manifestations of the weak interactions can be described and studied in high-energy collisions of hadrons, the search for the intermediate boson deservedly commands our attention. The specific prediction of intermediate boson properties is a key feature of the Weinberg–Salam model. It tests in an important fashion the choice of the electroweak gauge group and the particular mechanism for spontaneous symmetry breaking. To verify or convincingly to disprove the predictions of the model will be an achievement of the first importance.

7.5   Reflections

Despite the combined length of this and the preceding chapter, we have only discussed in a limited way the phenomena and theoretical descriptions of the weak and electromagnetic interactions. Among spontaneously broken gauge theories we have elaborated only the structure and predictions of the Weinberg–Salam model extended to hadrons by the Glashow–Iliopoulos–Maiani construction. The focus upon a single model is justified on at least two grounds. The standard model has the pedagogical advantage of being the simplest viable theory as well as one that exhibits all the main features of spontaneously broken gauge theories. Its experimental successes, which include the description of the known neutral-current phenomena with a common value of the weak mixing angle[14]

$$x_{\mathrm{W}} = \sin^2\theta_{\mathrm{W}} = 0.233 \pm 0.009(\pm 0.005) \tag{7.5.1}$$

are impressive indeed, and a decisive test of its predictions for the intermediate bosons may be imminent.

In spite of—indeed because of—the successes of the standard model, many questions present themselves. It is appropriate to close this chapter with a brief list.

Is the standard model correct? Is it complete?

Will the intermediate bosons be found with the desired properties?

Will a Higgs boson be found? What are its properties?

Are weak and electromagnetic interactions truly described by a gauge theory, or is it merely that the low-energy phenomenology is characterized by the point-coupling limit of the standard model?

Why are the charged-current weak interactions left-handed?

What is the origin of fermion masses?

How does the mixing of quark flavors arise?

Are the neutrinos massless?

What, if any, is the pattern of lepton mixing, and how does it arise?

What is the mechanism of CP violation?

What is the origin of quark and lepton generations?

Is there a relation between the quark and lepton spectra, as suggested by the anomaly cancellation?

The reader will be able to add many more.

Problems

7-1.   Calculate the angular distribution $d\Gamma/d\Omega$ for the decay $Z^0 \to e^+e^-$.

7-2.   Compute the decay rate for the disintegration $Z^0 \to H\mu^+\mu^-$, which proceeds through the intermediate state HZ^0_{virtual}. [Reference: J. D. Bjorken, in *Weak Interactions at High Energy and the Production of New Particles*, 1976 SLAC Summer Institute, SLAC Report No. 198, November, 1976, p. 1. A factor of π is missing from the denominator of equation (4.30), by virtue of a typographical error.]

7-3.   Show that the $n \times n$ unitary matrix U defined in (7.1.27–29) to describe the mixing of quarks of different flavors can be parametrized in terms of $n(n-1)/2$ real mixing angles and $(n-1)(n-2)/2$ complex phases, after the freedom to redefine the phases of quark fields has been taken into account. For the specific case of three quark doublets ($n = 3$), show that U may be written in the form

$$U = \begin{pmatrix} 1 & 0 & 0 \\ 0 & c_2 & s_2 \\ 0 & -s_2 & c_2 \end{pmatrix} \cdot \begin{pmatrix} c_1 & s_1 & 0 \\ -s_1 & c_1 & 0 \\ 0 & 0 & 1 \end{pmatrix} \cdot \begin{pmatrix} 1 & 0 & 0 \\ 0 & 1 & 0 \\ 0 & 0 & e^{i\delta} \end{pmatrix} \cdot \begin{pmatrix} 1 & 0 & 0 \\ 0 & c_3 & s_3 \\ 0 & -s_3 & c_3 \end{pmatrix},$$

where $s_i = \sin\theta_i$ and $c_i = \cos\theta_i$. Discuss the implications of the phase δ for CP invariance. [Reference: M. Kobayashi and T. Maskawa, *Prog. Theoret. Phys. (Kyoto)* **49**, 652 (1973).]

7-4.   Compute the cross section for the reaction $e^+e^- \to \nu_\mu\bar\nu_\mu$ in the point-coupling limit, and compare it with that for $\nu_\mu e \to \nu_\mu e$. For the special case $x_W = 1/4$, use angular momentum arguments to explain the ratio of the two cross sections.

7-5.   Verify the general expression (7.2.3) for the parton-model reaction $e^+e^- \to q\bar{q}$.

7-6.   Compute the cross section for the reaction $e^+e^- \to \nu_e\bar{\nu}_e$. What, if any, is the importance of the W–exchange diagram in the neighborhood of the Z^0 peak?

7-7.   Reconsider the momentum sum rule (7.3.66), taking into consideration the contribution of strange quarks. How large a strange-quark sea would be required to account for the "missing" momentum? What implications would such a sea have for neutrino-induced production of charm?

7-8.   Derive the general form (7.3.70) of the cross section for the inclusive reaction $\nu N \to \mu^- +$ anything, and place the result in the scaling form (7.3.71). For scattering on a structureless target, show that $x\mathscr{F}_3(x)/\mathscr{F}_2(x) = \pm 1$ for a (fermion, antifermion) target.

7-9.   Neglecting the antiquarks within a nucleon, compute the ratios of neutral-current to charged-current cross sections given in (7.3.90, 91).

7-10.   Working in the framework of the parton model, derive the general form (7.3.95) of the cross section for the reaction $e^{\mp}p \to e^{\mp} +$ anything.

For Further Reading

INTERMEDIATE BOSONS. The conventional expectations of the standard model are summarized in

C. Quigg, *Rev. Mod. Phys.* **49**, 297 (1977).

J. Ellis, M. K. Gaillard, G. Girardi, and P. Sorba, *Ann. Rev. Nucl. Part. Sci.* **32**, 443 (1982). Radiative corrections to the intermediate boson mass and to other observables have been computed to one-loop order in the standard model by

F. Antonelli, M. Consoli, and G. Corbo, *Phys. Lett.* **91B**, 90 (1980).

C. H. Llewellyn Smith and J. Wheater, *Phys. Lett.* **105B**, 486 (1981).

J. Wheater and C. H. Llewellyn Smith, *Nucl. Phys.* **B208**, 27 (1982).

A. Sirlin and W. J. Marciano, *Nucl. Phys.* **B189**, 442 (1981).

M. Veltman, *Phys. Lett.* **91B**, 95 (1980).

Additional references for specific processes may be found in the papers by Llewellyn Smith and Wheater. Rather general bounds on the intermediate boson masses in spontaneously broken gauge theories have been derived by

J. D. Bjorken, *Phys. Rev.* **D15**, 1330 (1977).

The comportment of theories with several neutral intermediate bosons and bounds on the Z^0 masses are investigated in

H. Georgi and S. Weinberg, *Phys. Rev.* **D17**, 275 (1978).

The search for intermediate bosons is the subject of

D. B. Cline, C. Rubbia, and S. van der Meer, *Sci. Am.* **246**, 38 (March, 1982).

PARTON MODEL. An important early paper is

J. D. Bjorken and E. A. Paschos, *Phys. Rev.* **185**, 1975 (1969).

Thorough developments of the quark-parton model are given by

F. E. Close, *An Introduction to Quarks and Partons*, Academic, New York, 1979.

R. P. Feynman, *Photon–Hadron Interactions*, Benjamin, Reading, Massachusetts, 1972.

T. M. Yan, *Ann. Rev. Nucl. Sci.* **26**, 199 (1976).

A somewhat broader perspective on deeply inelastic scattering is provided by

P. Roy, *Theory of Lepton–Hadron Processes at High Energies*, Clarendon, Oxford, 1975.

Many applications of the parton-model philosophy are treated in

S. M. Berman, J. D. Bjorken, and J. B. Kogut, *Phys. Rev.* **D4**, 3388 (1971).

OPERATOR PRODUCT EXPANSION. This fruitful alternative to the heuristic parton-model formulation was propounded by

K. G. Wilson, *Phys. Rev.* **179**, 1499 (1969).

Many early references to the concepts of broken scale invariance and the light-cone expansion are to be found in Roy's book. Some later work will be mentioned in Chapter 8.

CHARMED PARTICLES. A general survey appears in the prospectus by

M. K. Gaillard, B. W. Lee, and J. L. Rosner, *Rev. Mod. Phys.* **47**, 277 (1975).

Good recent experimental surveys are

G. Goldhaber and J. E. Wiss, *Ann. Rev. Nucl. Part. Sci.* **30**, 337 (1980).

G. H. Trilling, *Phys. Rep.* **75**, 57 (1981).

For a limit on charm-changing neutral currents, see

A. Bodek *et al.*, *Phys. Lett.* **113B**, 82 (1982).

Restrictions on spontaneously broken gauge theories imposed by limits on strangeness-changing neutral currents and on $|\Delta S| = 2$ transitions are reviewed in the Prologue of Gaillard, Lee, and Rosner.

ALTERNATIVES TO GAUGE THEORIES. Means of reproducing the low-energy phenomenology of the standard model without appealing to local gauge invariance have been investigated by

J. D. Bjorken, *Phys. Rev.* **D19**, 335 (1979).

P. Q. Hung and J. J. Sakurai, *Nucl. Phys.* **B143**, 81 (1978).

MASSIVE NEUTRINOS AND NEUTRINO OSCILLATIONS. The general phenomenology is set out in

S. M. Bilenky and B. Pontecorvo, *Phys. Rep.* **41C**, 225 (1978).

I. Yu. Kobzarev, B. V. Martem'yakov, L. B. Okun, and M. G. Shchepin, *Yad. Fiz.* **32**, 1599 (1980) [English translation: *Sov. J. Nucl. Phys.* **32**, 823 (1980)].

Recent experimental results and the possible connection with grand unified theories are reviewed by

P. H. Frampton and P. Vogel, *Phys. Rep.* **82**, 339 (1982).

CP VIOLATION. The discovery of CP nonconservation in the neutral kaon system and the search for its origin are described in the Nobel lectures by

V. L. Fitch, *Rev. Mod. Phys.* **53**, 367 (1981).

J. W. Cronin, *Rev. Mod. Phys.* **53**, 373 (1981).

A review of the experimental situation has been given by

K. Kleinknecht, *Ann. Rev. Nucl. Sci.* **26**, 1 (1976).

The theoretical situation is unsettled. Within the gauge theory framework, two approaches have been distinguished. The possibility that CP violation arises from complex elements of the quark mass matrix was raised by

M. Kobayashi and T. Maskawa, *Prog. Theoret. Phys.* (*Kyoto*) **49**, 652 (1973).

for theories with more than four quarks. Alternatively, CP violation may be due to a relative phase among several Higgs doublets. A realization for the (otherwise) standard four-quark model is due to

S. Weinberg, *Phys. Rev. Lett.* **37**, 657 (1976).

Present experiments remain consistent with the "superweak" theory of

L. Wolfenstein, *Phys. Rev. Lett.* **13**, 562 (1964).

NONLEPTONIC WEAK DECAYS. The theoretical situation is rather unsatisfactory. A partial understanding of the phenomenon of nonleptonic enhancement has been achieved by

M. K. Gaillard and B. W. Lee, *Phys. Rev. Lett.* **33**, 108 (1974).

G. Altarelli and L. Maiani, *Phys. Lett.* **52B**, 351 (1974).

A further elaboration of the possible connection between quantum chromodynamics and weak decays was initiated by

M. Shifman, A. I. Vainshtein, and V. I. Zakharov, *Nucl. Phys.* **B120**, 316 (1977); *Zh. Eksp. Teor. Fiz.* **72**, 1275 (1977) [English translation: *Sov. Phys.-JETP* **45**, 670 (1977)].

References

[1] R. E. Shrock and L.-L. Wang, *Phys. Rev. Lett.* **41**, 1692 (1978).

[2] Y. Asano *et al.*, *Phys. Lett.* **107B**, 159 (1981).

[3] S. L. Glashow, J. Iliopoulos, and L. Maiani, *Phys. Rev.* **D2**, 1285 (1970).

[4] R. N. Cahn, M. S. Chanowitz, and N. Fleishon, *Phys. Lett.* **82B**, 113 (1979).

[5] Earlier we denoted charges by Q_i, to avoid confusion with the electron spinor e; here we change notation to e_i, to avoid confusion with the momentum transfer variable Q^2.

[6] J. D. Bjorken, *Phys. Rev.* **179**, 1547 (1969).

[7] A convenient summary of the early SLAC experiments appears in J. I. Friedman and H. W. Kendall, *Ann. Rev. Nucl. Sci.* **22**, 203 (1972).

[8] See ref. 9 of Chapter 1. The corresponding measurement for neutrino–nucleon scattering yields $\sigma_S/\sigma_T = 0.10 \pm 0.07$ for $v \simeq 50$ GeV: H. Abramowicz *et al.*, *Phys. Lett.* **107B**, 141 (1981). In high-energy muon-nucleon scattering with 60 GeV $< v <$ 160 GeV and $\langle Q^2 \rangle = 22.5$ GeV^2, the result is $\sigma_S/\sigma_T = 0.0 \pm 0.1$: J. J. Aubert, *et al.*, *Phys. Lett.* **121B**, 87 (1983).

[9] J. Drees, in *Proceedings of the 1981 International Symposium on Lepton and Photon Interactions at High Energies*, edited by W. Pfeil, Physikalisches Institut Universität Bonn, Bonn, 1981, p. 474.

[10] See, for example, the comparison in Fig. 9 of G. Smadja, *Proceedings 1981 International Symposium on Lepton and Photon Interactions at High Energies*, edited by W. Pfeil, Physikalisches Institut Universität Bonn, Bonn, 1981, p. 444.

[11] C. Prescott *et al.*, *Phys. Lett.* **77B**, 347 (1978); **84B**, 524 (1979).

[12] C. Quigg, in *Physics in Collison*, edited by W. P. Trower and Gianpaolo Bellini, Plenum, New York, 1982, p. 345.

[13] S. D. Drell and T. M. Yan, *Phys. Rev. Lett.* **25**, 316 (1970); *Ann. Phys. (NY)* **66**, 578 (1971).

[14] This is taken from a global fit by J. E. Kim, P. Langacker, M. Levine, and H. H. Williams, *Rev. Mod. Phys.* **53**, 211 (1981).

CHAPTER 8

STRONG INTERACTIONS AMONG QUARKS

Having learned that local gauge invariance provides the key to understanding the weak and electromagnetic interactions, we turn our attention once again to the strong interactions. The work of many people, from Yang and Mills[1] through Sakurai,[2] Ne'eman,[3] Englert and Brout,[4] to 't Hooft,[5] among others, showed that it is unlikely that a flavor symmetry such as isospin or $SU(3)$ will be the basis of a successful gauge theory of the hadronic interactions. Furthermore, at what we currently perceive to be the constituent level of quarks and leptons, flavor has been seen to be an attribute tied more directly to the weak interactions than to the strong. The property that distinguishes quarks from leptons is color, so it is natural to attempt to construct a theory of the strong interactions among quarks based upon a local color gauge symmetry. The resulting theory has acquired the name quantum chromodynamics, or QCD.

For reasons connected with the nonobservation of free quarks, it is appealing to have long-range forces among the quarks, as would be mediated by massless gauge bosons in an unbroken gauge theory. In this respect, the absence of spontaneous symmetry breaking, the formulation of QCD is technically simpler than that of the $SU(2)_L \otimes U(1)_Y$ electroweak theory. The analysis of the theory is considerably complicated by the fact that the strong interactions are strong. There is thus at first sight no guarantee that perturbation theory—our most highly developed tool—will be at all useful. In a general field theory, the effective coupling constant is not a constant, but depends on a momentum or distance scale because of renormalization effects. It is a particular property of non-Abelian gauge theories that the effective coupling constant decreases at short distances, or equivalently at high momenta. Such theories are said to be asymptotically free. This raises

193

the hope that, in the regime of short-distance phenomena, perturbative methods may indeed be reliable. Furthermore, it suggests the beginning of a reconciliation of the quasifreedom of quarks, embodied in the parton model, with the observed low density of isolated quarks.

If QCD is indeed the correct theory of the strong interactions, it must describe an enormous range of phenomena, from the spectrum of light hadrons to deeply inelastic scattering. We shall investigate only a small sample of these, in the interest of restricting the book to finite size, because of the absence of complete or graceful solutions to certain problems, and because the limited techniques we have used are ill-suited to some others. Nevertheless, we shall be able to expose the essential features of QCD, to see why it appears so promising as a fundamental theory of the strong interactions, and to prepare the reader to confront the research literature. The emphasis will be not so much on the shortcomings of elementary methods as on the considerable amount that may be done with so little formal apparatus.

We shall begin by reviewing the motivation for hadronic color and for the choice of $SU(3)_{color}$ as the strong-interaction gauge group. Next we construct the QCD Lagrangian and extract the Feynman rules for tree diagrams. An analysis of the spectroscopic consequences of the theory in the crude approximation of lowest-order perturbation theory reveals no obvious contradictions with experiment. Thereafter we turn to a study of the coupling constant evolution in the context of charge renormalization in electrodynamics. This provides the occasion to introduce two useful techniques for the evaluation of the integrals over loop momenta, which occur in the computation of Feynman diagrams. Turning to QCD we derive the celebrated property of asymptotic freedom. Having found this signal of the existence of a regime in which perturbation theory may be expected to apply, we consider three concrete applications: electron–positron annihilation into hadrons, deeply inelastic lepton–nucleon scattering, and deeply inelastic lepton–photon scattering. To deal with the Q^2-evolution of the nucleon and photon structure functions, we develop the intuitive method applied to this problem by Altarelli and Parisi.[6] At the end of the chapter we discuss very briefly some applications for which perturbative methods do not suffice, and assess the current status and open problems in QCD.

8.1   A Color Gauge Theory

Several pieces of evidence have been arrayed in favor of the hypothesis that the quarks of common experience u, d, s, c, b are color triplets. These include the resolution of the spin-statistics problem for baryons, the magnitude of the cross section for electron–positron annihilation into hadrons, the branching ratios for τ-decay, the π^0 lifetime, and the requirement of anomaly cancellation in the standard model of weak and electromagnetic interactions. The known leptons, in contrast, are all colorless states. The distinction suggests the possibility that color may play the part of a charge of the strong interactions. We attempt, therefore, to formulate a dynamical theory based on color symmetry, which will evidently have to be a color gauge theory. Early steps in this direction were taken by Nambu.[7] It was soon recognized[8] that such a theory might provide a basis for the simple rules that mesons are quark–antiquark states and baryons are three-quark states. Let us see why $SU(3)$ is a promising choice for the color symmetry group.

When we entertain the possibility that the color quantum number reflects a continuous symmetry of the strong interaction Lagrangian, rather than merely a discrete degree of freedom, three candidates for the symmetry come immediately to mind: $SO(3)$, $SU(3)$, and $U(3)$. The choice of a gauge group is to be governed by the two empirical facts—that the familiar quarks are color triplets, but all the known hadrons are color singlets. Simple arguments discourage the use of $SO(3)$ and $U(3)$, as we shall now see.

In $SO(3)$ no distinction is made between color and anticolor, so in the computation of forces there will be no distinction between quarks and antiquarks. The existence of $(q\bar{q})$ mesons thus implies the existence of (qq) diquark states, which will be fractionally charged. Because fractionally charged matter is less commonplace than ordinary mesons, $SO(3)$ does not seem to be an apt choice for the color symmetry group. (A similar shortcoming of the original Yang–Mills theory was already seen in Section 4.3 and Problem 4-4.) We shall also see below, in Problem 8-10, that an $SO(3)$ gauge theory with the known quark flavors would not be asymptotically free.

In $U(3)$ color gauge theory, the color singlet gauge boson that occurs in the product

$$\mathbf{3} \otimes \mathbf{3} = \mathbf{1} \oplus \mathbf{8} \tag{8.1.1}$$

(in $SU(3)$ notation) cannot be dispensed with. It would mediate long-range strong interactions between color singlet hadrons and is thus ruled out by the same experimental facts that eliminate the Yang–Mills theory. Thus $U(3)$ is excluded, and we are left with $SU(3)$ as the candidate gauge group.

To construct the $SU(3)$ color gauge theory for the interactions of color triplet quarks, we follow the procedure used in Section 4.2 for the Yang–Mills theory. The color octet gauge bosons that will emerge in the theory are called gluons, because of their role in binding quarks together within hadrons. The Lagrangian for the theory will have the standard form

$$\mathscr{L} = \bar{\psi}(i\gamma^\mu \mathscr{D}_\mu - m)\psi - \tfrac{1}{2}\operatorname{tr}(G_{\mu\nu}G^{\mu\nu}), \tag{8.1.2}$$

where the composite spinor for the color triplet quarks is

$$\psi = \begin{pmatrix} q_{red} \\ q_{blue} \\ q_{green} \end{pmatrix}, \tag{8.1.3}$$

and the covariant derivative is

$$\mathscr{D}_\mu = \partial_\mu + igB_\mu, \tag{8.1.4}$$

where the object B_μ is a three-by-three matrix in color space formed from the eight color gauge fields b_μ^l and the generators $\lambda^l/2$ of the $SU(3)$ gauge group as

$$B_\mu = \tfrac{1}{2}\lambda \cdot \mathbf{b}_\mu = \tfrac{1}{2}\lambda^l b_\mu^l. \tag{8.1.5}$$

The gluon field-strength tensor is

$$G_{\mu\nu} = \tfrac{1}{2}\mathbf{G}_{\mu\nu} \cdot \lambda = \tfrac{1}{2}G_{\mu\nu}^l \lambda^l$$
$$= (ig)^{-1}[\mathscr{D}_\nu, \mathscr{D}_\mu] = \partial_\nu B_\mu - \partial_\mu B_\nu + ig[B_\nu, B_\mu].$$

The λ-matrices are familiar[9] from the study of flavor-$SU(3)$ symmetry. They have a number of simple properties, including

$$\text{tr}(\lambda^l) = 0, \tag{8.1.7}$$
$$\text{tr}(\lambda^k \lambda^l) = 2\delta^{kl}, \tag{8.1.8}$$

and

$$[\lambda^j, \lambda^k] = 2if^{jkl}\lambda^l, \tag{8.1.9}$$

which parallel those of the Pauli isospin matrices given in (4.2.18) and (4.2.25). Indeed, in the canonical basis

$$\lambda_1 = \begin{pmatrix} 0 & 1 & 0 \\ 1 & 0 & 0 \\ 0 & 0 & 0 \end{pmatrix} \begin{matrix} \bar{R} \\ \bar{B}, \\ \bar{G} \end{matrix} \qquad \lambda_2 = \begin{pmatrix} 0 & -i & 0 \\ i & 0 & 0 \\ 0 & 0 & 0 \end{pmatrix},$$
$$\;\;\; R \quad\; B \quad\; G$$

$$\lambda_3 = \begin{pmatrix} 1 & 0 & 0 \\ 0 & -1 & 0 \\ 0 & 0 & 0 \end{pmatrix}, \qquad \lambda_4 = \begin{pmatrix} 0 & 0 & 1 \\ 0 & 0 & 0 \\ 1 & 0 & 0 \end{pmatrix}, \tag{8.1.10}$$

$$\lambda_5 = \begin{pmatrix} 0 & 0 & -i \\ 0 & 0 & 0 \\ i & 0 & 0 \end{pmatrix}, \qquad \lambda_6 = \begin{pmatrix} 0 & 0 & 0 \\ 0 & 0 & 1 \\ 0 & 1 & 0 \end{pmatrix},$$

$$\lambda_7 = \begin{pmatrix} 0 & 0 & 0 \\ 0 & 0 & -i \\ 0 & i & 0 \end{pmatrix} \qquad \lambda_8 = \frac{1}{\sqrt{3}}\begin{pmatrix} 1 & 0 & 0 \\ 0 & 1 & 0 \\ 0 & 0 & -2 \end{pmatrix},$$

TABLE 8.1:   ANTISYMMETRIC STRUCTURE
CONSTANTS OF $SU(3)$

$$f_{123} = 1$$
$$f_{147} = f_{246} = f_{257} = f_{345} = f_{516} = f_{637} = \tfrac{1}{2}$$
$$f_{458} = f_{678} = \sqrt{3}/2$$

the matrices λ_1, λ_2, λ_3 correspond to the Pauli isospin matrices when the colors (R, B, G) are replaced by the flavors (u, d, s). Using (8.1.8) and (8.1.9) it is easy to compute the antisymmetric structure constants of the Lie group as

$$f^{jkl} = (4i)^{-1} \operatorname{tr}(\lambda^l [\lambda^j, \lambda^k]). \tag{8.1.11}$$

The nonzero elements are given by the entries in Table 8.1 and their permutations. The field-strength tensor may be rewritten in component form as

$$G_{\mu\nu}^l = \partial_\nu b_\mu^l - \partial_\mu b_\nu^l + g f^{jkl} b_\mu^j b_\nu^k. \tag{8.1.12}$$

Knowing the QCD Lagrangian, we may now study the interactions between quarks in a very simplified fashion, just as we did the interactions among nucleons in the Yang–Mills theory in Section 4.3. The point of the exercise will be to verify that color singlets enjoy a preferred status. This encourages the hope that the spectrum of QCD, when it is computed, will display the systematics that inspired the invention of the theory.

The quark–gluon interaction term in the QCD Lagrangian is

$$\mathscr{L}_{\text{int}} = -\frac{g}{2} b_\mu^a \bar{\psi} \gamma^\mu \lambda^a \psi, \tag{8.1.13}$$

which leads at once to the Feynman rule for the quark–antiquark–gluon vertex. For the transition depicted in Fig. 8-1 of a quark with color index α ($= R, B, G$ or $1, 2, 3$) into a quark with color index β and a gluon with Lorentz index μ and color label a ($= 1, 2, \ldots, 8$), the vertex factor is simply

$$-\frac{ig}{2} \lambda_{\alpha\beta}^a \gamma_\mu. \tag{8.1.14}$$

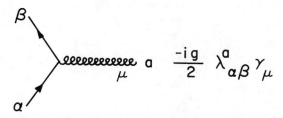

FIG. 8-1. Feynman rule for the quark–antiquark–gluon vertex in QCD.

Thus, the one-gluon-exchange force between quarks is proportional to

$$\mathcal{E} = \frac{g^2}{4} \sum_a \lambda^a_{\alpha\beta} \lambda^a_{\gamma\delta} \qquad (8.1.15)$$

for the transition $\alpha + \gamma \to \beta + \delta$. We shall take the quantity (8.1.15) as representative of the interaction energy between quarks and proceed to develop the consequences of QCD for the hadron spectrum, according to that measure.

To compute the interaction energy it is necessary to evaluate the expectation value of products such as $(1/4)\lambda^{(1)} \cdot \lambda^{(2)}$, where the superscripts label the interacting quarks and the $\lambda^{(i)}$ are 8-vectors in color space. The $SU(N)$ techniques are quite standard.[10] It will save writing to define the $SU(3)$ generators

$$\mathbf{T} \equiv \frac{1}{2}\lambda \qquad (8.1.16)$$

and to evaluate the expectation value $\langle \mathbf{T}^2 \rangle$ in various representations of interest.

In $SU(N)$, it is equivalent to average the square of any single generator over the representation, or to perform the sum over all the generators. The former tactic is simpler, and it is particularly convenient to choose I_3, the third component of isospin in the flavor analogy, as the designated generator. Consequently the expectation value in a representation of dimension d is

$$\langle \mathbf{T}^2 \rangle_d = (N^2 - 1) \sum_{\substack{\text{members} \\ \text{of rep.}}} I_3^2/d, \qquad (8.1.17)$$

where $N^2 - 1$ is the number of generators of $SU(N)$. Results for the low-dimensioned representations of $SU(3)$ are given in Table 8.2. To evaluate

TABLE 8.2: VALUE OF
THE COLOR CASIMIR
OPERATOR IN SMALL
REPRESENTATIONS OF $SU(3)$

Representation	$\langle \mathbf{T}^2 \rangle$
[1]	0
[3] or [3*]	4/3
[6] or [6*]	10/3
[8]	3
[10] or [10*]	6
[27]	8

$\langle \mathbf{T}^{(1)} \cdot \mathbf{T}^{(2)} \rangle$, we use the familiar identity

$$\langle \mathbf{T}^{(1)} \cdot \mathbf{T}^{(2)} \rangle = \frac{\langle \mathbf{T}^2 \rangle - \langle \mathbf{T}^{(1)2} \rangle - \langle \mathbf{T}^{(2)2} \rangle}{2}. \tag{8.1.18}$$

The "interaction energies" for two-body systems composed of quark–quark and quark–antiquark are given in Table 8.3. For the $(q\bar{q})$ systems, the one-gluon-exchange contribution is attractive for the color singlet but repulsive for the color octet. Similarly for diquark systems, the color triplet is attracted but the color sextet is repelled. Of all the two-body channels, the color singlet $(q\bar{q})$ is the most attractive. On the basis of this analysis, one may choose to believe that colored mesons should not exist, whereas color singlets should be found.

To analyze three(or more)-body systems, let us assume that the interaction is merely the sum of two-body forces, so that

$$\mathscr{E} = \sum_{i<j} \langle \mathbf{T}^{(i)} \cdot \mathbf{T}^{(j)} \rangle, \tag{8.1.19}$$

which is easily computed as

$$2 \sum_{i<j} \langle \mathbf{T}^{(i)} \cdot \mathbf{T}^{(j)} \rangle = \langle \mathbf{T}^2 \rangle - \sum_i \langle \mathbf{T}^{(i)2} \rangle. \tag{8.1.20}$$

For three-quark systems, the results in Table 8.3 show that the color singlet is again the most attractive channel. This is as desired.

Several potentially important effects have been neglected in these calculations:

1. Multiple gluon exchanges between quarks have been ignored, and we have put forward no arguments for faith in lowest-order perturbation theory.

TABLE 8.3: "Interaction Energies" for Few-Quark Systems

Configuration	$\left\langle \sum\limits_{i<j} \right\rangle \mathbf{T}^{(i)} \cdot \mathbf{T}^{(j)}$
$(q\bar{q})_{[1]}$	$-4/3$
$(q\bar{q})_{[8]}$	$+1/6$
$(qq)_{[3^*]}$	$-2/3$
$(qq)_{[6]}$	$1/3$
$(qqq)_{[1]}$	-2
$(qqq)_{[8]}$	$-1/2$
$(qqq)_{[10]}$	$+1$
$(qqqq)_{[3]}$	-2

FIG. 8-2. A baryon configuration that is not considered in the sum over two-body forces.

2. Configurations involving the three-gluon vertex, such as the (qqq) color singlet shown in Fig. 8-2, have not been taken into account. This is related to the incompleteness of the calculation noted in (1).

3. What may be the most serious shortcoming of the toy calculation is its neglect of the energetics associated with the creation of an isolated color nonsinglet state. In Section 8.8 we shall review a plausibility argument that implies an infinite cost in energy to isolate a colored system. If that is so, the attraction provided by one-gluon exchange will be insufficient to bind colored states.

In spite of these shortcomings, the elementary "maximally attractive channel" calculation we have just completed does make it plausible that color singlets are energetically favored states. In addition, it is easy to see that there is no long-range interaction with color singlets. As an example, consider whether a quark is bound to a baryon. The final entry in Table 8.3 shows that the interaction energy of the quark-plus-baryon system is precisely that which binds the baryon, with no additional attraction.

It is quite generally believed, but not proved, that QCD is in fact a confining theory, and that colored objects cannot be liberated. Seeking proofs, loopholes, and interpretations of the experimental indications[11] for fractionally charged matter is of obvious importance.

Before proceeding to a specific study of the predictions of QCD, it will be useful to present a qualitative discussion of the implications of an interacting field theory of quarks and gluons. A convenient setting for this discussion is the deeply inelastic scattering of leptons from a proton target, in which a virtual photon (or intermediate boson) of (mass)$^2 = -Q^2$ probes the target structure on a length scale characterized by $1/\sqrt{Q^2}$.

Viewed at very long wavelengths, the proton appears structureless, but as Q^2 increases and the resolution becomes finer, the proton is revealed as a composite object characterized, for example, by rapidly falling elastic form factors that decrease as $1/Q^4$. According to the parton model, which ignores interactions among the constituents of the protons, the picture for deeply inelastic scattering then becomes exceedingly simple, as sketched in Fig. 8-3. Once Q^2 has become large enough for the quark constituents to be resolved,

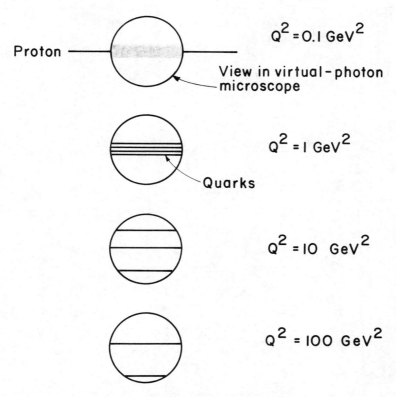

Proton

$Q^2 = 0.1 \text{ GeV}^2$

View in virtual-photon
microscope

$Q^2 = 1 \text{ GeV}^2$

Quarks

$Q^2 = 10 \text{ GeV}^2$

$Q^2 = 100 \text{ GeV}^2$

FIG. 8-3. Parton model view of the proton.

no finer structure is seen. The quarks are structureless, have no size, and thus introduce no length scale. When Q^2 exceeds a few GeV^2, all fixed mass scales become irrelevant and the Q^2-dependence of structure functions can be determined by dimensional analysis. We have seen in Section 7.3 that this description is not only simple, but also extraordinarily successful in reproducing the essential features of the data. It is also a description that appears difficult to reconcile with the idea of an underlying quantum field theory, for reasons that will immediately become apparent.

In an interacting field theory, a more complex picture of hadron structure emerges. As Q^2 increases beyond the magnitude required to resolve quarks, the quarks themselves are found to have an apparent structure, which arises from the interactions mediated by the boson fields. This is indicated in Fig. 8-4. The quantum fluctuations shown there lead to deviations from Bjorken scaling in deeply inelastic scattering. As we have seen in Section 7.3, the structure functions $F_i(x, Q^2)$ measure the distribution of quarks in a

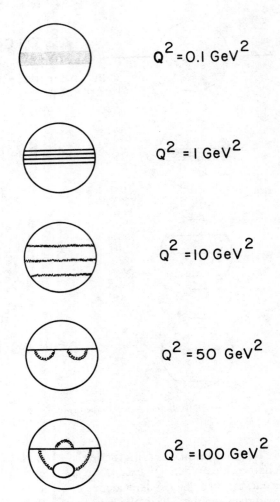

FIG. 8-4. Interacting field theory view of the proton.

fast-moving proton as a function of the momentum fraction

$$x = p_{\text{quark}}/P_{\text{proton}}. \qquad (8.1.21)$$

As Q^2 grows, the structure functions undergo a characteristic evolution. At large values of x ($0.3 \lesssim x < 1$), uncertainty principle considerations make it increasingly likely with increasing Q^2 that a quark with momentum fraction x will be caught in mid-dissociation into components with momentum fractions $x_1 + x_2 = x$, as shown in Fig. 8-5. For small values of x, the population of both quarks and antiquarks will be enhanced by

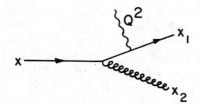

FIG. 8-5. Virtual dissociation of a quark into a quark and a gluon. A high-Q^2 probe may catch the system in mid-fluctuation.

the virtual dissociation of the boson fields, such as the second-order fluctuations illustrated in Fig. 8-6.

It is therefore plausible to expect, in any interacting field theory, that as Q^2 increases the structure function will fall at large values of x and rise at small values of x, as sketched in Fig. 8-7. It remains for a quantitative

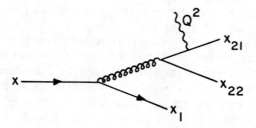

FIG. 8-6. Virtual dissociation of a gluon into a quark and an antiquark, which enhances the population of low-x quarks and antiquarks seen by a probe of high Q^2.

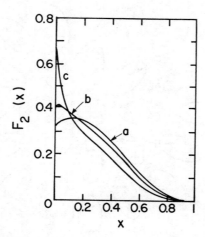

FIG. 8-7. Schematic evolution of the parton structure function in an interacting field theory: (a) low, (b) medium, (c) high Q^2.

analysis to show whether these effects are calculable in a particular field theory, specifically in QCD, using perturbative techniques. In the general case, we have no reason to anticipate that the effects will be small, as they are in Nature. Indeed, the straightforward inference is that structure functions should decrease as inverse powers of Q^2. It therefore would appear that the parton model cannot be accomodated in an interacting field theory. In the next two sections, we shall begin to see why non-Abelian gauge theories—QCD among them—constitute an important, and indeed unique, exceptional case.

8.2   Charge Renormalization in Electrodynamics

In quantum field theories, observables such as scattering amplitudes may be sensitive to higher-order corrections, in addition to Born diagrams. The modifications to lowest-order contributions are in general dependent upon kinematic variables. A convenient way of representing an important class of these modifications is to introduce a so-called "running coupling constant," which is to say an effective coupling strength that depends upon the kinematic circumstances. A phenomenon of this sort is in fact familiar in the classical electrodynamics of a polarizable substance. A test charge placed in a dielectric will polarize the medium as indicated in Fig. 8-8. At any distance r (larger than the molecular scale) from the test charge, the total charge enclosed within a sphere centered on the test charge will be smaller

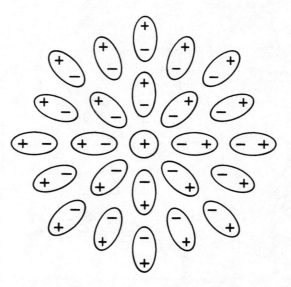

FIG. 8-8. Polarization of a dielectric medium by a test charge.

than the test charge itself, because of the opposite charge attracted by the test charge. By Gauss's law, the effective charge will be smaller in magnitude than the test charge,

$$Q_{\text{eff}} = Q_{\text{test}}/\varepsilon, \qquad (8.2.1)$$

where $\varepsilon \geq 1$ is the dielectric constant of the medium, in the sense that the electric field due to the test charge is related to what the field would have been in vacuo by

$$\mathbf{E}_{\text{medium}} = \mathbf{E}_{\text{vacuo}}/\varepsilon. \qquad (8.2.2)$$

At very short distances, closer to the test charge than the molecular size, screening cannot occur, and the effective charge is equal to the full magnitude of the test charge. Thus we may say that the effective electric charge increases at short distances. In quantum electrodynamics, the vacuum behaves as a polarizable medium in which virtual electron–positron pairs are polarized by the test charge. Consequently the effective charge shows a similar dependence on the distance scale, as we shall now derive.

The lowest-order manifestation of the vacuum polarization phenomenon occurs in the one-loop corrections to Coulomb scattering. For definiteness, we consider the scattering of an electron from an infinitely massive "proton," for which the Born diagram is shown in Fig. 8-9(a). The invariant amplitude is (see the Feynman rules in Appendix B)

$$\mathcal{M} = -ie^2\bar{u}(p')\gamma_\mu u(p)\frac{g^{\mu\nu}}{q^2}\,\bar{U}(P')\gamma_\nu U(P), \qquad (8.2.3)$$

where the momentum transfer is

$$q = p - p' \qquad (8.2.4)$$

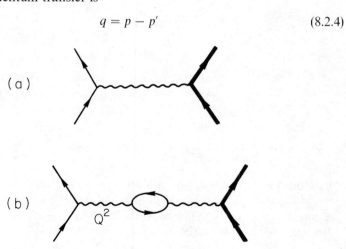

FIG. 8-9. Feynman diagrams for Coulomb scattering. (a) Born term; (b) one-loop correction to the photon propagator.

and U has been used to denote proton spinors, which, in the limit of infinite proton mass ($M \to \infty$), become

$$
\left.\begin{aligned}
U_\lambda(P) &= \sqrt{2M}\begin{pmatrix}\chi_\lambda\\0\end{pmatrix}\\
\bar{U}_{\lambda'}(P') &= \sqrt{2M}(\bar{\chi}_{\lambda'},0)
\end{aligned}\right\}
$$
(8.2.5)

so that

$$
\bar{U}_{\lambda'}(P')\gamma^\nu U_\lambda(P) = 2M\delta_{\lambda\lambda'}\delta^{\nu 0}.
$$
(8.2.6)

The last result follows because the matrices γ mix large and small components. In this limit, the invariant amplitude takes the form

$$
\mathcal{M} = -\frac{ie^2}{q^2}\bar{u}(p')\gamma_0 u(p)\cdot 2M,
$$
(8.2.7)

which is precisely the form of a Coulomb interaction, namely the interaction of the lepton current with a purely timelike potential.

Next we examine the contribution of the one-loop vacuum polarization graph shown in Fig. 8-9(b). The invariant amplitude is given by

$$
\mathcal{M}' = -\mathrm{Tr}\int\frac{d^4k}{(2\pi)^4}\left[(+ie)\bar{u}(p')\gamma_\mu u(p)\frac{(-ig^{\mu\rho})}{q^2}(ie)\gamma_\rho\frac{i(\slashed{k}+m)}{k^2-m^2+i\varepsilon}\right.
$$
$$
\left.\times (ie)\gamma_\sigma\frac{i(\slashed{k}-\slashed{q}+m)}{(q-k)^2-m^2+i\varepsilon}\frac{(-ig^{\nu\sigma})}{q^2}(-ie)\bar{U}(P')\gamma_\nu U(P)\right],
$$
(8.2.8)

corresponding to the replacement in (8.2.3) of $-ig^{\mu\nu}/q^2$ by

$$
\left(\frac{-i}{q^2}\right)I^{\mu\nu}\left(\frac{-i}{q^2}\right),
$$
(8.2.9)

where

$$
I^{\mu\nu} = -\mathrm{Tr}\int\frac{d^4k}{(2\pi)^4}(ie)^2\gamma^\mu\frac{i(\slashed{k}+m)}{k^2-m^2+i\varepsilon}\gamma^\nu\frac{i(\slashed{k}-\slashed{q}+m)}{(q-k)^2-m^2+i\varepsilon}
$$
$$
= e^2\,\mathrm{Tr}\int\frac{d^4k}{(2\pi)^4}\gamma^\mu\frac{i(\slashed{k}+m)}{k^2-m^2+i\varepsilon}\gamma^\nu\frac{i(\slashed{k}-\slashed{q}+m)}{(q-k)^2-m^2+i\varepsilon}.
$$
(8.2.10)

The closed-loop integral diverges for $k \to \infty$, and its evaluation thus will require some cutoff procedure. To proceed further, it is convenient to bring the integrand to an exponential form, in order that gaussian methods may eventually be used. With the aid of the identity

$$
\frac{i(\slashed{k}+m)}{k^2-m^2+i\varepsilon} = (\slashed{k}+m)\int_0^\infty dz\, e^{iz(k^2-m^2+i\varepsilon)},
$$
(8.2.11)

we may write

$$I^{\mu\nu} = e^2 \int \frac{d^4k}{(2\pi)^4} \int_0^\infty dz_1 \int_0^\infty dz_2 \, (\text{Tr}[\gamma^\mu(\not{k} + m)\gamma^\nu(\not{k} - \not{q} + m)]$$

$$\times \exp\{iz_1(k^2 - m^2 + i\varepsilon) + iz_2[(k - q)^2 - m^2 + i\varepsilon]\}) . \quad (8.2.12)$$

A convenient, and gauge-invariant, cutoff procedure is the method of regulators due to Pauli and Villars,[12] which consists in adding to the Lagrangian unphysical fields with adjustable masses, which are taken to infinity at the end of the calculation. The procedure succeeds if renormalized observables are finite and independent of the cutoff masses. In cases such as the one before us, this amounts to the replacement

$$I^{\mu\nu}(q^2, m^2) \to \bar{I}^{\mu\nu}(q^2) = I^{\mu\nu}(q^2, m^2) + \sum_{i=1}^N a_i I^{\mu\nu}(q^2, M_i^2)$$

$$= \sum_{i=0}^N a_i I^{\mu\nu}(q^2, M_i^2), \quad (8.2.13)$$

where the number and coefficients of the regulator terms are chosen to render the loop integral convergent, and evidently $a_0 = 1$, $M_0^2 = m^2$. Evaluation of integrals of the general form of (8.2.12) will be facilitated by completing the square in the exponential. Because the final expression will be convergent, we are free to introduce the new integration variable

$$l = \frac{kz_1 + (k - q)z_2}{z_1 + z_2}, \quad (8.2.14)$$

in terms of which

$$k = l + qz_2/(z_1 + z_2),$$
$$k - q = l - qz_1/(z_1 + z_2). \quad (8.2.15)$$

Carrying out the trace, completing the square in the exponential, and inverting the order of integration, we have

$$I^{\mu\nu}(q^2, m^2) = 4e^2 \int_0^\infty dz_1 \int_0^\infty dz_2$$

$$\times \exp\left\{i\left[\frac{q^2 z_1 z_2}{z_1 + z_2} - (m^2 - i\varepsilon)(z_1 + z_2)\right]\right\} \int \frac{d^4l}{(2\pi)^4}$$

$$\times e^{il^2(z_1 + z_2)} \left(2l^\mu l^\nu - (l^\mu q^\nu + l^\nu q^\mu)\frac{z_1 - z_2}{z_1 + z_2} - 2q^\mu q^\nu \frac{z_1 z_2}{(z_1 + z_2)^2}\right.$$

$$\left. + g^{\mu\nu}\left\{m^2 - \left[l^2 - l \cdot q\frac{z_1 - z_2}{z_1 + z_2} - q^2 \frac{z_1 z_2}{(z_1 + z_2)^2}\right]\right\}\right). \quad (8.2.16)$$

The momentum integrals are conveniently organized according to powers of l as

$$\mathscr{I}_2^{\mu\nu} = \int \frac{d^4 l}{(2\pi)^4} e^{il^2(z_1 + z_2)}(2l^\mu l^\nu - g^{\mu\nu}l^2), \tag{8.2.17}$$

$$\mathscr{I}_1^{\mu\nu} = \int \frac{d^4 l}{(2\pi)^4} e^{il^2(z_1 + z_2)}(g^{\mu\nu}l\cdot q - l^\mu q^\nu - l^\nu q^\mu)\frac{z_1 - z_2}{z_1 + z_2}, \tag{8.2.18}$$

$$\mathscr{I}_0^{\mu\nu} = \int \frac{d^4 l}{(2\pi)^4} e^{il^2(z_1 + z_2)}\left[(g^{\mu\nu}q^2 - 2q^\mu q^\nu)\frac{z_1 z_2}{(z_1 + z_2)^2} + g^{\mu\nu}m^2 \right]. \tag{8.2.19}$$

On symmetric integration [cf. equation (B.3.11a, 17)] $\mathscr{I}_1^{\mu\nu}$ vanishes and the other pieces become

$$\mathscr{I}_2^{\mu\nu} = \frac{1}{16\pi^2 i(z_1 + z_2)^2} \cdot \frac{ig^{\mu\nu}}{2(z_1 + z_2)}(2 - 4)$$

$$= -g^{\mu\nu}/16\pi^2(z_1 + z_2)^3 \tag{8.2.20}$$

and

$$\mathscr{I}_0^{\mu\nu} = \frac{1}{16\pi^2 i(z_1 + z_2)^2} \left\{ 2(g^{\mu\nu}q^2 - q^\mu q^\nu)\frac{z_1 z_2}{(z_1 + z_2)^2} + g^{\mu\nu}\left[m^2 - \frac{q^2 z_1 z_2}{(z_1 + z_2)^2} \right] \right\}. \tag{8.2.21}$$

We therefore obtain

$$I^{\mu\nu}(q^2, m^2) = \frac{-i\alpha}{\pi} \int_0^\infty dz_1 \int_0^\infty dz_2 \frac{1}{(z_1 + z_2)^2} \exp\left\{ i\left[\frac{q^2 z_1 z_2}{z_1 + z_2} - (m^2 - i\varepsilon)(z_1 + z_2) \right] \right\}$$

$$\times \left\{ 2(g^{\mu\nu}q^2 - q^\mu q^\nu)\frac{z_1 z_2}{(z_1 + z_2)^2} + g^{\mu\nu}\left[m^2 - \frac{q^2 z_1 z_2}{(z_1 + z_2)^2} - \frac{i}{z_1 + z_2} \right] \right\}. \tag{8.2.22}$$

The first term in braces has been constructed to satisfy the current conservation requirement

$$q_\mu I^{\mu\nu} = 0 = q_\nu I^{\mu\nu}. \tag{8.2.23}$$

The term proportional to $g^{\mu\nu}$ does not obey this condition but can readily be shown to vanish identically, if we note that its contribution to $I^{\mu\nu}(q^2)$ may be rewritten as

$$\Delta^{\mu\nu} = \frac{-i\alpha}{\pi} g^{\mu\nu} \int_0^\infty dz_1 \int_0^\infty dz_2 \frac{1}{(z_1 + z_2)^3} i\sigma \frac{\partial}{\partial\sigma} \frac{1}{\sigma}$$

$$\times \sum_{i=0}^N a_i \exp\left\{ i\sigma\left[\frac{q^2 z_1 z_2}{z_1 + z_2} - (M_i^2 - i\varepsilon)(z_1 + z_2) \right] \right\}_{\sigma = 1}. \tag{8.2.24}$$

Because the integrals are convergent, we may interchange the order of the integrations and the differentiation $\partial/\partial\sigma$, so that

$$\Delta^{\mu\nu} = \frac{\alpha}{\pi} g^{\mu\nu} \sigma \frac{\partial}{\partial\sigma} \left(\int_0^\infty dz_1 \int_0^\infty dz_2 \frac{1}{(z_1 + z_2)^3 \sigma} \right.$$

$$\left. \times \sum_{i=0}^N a_i \exp\left\{ i\sigma\left[\frac{q^2 z_1 z_2}{z_1 + z_2} - (M_i^2 - i\varepsilon)(z_1 + z_2) \right] \right\} \right)_{\sigma=1} \quad (8.2.25)$$

If we now rescale the variables of integration as

$$z_i \to z_i/\sigma, \quad (8.2.26)$$

the expression in bold-face parentheses is seen to be independent of σ, and so the contribution $\Delta^{\mu\nu}$ vanishes.

We are left with the task of evaluating

$$\bar{I}^{\mu\nu}(q^2) = -\frac{2i\alpha}{\pi}(g^{\mu\nu}q^2 - q^\mu q^\nu) \int_0^\infty dz_1 \int_0^\infty dz_2 \frac{z_1 z_2}{(z_1 + z_2)^4}$$

$$\times \sum_{i=0}^N a_i \exp\left\{ i\left[\frac{q^2 z_1 z_2}{z_1 + z_2} - (M_i^2 - i\varepsilon)(z_1 + z_2) \right] \right\}. \quad (8.2.27)$$

This is conveniently accomplished by rescaling the integration variables by introducing the factor

$$1 = \int_0^\infty d\sigma \, \delta(\sigma - z_1 - z_2), \quad (8.2.28)$$

under the integral signs and letting $z_i \to \sigma z_i$, so that

$$\bar{I}^{\mu\nu}(q^2) = \frac{2i\alpha}{\pi}(q^\mu q^\nu - g^{\mu\nu}q^2) \int_0^1 dz_1 \int_0^1 dz_2 \, z_1 z_2 \, \delta(1 - z_1 - z_2)$$

$$\times \int_0^\infty \frac{d\sigma}{\sigma} \sum_{i=0}^N a_i \exp[i\sigma(q^2 z_1 z_2 - M_i^2 + i\varepsilon)], \quad (8.2.29)$$

which appears logarithmically divergent at the point $\sigma = 0$. This mild divergence can be removed with a single regulator, by the choice

$$a_1 = -1, \qquad M_1^2 = \Lambda^2 \gg m^2, q^2, \quad (8.2.30)$$

whereupon for $q^2 < 4m^2$

$$\bar{I}^{\mu\nu}(q^2) = I^{\mu\nu}(q^2, m^2) - I^{\mu\nu}(q^2, \Lambda^2)$$

$$= \frac{2i\alpha}{\pi}(q^\mu q^\nu - g^{\mu\nu}q^2) \int_0^1 dz \, z(1-z) \log\left[\frac{\Lambda^2}{m^2 - q^2 z(1-z)} \right]$$

$$= \frac{i\alpha}{3\pi}(q^\mu q^\nu - g^{\mu\nu}q^2)\left\{ \log\left(\frac{\Lambda^2}{m^2}\right) \right.$$

$$\left. - 6\int_0^1 dz \, z(1-z) \log\left[1 - \frac{q^2 z(1-z)}{m^2} \right] \right\}. \quad (8.2.31)$$

Returning to the problem of original interest, the scattering of an electron in a Coulomb field, we note that the $q^\mu q^\nu$ factor does not contribute to the scattering amplitude by virtue of current conservation at the electron vertex, or, more operationally, because of the Dirac equation. Hence the modified photon propagator is

$$\frac{-ig^{\mu\nu}}{q^2} + \left(\frac{-i}{q^2}\right)\overline{I}^{\mu\nu}\left(\frac{-i}{q^2}\right) = \frac{-ig^{\mu\nu}}{q^2}\left\{1 - \frac{\alpha}{3\pi}\log\left(\frac{\Lambda^2}{m^2}\right)\right.$$
$$\left. + \frac{2\alpha}{\pi}\int_0^1 dz\, z(1-z)\log\left[1 - \frac{q^2 z(1-z)}{m^2}\right]\right\}.$$

$$(8.2.32)$$

A number of limiting cases are worth comment. For $q^2 \to 0$, the long-wavelength limit, the propagator is modified by the factor

$$1 - \frac{\alpha}{3\pi}\log\left(\frac{\Lambda^2}{m^2}\right) \equiv Z_3, \qquad (8.2.33)$$

so that the Coulomb scattering amplitude is changed by the same factor. In other words, the electric charge measured in Coulomb scattering is not the "bare" charge e that appears in the Lagrangian, but a renormalized quantity e_R that differs by a factor $\sqrt{Z_3}$. The term $(-\alpha/3\pi)\log(\Lambda^2/m^2)$ is but the first term in the renormalization of the electric charge, which can be carried out systematically to all orders in perturbation theory. We accommodate this by replacing the bare charge e by the renormalized charge e_R in all our expressions for observables, though it will usually be convenient to omit the subscript. It is thus the renormalized charge that is measured in experiments as $e_R^2 \simeq 4\pi/137$.

Of greater interest to us for what follows is the modification to the Coulomb scattering amplitude for large spacelike values of q^2, such that $-q^2 \gg m^2$. In this circumstance the remaining integral in (8.2.32) may be approximated by

$$\frac{2\alpha}{\pi}\int_0^1 dz\, z(1-z)\log\left(\frac{-q^2}{m^2}\right) = \frac{\alpha}{3\pi}\log\left(\frac{-q^2}{m^2}\right). \qquad (8.2.34)$$

As a result, the photon propagator is given by

$$\frac{-ig^{\mu\nu}}{q^2}\left[1 - \frac{\alpha}{3\pi}\log\left(\frac{\Lambda^2}{m^2}\right) + \frac{\alpha}{3\pi}\log\left(\frac{-q^2}{m^2}\right)\right], \qquad (8.2.35)$$

which amounts to the replacement of α_R in the scattering amplitude by the "running coupling constant"

$$\alpha_R(q^2) = \alpha_R(m^2)\left[1 + \frac{\alpha_R(m^2)}{3\pi}\log\left(\frac{-q^2}{m^2}\right) + O(\alpha_R^2)\right]. \qquad (8.2.36)$$

We see that to this approximation the effective electric charge increases logarithmically with $-q^2$, which is to say with improving resolution. This is the quantum-mechanical analog of the macroscopic charge screening discussed at the beginning of this section.

For small values of $-q^2 \ll m^2$, it is straightforward (see Problem 8-8) to derive the modification to Coulomb's law in position space, which results from vacuum polarization. The additional term in the force law first calculated by Uehling[13] in 1935 induces a shift in the s-wave energy levels of atoms that is especially significant for high-Z muonic atoms.

It is possible to continue the study of modifications to Coulomb's law in higher orders in perturbation theory. For this purpose, the most important class of Feynman diagrams is the "sum of bubbles" indicated in Fig. 8-10, which correspond to the most divergent set of logarithms. It does not require detailed calculation to see that for $-q^2 \gg m^2$ the running coupling constant will be given by

$$
\begin{aligned}
\alpha_R(q^2) &= \alpha_R(m^2)\left\{1 + \frac{\alpha_R(m^2)}{3\pi}\log\left(\frac{-q^2}{m^2}\right) + \left[\frac{\alpha_R(m^2)}{3\pi}\log\left(\frac{-q^2}{m^2}\right)\right]^2 + \cdots\right\} \\
&\rightarrow \frac{\alpha_R(m^2)}{\left[1 - \dfrac{\alpha_R(m^2)}{3\pi}\log\left(\dfrac{-q^2}{m^2}\right)\right]},
\end{aligned}
\tag{8.2.37}
$$

where we have identified

$$
\sum_{n=0}^{\infty} x^n = 1/(1 - x),
\tag{8.2.38}
$$

without worrying about convergence issues. To this approximation, it is convenient to write

$$
1/\alpha_R(q^2) = 1/\alpha_R(m^2) - (1/3\pi)\log(-q^2/m^2),
\tag{8.2.39}
$$

FIG. 8-10. Leading logarithmic corrections to the photon propagator in spinor electrodynamics.

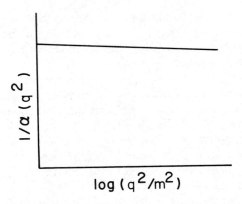

FIG. 8-11. Evolution of the running coupling constant in QED.

and to represent the q^2-evolution of the coupling constant by the graph shown in Fig. 8-11.

This brief calculation has shown how a running coupling constant may arise in an interacting field theory. We have not, in this limited discussion, shown how the renormalization program can be carried through in detail, nor have we justified the implicit claim of the foregoing discussion that the effects of vacuum polarization can be systematically accounted for by replacing α by $\alpha(q^2)$ as computed from the bubble sum of the photon propagator. These issues are well treated in the standard textbooks on QED, and the reader who has not yet experienced the renormalization program is counseled to do so, pen in hand, although it will not be required for what follows.

Although the principal interest in this chapter is QCD, and among other things the comportment of the running coupling constant in that theory, it will be useful to spend a little more time on electrodynamics. We consider the evolution of the running coupling constant in scalar electrodynamics, both because the result is interesting and instructive, and also in order to introduce a second renormalization technique that is better suited to non-Abelian gauge theories than the method of Pauli and Villars.

Applying the Feynman rules for scalar electrodynamics given in Appendix B.5, we find that the lowest-order corrections to the photon propagator are the two diagrams shown in Fig. 8-12. In the same notation employed previously in this section, we wish to evaluate

$$I^{\mu\nu}(q^2, m^2) = e^2 \int \frac{d^4k}{(2\pi)^4} \left\{ \frac{(2k-q)^\mu (2k-q)^\nu}{(k^2 - m^2 + i\varepsilon)[(k-q)^2 - m^2 + i\varepsilon]} - \frac{2g^{\mu\nu}}{k^2 - m^2 + i\varepsilon} \right\},$$

$$(8.2.40)$$

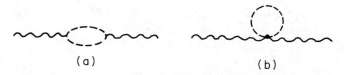

FIG. 8-12. One-loop corrections to the photon propagator in scalar electrodynamics.

where the first term corresponds to the bubble of Fig. 8-12(a) and the second to the contact term of Fig. 8-12(b). We combine the terms over a common denominator as

$$I^{\mu\nu}(q^2, m^2) = e^2 \int \frac{d^4k}{(2\pi)^4} \frac{(2k-q)^\mu (2k-q)^\nu - 2g^{\mu\nu}[(k-q)^2 - m^2 + i\varepsilon]}{(k^2 - m^2 + i\varepsilon)[(k-q)^2 - m^2 + i\varepsilon]},$$

(8.2.41)

and promote the denominators to an exponent by means of the identity

$$i(k^2 - m^2 + i\varepsilon)^{-1} = \int_0^\infty dz\, e^{iz(k^2 - m^2 + i\varepsilon)}.$$

(8.2.42)

The result is

$$I^{\mu\nu}(q^2, m^2) = -e^2 \int \frac{d^4k}{(2\pi)^4} \int_0^\infty dz_1 \int_0^\infty dz_2 \{(2k-q)^\mu (2k-q)^\nu$$

$$- 2g^{\mu\nu}[(k-q)^2 - m^2]\}$$

$$\times \exp\{iz_1(k^2 - m^2 + i\varepsilon) + iz_2[(k-q)^2 - m^2 + i\varepsilon]\}. \quad (8.2.43)$$

We complete the square in the exponential on introducing the new variable of integration

$$l = \frac{kz_1 + (k-q)z_2}{z_1 + z_2},$$

(8.2.14)

and regularize the integral by changing the dimensionality of space-time to n. We are left with

$$I^{\mu\nu}(q^2, m^2) = -e^2 \int_0^\infty dz_1 \int_0^\infty dz_2 \exp\left\{i\left[\frac{q^2 z_1 z_2}{z_1 + z_2} - (m^2 - i\varepsilon)(z_1 + z_2)\right]\right\}$$

$$\times \int \frac{d^n l}{(2\pi)^n} e^{il^2(z_1 + z_2)} \left\{4l^\mu l^\nu + \frac{(z_1 - z_2)^2}{(z_1 + z_2)^2} q^\mu q^\nu\right.$$

$$\left. - 2g^{\mu\nu}\left[l^2 - m^2 + \frac{q^2 z_1^2}{(z_1 + z_2)^2}\right]\right\},$$

(8.2.44)

where odd powers of l have been omitted in anticipation of symmetric integration and the order of integrations has been interchanged. As we did for spinor QED, we may rearrange the momentum integral as

$$\int \frac{d^n l}{(2\pi)^n} e^{il^2(z_1+z_2)} \left(\frac{(z_1-z_2)^2}{(z_1+z_2)^2}(q^\mu q^\nu - g^{\mu\nu}q^2) \right.$$
$$\left. + 4l^\mu l^\nu - 2g^{\mu\nu} \left\{ l^2 - m^2 + q^2 \frac{[z_1^2-(z_1-z_2)^2/2]}{(z_1+z_2)^2} \right\} \right). \quad (8.2.45)$$

The first term in bold-face parentheses satisfies the requirements of current conservation, and we shall evaluate its contribution presently. First we show by the 't Hooft–Veltman method of dimensional regulation[14] that the remaining terms do not in fact contribute. Using the expressions for n-dimensional gaussian integrals given in Appendix B.4, we may write

$$\int \frac{d^n l}{(2\pi)^n} e^{il^2(z_1+z_2)} \left(4l^\mu l^\nu - 2l^2 g^{\mu\nu} + g^{\mu\nu} \left\{ 2m^2 - \frac{2[z_1^2-(z_1-z_2)^2/2]q^2}{(z_1+z_2)^2} \right\} \right)$$
$$= \frac{ie^{-i\pi n/4}g^{\mu\nu}}{[4\pi(z_1+z_2)]^{n/2}} \left\{ \frac{in(4/n-2)}{2(z_1+z_2)} + 2m^2 - \frac{2[z_1^2-(z_1-z_2)^2/2]}{(z_1+z_2)^2}q^2 \right\}. \quad (8.2.46)$$

The contribution to $I^{\mu\nu}$ is therefore

$$\Delta^{\mu\nu} = \frac{ie^2 e^{-i\pi n/4}}{(4\pi)^{n/2}} g^{\mu\nu} \int_0^\infty dz_1 \int_0^\infty dz_2 \exp\left\{ i\left[\frac{q^2 z_1 z_2}{z_1+z_2} - (m^2-i\varepsilon)(z_1+z_2) \right] \right\}$$
$$\times \frac{2}{(z_1+z_2)^{1+n/2}} \left\{ i\left(1-\frac{n}{2}\right) + m^2(z_1+z_2) - \frac{[z_1^2-(z_1-z_2)^2/2]q^2}{z_1+z_2} \right\}. \quad (8.2.47)$$

Introducing a scaling parameter σ as before, we have

$$\Delta^{\mu\nu} = \frac{-2e^2 e^{-i\pi n/4}}{(4\pi)^{n/2}} g^{\mu\nu} \int_0^1 dz_1 \int_0^1 dz_2\, \delta(1-z_1-z_2) \int_0^\infty \frac{d\sigma}{\sigma^{n/2-1}}$$
$$\times \exp[i\sigma(q^2 z_1 z_2 - m^2 + i\varepsilon)] \times \left[\frac{(1-n/2)}{\sigma} + i(z_1 z_2 q^2 - m^2) \right], \quad (8.2.48)$$

in which, in anticipation of the restrictions on the z_1 and z_2 integrations, we have replaced $z_1 + z_2$ by 1 and dropped terms that are odd in $(1-2z_i)$. The integral over σ may be expressed as

$$\int_0^\infty d\sigma \frac{\partial}{\partial \sigma} \{\sigma^{1-n/2} \exp[i\sigma(q^2 z_1 z_2 - m^2 + i\varepsilon)]\} \quad (8.2.49)$$

which is convergent and equal to zero for sufficiently small values of n, namely $n < 2$. Thus we obtain by dimensional regularization the anticipated gauge-invariant result,

$$\Delta^{\mu\nu} = 0. \tag{8.2.50}$$

Let us return to the gauge-invariant contribution. We have

$$I^{\mu\nu} = \frac{ie^2 e^{-i\pi n/4}}{(4\pi)^{n/2}} (q^\mu q^\nu - q^2 g^{\mu\nu}) \int_0^\infty dz_1 \int_0^\infty dz_2$$

$$\times \exp\left\{i\left[\frac{q^2 z_1 z_2}{z_1 + z_2} - (m^2 - i\varepsilon)(z_1 + z_2)\right]\right\} \times \frac{(z_1 - z_2)^2}{(z_1 + z_2)^{2 + n/2}}$$

$$= \frac{ie^2 e^{-i\pi n/4}}{(4\pi)^{n/2}} (q^\mu q^\nu - q^2 g^{\mu\nu}) \int_0^1 dz_1 \int_0^1 dz_2 \, \delta(1 - z_1 - z_2) \int_0^\infty \frac{d\sigma}{\sigma^{n/2 - 1}}$$

$$\times \exp[i\sigma(q^2 z_1 z_2 - m^2 + i\varepsilon)](z_1 - z_2)^2$$

$$= \frac{ie^2 e^{-i n\pi/4}}{(4\pi)^{n/2}} (q^\mu q^\nu - q^2 g^{\mu\nu}) \int_0^1 dz(1 - 2z)^2$$

$$\times \int_0^\infty \frac{d\sigma}{\sigma^{n/2 - 1}} \exp\{i\sigma[q^2 z(1 - z) - m^2 + i\varepsilon]\}. \tag{8.2.51}$$

The σ-integration produces a gamma function, so that finally we have

$$I^{\mu\nu} = \frac{ie^2 e^{-i\pi n/4}}{(4\pi)^{n/2}} (q^\mu q^\nu - q^2 g^{\mu\nu})\Gamma\left(2 - \frac{n}{2}\right)\int_0^1 \frac{dz(1 - 2z)^2}{[q^2 z(1 - z) - m^2 + i\varepsilon]^{2 - n/2}}, \tag{8.2.52}$$

which has simple poles at $n = 4, 6, 8, \ldots$. To regularize the integral, we subtract from (8.2.52) the pole at $n = 4$ with its residue, given by

$$I^{\mu\nu}_{\text{pole}} = \frac{-ie^2}{16\pi^2} (q^\mu q^\nu - q^2 g^{\mu\nu})\frac{2}{4 - n} \int_0^1 dz(1 - 2z)^2. \tag{8.2.53}$$

Subtracting the pole term, and taking the limit $n \to 4$ in the regularized expression, we obtain

$$\bar{I}^{\mu\nu} = \lim_{n \to 4} (I^{\mu\nu} - I^{\mu\nu}_{\text{pole}}) = \frac{-ie^2}{16\pi^2} (q^\mu q^\nu - q^2 g^{\mu\nu})$$

$$\times \left\{\int_0^1 dz(1 - 2z)^2 \ln\left[1 - \frac{q^2 z(1 - z)}{m^2}\right] + K\right\}, \tag{8.2.54}$$

where the constant K arises from the n-dependence other than in the denominator of the integrand in (8.2.52). It is arbitrary in the same way as the Pauli–Villars cutoff is arbitrary, and contributes to the overall charge renormalization.

Let us compare the q^2-evolution of the coupling constant with the result we found in spinor electrodynamics. For $-q^2 \gg m^2$, we have that

$$
\begin{aligned}
\bar{I}^{\mu\nu} &= \frac{-i\alpha}{4\pi}(q^\mu q^\nu - q^2 g^{\mu\nu})\left[K + \log\left(\frac{-q^2}{m^2}\right)\int_0^1 dz(1-2z)^2 \right] \\
&= \frac{-i\alpha}{4\pi}(q^\mu q^\nu - q^2 g^{\mu\nu})\left[K + \log\left(\frac{-q^2}{m^2}\right)\left(\frac{1}{3}\right) \right].
\end{aligned}
\tag{8.2.55}
$$

Forming the photon propagator as in (8.2.32) and using current conservation at the scalar vertex, we find for the running renormalized coupling constant in scalar electrodynamics

$$
\alpha_R(q^2) = \alpha_R(m^2)\left[1 + \frac{\alpha_R(m^2)}{12\pi}\log\left(\frac{-q^2}{m^2}\right) + \cdots \right],
\tag{8.2.56}
$$

which is to be compared with the expression (8.2.36) obtained in spinor electrodynamics. The rate of change of the coupling constant in scalar electrodynamics is one-fourth of that in spinor electrodynamics. This is precisely the ratio of the cross section for the reactions $e^+e^- \to \sigma^+\sigma^-$ and $e^+e^- \to \mu^+\mu^-$ evaluated in Problems 1-4 and 1-5, which are directly related to the same polarization tensor.

It is of some importance that, whereas the variation in the effective coupling constant is logarithmically divergent, the quantity $\partial\alpha(q^2)/\partial(\log q^2)$ is finite. This tame behavior, when combined with the observation that the effects of renormalization are multiplicative in QED, permits the introduction of the powerful general techniques of the renormalization group.

We now proceed to investigate the running coupling constant in QCD. This will lead us to the notion of asymptotic freedom and the prospect of reliable perturbative calculations.

8.3   The Running Coupling Constant in QCD

In order to discuss the effects of higher-order corrections in QCD, we must first develop the Feynman rules for the theory in a way that leads to consistent results for diagrams containing closed gluon loops. There is no particular difficulty in evaluating the multigluon vertex factors shown[15] in Fig. 8-13, but there are some technical matters to be confronted. These arise from the familiar problem in gauge theories that the gauge-invariant Lagrangian does not uniquely determine the gauge field in terms of a source. In our treatment of electrodynamics in Section 3.6 we resolved this ambiguity in the definition of the photon propagator by imposing the covariant gauge condition

$$
\partial_\mu A^\mu = 0.
\tag{8.3.1}
$$

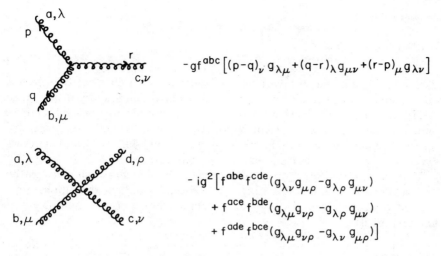

$$-gf^{abc}\left[(p-q)_\nu\, g_{\lambda\mu}+(q-r)_\lambda g_{\mu\nu}+(r-p)_\mu g_{\lambda\nu}\right]$$

$$-ig^2\left[f^{abe}f^{cde}(g_{\lambda\nu}g_{\mu\rho}-g_{\lambda\rho}g_{\mu\nu})\right.$$
$$+f^{ace}f^{bde}(g_{\lambda\mu}g_{\nu\rho}-g_{\lambda\rho}g_{\mu\nu})$$
$$\left.+f^{ade}f^{bce}(g_{\lambda\mu}g_{\nu\rho}-g_{\lambda\nu}g_{\mu\rho})\right]$$

FIG. 8-13. Feynman rules for multigluon interactions in non-Abelian gauge theories.

The resulting form of the photon propagator

$$-ig^{\mu\nu}/(q^2+i\varepsilon)\tag{8.3.2}$$

and associated Feynman rules are convenient for many, but by no means all, computations in electrodynamics. Equivalently, in the case of electrodynamics, we may modify the source-free Lagrangian by adding to it a "gauge-fixing" term $(-1/2\xi)(\partial_\mu A^\mu)^2$, so that

$$\mathscr{L}_{\text{photon}}=-\tfrac{1}{4}F_{\mu\nu}F^{\mu\nu}-(1/2\xi)(\partial_\mu A^\mu)^2.\tag{8.3.3}$$

The Euler–Lagrange equations then lead to the equations of motion

$$\square A^\nu-(1-\xi^{-1})\partial^\nu(\partial_\mu A^\mu)=J^\nu,\tag{8.3.4}$$

from which the photon propagator is

$$\frac{-i[g^{\mu\nu}+(\xi-1)q^\mu q^\nu/(q^2+i\varepsilon)]}{q^2+i\varepsilon}.\tag{8.3.5}$$

This is entirely parallel to the gauge-fixing procedure for spontaneously broken gauge theories such as the Abelian Higgs model mentioned in Section 5.3. Again we recognize the well-known special cases of the Feynman (or Lorentz) gauge, for $\xi=1$, and the Landau gauge for $\xi=0$. The parameter ξ is a reminder of the arbitrariness of the gauge-fixing condition. It must be absent from any physical quantity we calculate.

It is important to remark that the Lagrangian (8.3.3) remains gauge-invariant provided that the local phase rotation that generates the gauge transformation

$$A_\mu(x) \to A_\mu(x) - \partial_\mu\alpha(x) \qquad (3.5.14)$$

satisfies the condition

$$\Box\alpha(x) = 0. \qquad (8.3.6)$$

Maintaining the gauge invariance is important in order that the physical requirements of local current conservation and transversality of the gauge fields remain in force. The covariant gauges are not the only possible choices, nor are they the most convenient for every sort of calculation. They will prove adequate to our limited needs, however, and will lead us to acknowledge an important technical issue, as we shall now see.

The preceding discussion would seem to apply, *mutatis mutandis*, to the propagator for any massless gauge field. In particular for the gluon propagator we need only add a Kronecker delta in color space, so that the gluon propagator is

$$\frac{-i\delta_{jk}[g^{\mu\nu} + (\xi - 1)q^\mu q^\nu/(q^2 + i\varepsilon)]}{q^2 + i\varepsilon}. \qquad (8.3.7)$$

The use of this form indeed leads to consistent results for tree diagrams. Because of the non-Abelian nature of the color gauge transformations

$$b_\mu^l \to b_\mu'^l = b_\mu^l - (1/g)\partial_\mu\alpha^l - f_{jkl}\alpha^j b^k \qquad (8.3.8)$$

upon the gluon field, it is no longer possible to find a condition analogous to (8.3.6) that preserves the gauge-invariant appearance of the Lagrangian. For diagrams involving closed gluon loops, this leads to problems that, at the technical level of this volume, are best encountered by direct computation. In this way we shall also see how the general cure solves our specific problems.

Consider the one-loop modifications to the gluon propagator, the analog to the calculation discussed in Section 8.2 for the photon. In addition to the quark-loop diagram of Fig. 8-14(a), there is the tadpole term of Fig. 8-14(b) arising from the four-gluon vertex, and the gluon loop shown in Fig. 8-14(c). We work in the dimensional regularization scheme to maintain gauge invariance. The quark-loop diagram may be evaluated directly from the corresponding QED expressions (8.2.31, 34) simply by replacing

$$e^2 \to g^2 \, \text{tr}\left(\frac{\lambda^a}{2}\frac{\lambda^b}{2}\right) = g^2\frac{\delta^{ab}}{2} \qquad (8.3.9)$$

for each quark flavor, to account for the difference in Feynman rules. In the short-wavelength limit, we find for the logarithmically divergent part

$$\bar{I}^{\mu\nu}_{\text{quarks}}(q^2) = \frac{-ig^2}{16\pi^2}(q^\mu q^\nu - q^2 g^{\mu\nu}) \cdot \frac{4}{3}\log\left(\frac{-q^2}{\mu^2}\right)n_f\frac{\delta^{ab}}{2}, \qquad (8.3.10)$$

(a)

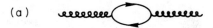

(b)

(c)

FIG. 8-14. One-loop corrections to the gluon propagator in QCD.

where n_f is the number of quark flavors appearing in the loops and $-\mu^2$ is a spacelike renormalization point.

Within the dimensional regularization scheme, massless tadpole graphs do not contribute.[14] Hence we have

$$\bar{I}^{\mu\nu}{}_{\text{tadpole}}(q^2) = 0. \tag{8.3.11}$$

Evaluation of the gluon loop is algebraically tedious, but entirely parallel to the methods used for scalar electrodynamics in the previous section. The result is

$$\bar{I}^{\mu\nu}{}_{\text{gluons}}(q^2) = \frac{ig^2}{16\pi^2} f^{acd} f^{bcd} \left[\frac{11}{6} q^\mu q^\nu - \frac{19}{12} q^2 g^{\mu\nu} \right.$$
$$\left. + \frac{(1-\xi)}{2} (q^\mu q^\nu - q^2 g^{\mu\nu}) \right] \log\left(\frac{-q^2}{\mu^2} \right), \tag{8.3.12}$$

which is evidently not transverse—i.e., not proportional to $(q^\mu q^\nu - q^2 g^{\mu\nu})$. This nontransversality is a reflection of the fact that the gauge-fixing term has interfered with the gauge invariance of the theory. A precursor of this effect is to be found in the results of Problem 8-5, where the reaction $q\bar{q} \to gg$ is considered. There it will be seen that, in contrast to electrodynamics, the current conservation conditions $k_{1\nu} T^{\mu\nu} = 0 = k_{2\mu} T^{\mu\nu}$, where $\varepsilon^*_{1\nu} \varepsilon^*_{2\mu} T^{\mu\nu}$ is the total amplitude corresponding to the tree diagrams, are satisfied only if the gluons are themselves transverse—i.e., on-mass-shell. This is a much weaker guarantee than exists in electrodynamics and suggests the delicacy of the comportment of diagrams with loops.

A systematic resolution of this problem is beyond the elementary means we have chosen to employ in this book. We shall simply state that such a

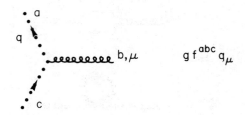

$$g\,f^{abc}\,q_\mu$$

FIG. 8-15. Feynman rule for the gluon–ghost–ghost interaction in QCD.

resolution exists and that it may be characterized by the introduction of a color octet of fictional scalar particles commonly known as Faddeev–Popov ghosts, which appear only in closed loops. These particles are known as ghosts because, although spinless, they obey Fermi statistics and thus each closed ghost loop appears multiplied by an explicit minus sign, like a fermion loop. They have the usual propagator for massless scalars, and their interactions with gluons are given by the Feynman rule shown in Fig. 8-15.

It is now straightforward (Problem 8-9) to evaluate the modification to the gluon propagator given by the ghost loop shown in Fig. 8-16, using the method of dimensional regularization. It is

$$\bar{I}^{\mu\nu}_{\text{ghosts}}(q^2) = \frac{-ig^2}{16\pi^2} f^{acd} f^{bcd} \left(\frac{1}{6} q^\mu q^\nu + \frac{1}{12} q^2 g^{\mu\nu} \right) \log\left(\frac{-q^2}{\mu^2} \right), \quad (8.3.13)$$

which is not transverse. The sum of gluon and ghost loop contributions,

$$\bar{I}^{\mu\nu}_{\text{gluons}}(q^2) + \bar{I}^{\mu\nu}_{\text{ghosts}}(q^2) = \frac{ig^2}{16\pi^2} f^{acd} f^{bcd} \left[\frac{5}{3} + \frac{(1 - \xi)}{2} \right]$$

$$\times (q^\mu q^\nu - q^2 g^{\mu\nu}) \log\left(\frac{-q^2}{\mu^2} \right), \quad (8.3.14)$$

is, however, transverse, as required. Thus we have verified that in this simplest case the Faddeev–Popov ghosts have performed the task for which they were introduced: the removal of unphysical polarizations from the contributions of gluon loops.

FIG. 8-16. Ghost-loop correction to the gluon propagator in QCD.

The modification to the gluon propagator in one-loop order is now given by

$$\bar{I}^{\mu\nu}(q^2) = \bar{I}^{\mu\nu}_{\text{quarks}}(q^2) + \bar{I}^{\mu\nu}_{\text{gluons}}(q^2) + \bar{I}^{\mu\nu}_{\text{ghosts}}(q^2)$$

$$= \frac{ig^2}{16\pi^2} \delta^{ab} \log\left(\frac{-q^2}{\mu^2}\right)(q^\mu q^\nu - q^2 g^{\mu\nu})\left\{ N\left[\frac{5}{3} + \frac{(1-\xi)}{2}\right] - \frac{2n_f}{3}\right\}, \quad (8.3.15)$$

which is explicitly gauge-dependent, and thus cannot alone describe the q^2-dependent evolution of observable quantities. In passing to (8.3.15) we have simplified the color factor for the boson loops using

$$f^{acd}f^{bcd} = N\delta^{ab}, \qquad (8.3.16)$$

where N is the number of colors—i.e., the dimension of the fundamental representation of the $SU(N)_{\text{color}}$ gauge group. It remains to evaluate the one-loop corrections to the quark propagator, shown in Fig. 8-17(a), and the one-loop vertex correction, drawn in Fig. 8-17(b). The modified quark propagator may be taken over directly from electrodynamics, upon inserting the required color factor. It is

$$G(p) = \frac{i\not{p}}{p^2}\left[\delta_{\alpha\beta} - \xi\frac{g^2}{16\pi^2}\log\left(\frac{-p^2}{\mu^2}\right)\sum_a \frac{\lambda^a_{\alpha\gamma}\lambda^a_{\gamma\beta}}{4}\right], \qquad (8.3.17)$$

where α, β, γ are color indices of the quarks, and we may replace

$$\frac{1}{4}\sum_a \lambda^a_{\alpha\gamma}\lambda^a_{\gamma\beta} = \frac{N^2 - 1}{2N}\delta_{\alpha\beta} \qquad (8.3.18)$$

to obtain

$$G(p) = \frac{i\not{p}}{p^2}\delta_{\alpha\beta}\left[1 - \xi\frac{(N^2 - 1)}{2N}\frac{g^2}{16\pi^2}\log\left(\frac{-p^2}{\mu^2}\right)\right]. \qquad (8.3.19)$$

(a)

(b)

FIG. 8-17. One-loop corrections to (a) the fermion propagator; (b) the quark–antiquark–gluon vertex in QCD.

The vertex modification includes a piece familiar from electrodynamics plus one involving the three-gluon vertex. The one-loop corrected vertex is

$$\Gamma_\mu^{\alpha\beta;a}(q^2) = -ig\gamma_\mu\left(\frac{\lambda_{\alpha\beta}^a}{2} - \frac{g^2}{16\pi^2}\log\left(\frac{-q^2}{\mu^2}\right)\right.$$

$$\left.\times\left\{\xi\frac{(N^2-1)}{2N}\frac{\lambda_{\alpha\beta}^a}{2} - \frac{i}{2}\left[1 - \frac{(1-\xi)}{4}\right]f^{abc}\lambda_{\alpha\gamma}^b\lambda_{\gamma\beta}^c\right\}\right) \qquad (8.3.20)$$

which may be simplified using the identity

$$f^{abc}\lambda^b\lambda^c = iN\lambda^a, \qquad (8.3.21)$$

whereupon

$$\Gamma_\mu^{\alpha\beta;a}(q^2) = -ig\frac{\lambda_{\alpha\beta}^a}{2}\gamma_\mu\left(1 - \frac{g^2}{16\pi^2}\log\left(\frac{-q^2}{\mu^2}\right)\left\{\xi\frac{(N^2-1)}{2N} + \left[1 - \frac{(1-\xi)}{4}\right]N\right\}\right).$$
$$(8.3.22)$$

Note that the QED results may be recovered by setting

$$\left.\begin{array}{r} (N^2-1)/2N \to 1 \\ N \to 0 \\ g^2/4\pi \to \alpha \end{array}\right\}, \qquad (8.3.23)$$

To assemble the pieces for a physical process, it is convenient to work in Landau gauge ($\xi = 0$), for which there is no wave function renormalization to this order. It is instructive to consider the reaction

$$qq \to qq, \qquad (8.3.24)$$

for which the one-gluon-exchange Born diagram is shown in Fig. 8-18. Denoting the Born amplitude as $\mathcal{M}_B$, we may write the amplitude modified by one-loop corrections as

$$\mathcal{M} = \mathcal{M}_B\left[1 + \frac{g^2}{16\pi^2}\log\left(\frac{-q^2}{\mu^2}\right)\left(\frac{2n_f}{3} - \frac{13N}{6} - 2\cdot\frac{3N}{4}\right) + O(g^4)\right], \qquad (8.3.25)$$

where the terms multiplying the logarithm correspond to the contribution of fermion and boson loops to the gluon propagator, and to the corrections

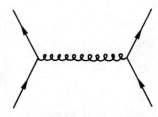

FIG. 8-18. Lowest-order contribution to quark–quark scattering in QCD.

at both vertices arising from the three-gluon interaction.[16] Combining terms and defining $\alpha_s \equiv g^2/4\pi$ as the strong-interaction coupling constant, we may evidently define for this process a running coupling constant

$$\alpha_s(q^2) = \alpha_s(\mu^2)\left[1 + \frac{\alpha_s(\mu^2)}{12\pi}\log\left(\frac{-q^2}{\mu^2}\right)(2n_f - 11N) + O(\alpha_s^2)\right], \quad (8.3.26)$$

which is similar in form to the running coupling constant (8.2.36) of QED. Although it has been camouflaged in our abbreviated discussion, this expression of course refers to the renormalized coupling constant.

Consider now the implications of the form (8.3.26). The fermion loop contribution has qualitatively the same effect as in QED; it is a vacuum polarization term that tends to enhance the effective coupling constant at short distances or large values of q^2. The contribution due to the three-gluon interaction is of the *opposite sign* and tends to decrease the strength of the interaction at short distances. It corresponds not to screening, but to anti-screening. The possibility of antiscreening can easily be understood in qualitative terms using the sketch in Fig. 8-19. Suppose our "test charge," in the language of the earlier electrodynamics discussion, is a blue quark at the coordinate origin, and that the probe we employ to measure its charge is a red–antiblue gluon. It may happen that, while the probe is en route to the origin, the blue quark radiates a virtual blue–antigreen gluon, and thus fluctuates into a green quark—to which the probe is blind. Rather than being concentrated at the origin, the net color charge will thus be dispersed throughout the gluon cloud. Therefore only by inspecting the test charge from long distances will one be able to measure its full effect.

The combined effect of quark loops and (roughly speaking) gluon loops is thus the result of a competition between screening and antiscreening. So

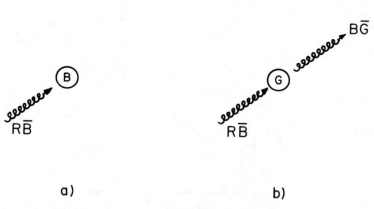

a)   b)

FIG. 8-19. (a) $R\bar{B}$ gluon probe incident on a blue quark may find (b) the blue charge dispersed as a result of vacuum fluctuations.

long as $11N$ exceeds $2n_f$, the antiscreening contribution will dominate and the running coupling constant will become small at short distances, or at large values of q^2. Our discussion has been lacking both in generality and in systematic rigor, but the result is indeed generally true. The summation of leading logarithms may be carried out as in the case of electrodynamics most conveniently using renormalization group techniques. As our expression (8.3.26) makes plausible, the running coupling constant takes the form for QCD ($N = 3$)

$$\frac{1}{\alpha_s(q^2)} = \frac{1}{\alpha_s(\mu^2)} + \frac{(33 - 2n_f)}{12\pi} \log\left(\frac{-q^2}{\mu^2}\right), \qquad (8.3.27)$$

the evolution of which is sketched in Fig. 8-20 for the case in which antiscreening prevails. The profound significance of this behavior for a calculable theory of the strong interactions was recognized[17] by Gross and Wilczek and by Politzer in 1973. The existence of a regime in which $\alpha_s(q^2) \ll 1$ implies a realm in which QCD perturbation theory should be valid. This property of non-Abelian gauge theories is known as *asymptotic freedom*. Although it by no means justifies all the hypotheses of the parton model, it does make it plausible that at very short distances (i.e., when examined by very-high-Q^2 probes) quarks may behave nearly as free particles within hadrons. By contrast, the growth of the coupling constant at large distances indicates the existence of a domain in which the strong interactions become formidable. This strong-coupling regime undoubtedly is of key importance for quark—or color—confinement.

As in electrodynamics,[18] one may calculate the "radiative corrections" to the effective coupling constant to higher orders in perturbation theory. The result is traditionally expressed in the notation of the renormalization

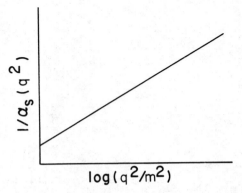

FIG. 8-20. Evolution of the running coupling constant in QCD.

group approach in terms of the function

$$\beta \equiv \frac{2\partial g}{\partial \log(-q^2/\mu^2)} \equiv -b_0 \frac{g^3}{16\pi^2} - b_1 \frac{g^5}{(16\pi^2)^2} + \cdots, \qquad (8.3.28)$$

where

$$b_0 = 11 - 2n_f/3, \qquad (8.3.29)$$

as we have just seen, and[19]

$$b_1 = 102 - 38n_f/3. \qquad (8.3.30)$$

8.4   Perturbative QCD: An Example

We now begin to consider specific applications of QCD perturbation theory. The general spirit of these calculations is that by carrying out a computation in some low order of perturbation theory, but replacing the strong coupling constant α_s consistently by the running coupling constant $\alpha_s(q^2)$, one will have included the effects of the strong interaction to all orders in perturbation theory. Such a program can be given a precise definition within the framework of the renormalization group approach, but clearly one cannot escape entirely from the requirement that a perturbation expansion make sense, in that the running coupling constant must be by an appropriate standard small. The approach is frequently called renormalization-group-improved perturbation theory.

The simplest illustration of QCD perturbation theory is the calculation of the cross section for electron–positron annihilation into hadrons. As we have discussed many times, this process is represented in the parton model by the elementary transition illustrated in Fig. 8-21(a), which yields

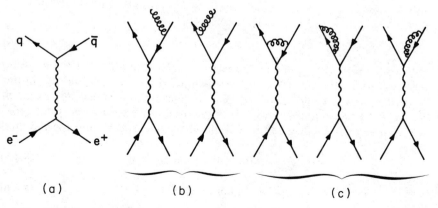

FIG. 8-21. Feynman diagrams for the reaction $e^+e^- \to q\bar{q}$ in order (a) α_s^0, (b) α_s^1, (c) α_s^2.

a cross section

$$\sigma_{\text{parton}}(s) = \frac{4\pi\alpha^2}{3s} \left[3 \sum_{\substack{\text{quark} \\ \text{flavors}}} e_q^2 \theta(s - 4m_q^2) \right], \tag{8.4.1}$$

where the theta function provides a crude representation of kinematic threshold effects.

To the extent that α_s is small, it makes sense to compute the strong-interaction corrections to the parton model in perturbation theory. To order α_s, this entails the calculation of the two gluon-radiation graphs shown in Fig. 8-21(b) and of the three virtual-gluon graphs of Fig. 8-21(c). In a schematic but self-evident notation, the cross section is given by

$$\sigma(e^+e^- \rightarrow \text{hadrons}) = \underbrace{|A|^2}_{\alpha_s^0} + \underbrace{|B|^2 + |A \otimes C|}_{\alpha_s^1} + \text{O}(\alpha_s)^2. \tag{8.4.2}$$

The ultraviolet divergences that appear separately in each loop diagram are eliminated in the usual renormalization program and introduce no particular problems. This cancellation is a consequence of the gauge invariance of the theory, or equivalently of the Ward–Takahashi identities, and may be seen to follow from the explicit results of the preceding section. The cancellation of infrared divergences is more subtle and has interesting consequences for the definition of experimental observables.

Consider the gluon-radiation graphs of Fig. 8-21(b). If we denote by p the final momentum of a quark that has radiated a gluon of momentum k, the virtual quark propagator is proportional to

$$1/[(p + k)^2 - m^2] = 1/(k^2 + 2p \cdot k) \tag{8.4.3}$$

which vanishes in the soft gluon limit as $k_\mu \rightarrow 0$ and, for the emission of massless gluons with $k^2 = 0$, also vanishes if

$$k \cdot p = 0. \tag{8.4.4}$$

The latter condition occurs, for massless quarks, when the quark and gluon momenta are collinear. These infrared singularities are canceled by similar terms in the vertex and quark-propagator corrections, which will occur in the $|A \otimes C|$ interference terms. This means that, whereas the total cross section given by (8.4.2) is infrared finite, the cross sections for specific final states

$$\sigma(e^+e^- \rightarrow q\bar{q}) = |A|^2 + |A \otimes C| + \cdots \tag{8.4.5}$$

and

$$\sigma(e^+e^- \rightarrow q\bar{q}g) = |B|^2 + \cdots \tag{8.4.6}$$

have infrared divergences and cannot meaningfully be calculated.

Once this is recognized, it is straightforward to evaluate the cross section, which is the same, modulo color factors, as that calculated in QED by Jost and Luttinger.[20] The result is simply[21]

$$\sigma(e^+e^- \to \text{hadrons}) = \sigma_{\text{parton}}(s)\left[1 + \frac{\alpha_s(s)}{\pi} + O(\alpha_s^2)\right]. \qquad (8.4.7)$$

We may therefore see that the strong-interaction corrections to the total cross section are

- calculable, and free of infrared problems;
- small, in the asymptotically free region, with $O(\alpha_s^2)$ corrections[22] that are not enormous;
- positive; and
- decreasing with increasing s.

The effect of these corrections to the parton model is shown schematically in Fig. 8-22, in which the quantity

$$R \equiv \frac{\sigma(e^+e^- \to \text{hadrons})}{\sigma(e^+e^- \to \mu^+\mu^-)} \qquad (8.4.8)$$

is plotted as a function of energy.

Although QCD prescribes the evolution of the strong coupling constant, it does not specify the value $\alpha_s(\mu^2)$ of the running coupling constant at the reference point $q^2 = -\mu^2$. We are thus left to wonder at what value of s

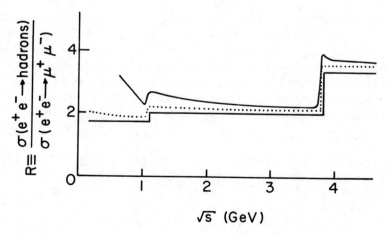

FIG. 8-22. Schematic behavior of the ratio $R = \sigma(e^+e^- \to \text{hadrons})/\sigma(e^+e^- \to \mu^+\mu^-)$ as a function of energy in the parton model (steps) and including lowest order strong interaction corrections (smooth curve). The dotted curve corresponds to a smaller value of $\alpha_s(0)$.

TABLE 8.4: Two Examples of
the Evolution of α_s

s (GeV^2)	$\alpha_s(s)$	
	A	B
1	0.30	2.9
5	0.22	0.67
10 ψ	0.2	0.5
16 $c\bar{c}$ threshold	0.19	0.43
100 Υ	0.15	0.28
1000 PETRA/PEP	0.12	0.20

(or q^2) the first-order calculation will become reliable. The value of $\alpha_s(\mu^2)$ must be determined experimentally, and the absence of a well-defined Thomson limit in QCD makes this a nontrivial task. Two estimates, which perhaps represent reasonable extremes, are shown in Table 8.4.

Experimental results are at present somewhat equivocal, because total cross-section measurements carry a systematic uncertainty of no less than 5% in the current generation of experiments. There is also the need to account exactingly for the contribution of the reaction

$$e^+e^- \rightarrow \tau^+\tau^- \tag{8.4.9}$$

to the apparent hadronic cross section, and for the contribution of weak-electromagnetic interference effects, which, as shown in Fig. 7-2, are not entirely negligible at the highest accessible energies. Typical of the state of the experimental art is a recent fit to the total cross-section data of the TASSO collaboration[23] working at the storage ring PETRA in Hamburg. Retaining QCD corrections through $O(\alpha_s^2)$ and the influence of $\gamma - Z^0$ interference parametrized according to the standard model, they find

$$\alpha_s \,(s = 1000 \text{ GeV}^2) = 0.18 \pm 0.03 \pm 0.14, \tag{8.4.10}$$

where the first error is of statistical origin, and the second reflects systematic uncertainties. Apparently the evidence for QCD corrections to the parton-model cross section is not yet strong. Because of the definiteness of the theoretical predictions, and the opportunity to measure in a direct way the evolution of α_s, improvement of the accuracy of total cross section measurements is an experimental imperative.

Although the definition of a three-jet cross section corresponding to the quark–antiquark–gluon final state is plagued by infrared difficulties, it is nevertheless apparent[24] that three-jet events are to be expected as a consequence of the gluon radiation diagrams. Indeed, it was the observation of three-jet events in high-energy e^+e^- annihilations that was interpreted as evidence for the gluon, as discussed in Section 1.3. The path toward a systematic confrontation of the complete content of the theory with experiment was indicated by Sterman and Weinberg,[25] who showed that it is possible to define infrared-finite energy-weighted cross sections that are calculable within QCD without explicit reference to the parton model and are, at least in principle, measurable.

At still higher orders in α_s, the cross section receives contributions from diagrams that have an obvious (though infrared dangerous) interpretation in terms of four (or more!) jet events. It is of considerable interest to devise means of confronting these expectations with experiment. Beyond the obvious problem of infrared finiteness, one must confront the problem of the hadronization of partons, which enters in the experimental definition of jet cross sections and in the determination of experimental efficiencies. This is an active and nontrivial area of research.

8.5   QCD Corrections to Deeply Inelastic Scattering

The evolution with Q^2 of the structure functions measured in deeply inelastic scattering may also be analyzed in QCD perturbation theory. To make clear the logical structure, it is helpful to revert to electrodynamics and to consider a *Gedankenexperiment* to measure the momentum spectrum of electrons in a "monochromatic" beam, by observing the scattering of virtual photons as shown in Fig. 8-23. A perfect calorimeter measures the energy of the backscattered photon and thus determines the energy of the electron from which the photon was scattered. If the momentum of the prepared beam is defined to be 1, then in zeroth order the momentum distribution of electrons in the beam is

$$d\mathcal{N}/dz = \mathcal{N}\,\delta(z-1),\qquad(8.5.1)$$

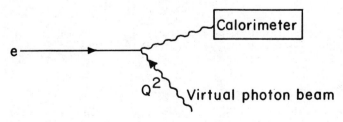

FIG 8-23. Conceptual experiment to measure the momentum spectrum of electrons in a beam prepared as monochromatic.

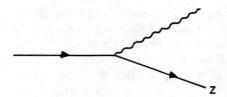

F $_{\text{IG}}$. 8-24. Fluctuation of an electron into an electron carrying momentum fraction z and a photon.

where
$$z \equiv \text{(measured momentum)/(prepared momentum)}. \qquad (8.5.2)$$

The virtual dissociation of an electron into an electron plus a photon, depicted in Fig. 8-24, induces in the beam a component with momentum fraction $z < 1$. The sensitivity of the apparatus to these fluctuations is a function of the invariant mass of the virtual photon, as uncertainty principle reasoning makes clear. The fluctuations are calculable in QED.

To see this explicitly, let us define the parameter
$$\tau \equiv \log(Q^2/q_0^2) \qquad (8.5.3)$$
and let the quantity
$$\frac{\alpha}{2\pi} P_{e \leftarrow e}(z)\, d\tau \qquad (8.5.4)$$

represent the probability of observing an electron carrying a fraction z of the momentum of the parent electron. Then if $e(z, \tau)$ is the number density of electrons observed with fractional momenta between z and $z + dz$ by a probe with resolving power characterized by τ, it follows at once that

$$\frac{de}{d\tau}(x, \tau) = \frac{\alpha(\tau)}{2\pi} \int_0^1 dy \int_0^1 dz\, \delta(zy - x) e(y, \tau) P_{e \leftarrow e}(z)$$

$$= \frac{\alpha(\tau)}{2\pi} \int_x^1 \frac{dy}{y} e(y, \tau) P_{e \leftarrow e}\left(\frac{x}{y}\right). \qquad (8.5.5)$$

For an initial distribution $e(y, \tau) = \mathcal{N}\delta(y - 1)$ corresponding to the beam prepared in (8.5.1), we recover

$$\frac{de}{d\tau}(x, \tau) = \mathcal{N} \frac{\alpha(\tau)}{2\pi} P_{e \leftarrow e}(x). \qquad (8.5.6)$$

By virtue of the same fluctuations, there are now photons in the beam as well. We let the probability of finding a photon carrying a fraction z of the

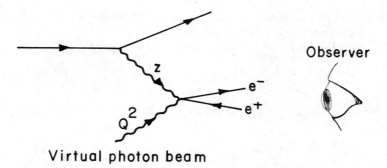

FIG. 8-25. Conceptual experiment to measure the momentum spectrum of photons in an electron beam prepared as monochromatic.

parent electron's momentum be represented by

$$\frac{\alpha}{2\pi} P_{\gamma \leftarrow e}(z)\, d\tau, \tag{8.5.7}$$

and let $\gamma(z, \tau)$ be the number density of photons observed with fractional momenta between z and $z + dz$ by a probe with resolving power characterized by τ. A schematic experiment to observe these photons is shown in Fig. 8-25. If the source of the virtual-photon probe is (for example) a nitrogen nucleus, the *Gedankenapparat* is recognized as a surrogate for the development of electromagnetic showers in the atmosphere. Indeed, the theory of cascade showers presented in the book by Rossi[26] has much in common with the present discussion. The evolution of the photon distribution is given by

$$\frac{d\gamma}{d\tau}(x, \tau) = \frac{\alpha(\tau)}{2\pi} \int_x^1 \frac{dy}{y}\, e(y, \tau) P_{\gamma \leftarrow e}\!\left(\frac{x}{y}\right). \tag{8.5.8}$$

Of course, photons may themselves fluctuate into electron–position pairs, as indicated in Fig. 8-26, so we are led to define

$$\frac{\alpha}{2\pi} P_{e \leftarrow \gamma}(z) \tag{8.5.9}$$

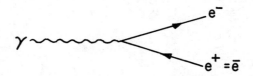

FIG. 8-26. Fluctuation of a photon into an electron–positron pair.

as the probability of finding an electron (or a positron) carrying a fraction z of the momentum of the parent photon.

The evolution of the electron distribution is now given by

$$\frac{de}{d\tau}(x, \tau) = \frac{\alpha(\tau)}{2\pi} \int_x^1 \frac{dy}{y} \left[e(y, \tau) P_{e \leftarrow e}\left(\frac{x}{y}\right) + \gamma(y, \tau) P_{e \leftarrow \gamma}\left(\frac{x}{y}\right) \right], \quad (8.5.10)$$

and the induced positron component will obey an equation identical in form,

$$\frac{d\bar{e}}{d\tau}(x, \tau) = \frac{\alpha(\tau)}{2\pi} \int_x^1 \frac{dy}{y} \left[\bar{e}(y, \tau) P_{e \leftarrow e}\left(\frac{x}{y}\right) + \gamma(y, \tau) P_{e \leftarrow \gamma}\left(\frac{x}{y}\right) \right]. \quad (8.5.11)$$

The photon in turn evolves according to

$$\frac{d\gamma}{d\tau}(x, \tau) = \frac{\alpha(\tau)}{2\pi} \int_x^1 \frac{dy}{y} (e(y, \tau) + \bar{e}(y, \tau)) P_{\gamma \leftarrow e}\left(\frac{x}{y}\right). \quad (8.5.12)$$

To solve equations of this kind, it is convenient to define moments of the distribution functions as

$$\left.\begin{array}{l} M_n(\tau) \equiv \int_0^1 dx \, x^{n-1} e(x, \tau) \\[2mm] \bar{M}_n(\tau) \equiv \int_0^1 dx \, x^{n-1} \bar{e}(x, \tau) \end{array}\right\}, \quad (8.5.13)$$

and similarly for the photon. The evolution of the moments is then easily computed. It takes a particularly simple form for the combination $M_n(\tau) - \bar{M}_n(\tau)$, for which

$$\frac{d}{d\tau}(M_n(\tau) - \bar{M}_n(\tau)) = \int_0^1 dx \, x^{n-1} \left(\frac{de}{d\tau}(x, \tau) - \frac{d\bar{e}}{d\tau}(x, \tau) \right)$$

$$= \frac{\alpha}{2\pi} \int_0^1 dx \, x^{n-1} \int_0^1 dy$$

$$\times \int_0^1 dz \, \delta(zy - x) P_{e \leftarrow e}(z)(e(y, \tau) - \bar{e}(y, \tau))$$

$$= \frac{\alpha}{2\pi} \int_0^1 dx \, x^{n-1} \int_x^1 \frac{dy}{y} P_{e \leftarrow e}\left(\frac{x}{y}\right)(e(y, \tau) - \bar{e}(y, \tau))$$

$$= \frac{\alpha}{2\pi} \int_0^1 dy \, y^{n-1}(e(y, \tau) - \bar{e}(y, \tau))$$

$$\times \int_0^1 d\left(\frac{x}{y}\right)\left(\frac{x}{y}\right)^{n-1} P_{e \leftarrow e}\left(\frac{x}{y}\right)$$

$$= \frac{\alpha}{2\pi}(M_n(\tau) - \bar{M}_n(\tau)) A_n, \quad (8.5.14)$$

where

$$A_n \equiv \int_0^1 dz \, z^{n-1} P_{e \leftarrow e}(z) \tag{8.5.15}$$

is the moment of the so-called splitting function $P_{e \leftarrow e}(z)$. Now writing

$$\Delta_n \equiv M_n(\tau) - \bar{M}_n(\tau), \tag{8.5.16}$$

we have that

$$\frac{d(\log \Delta_n(\tau))}{d\tau} = \frac{\alpha(\tau)}{2\pi} A_n. \tag{8.5.17}$$

If the effective coupling constant itself evolves as

$$\alpha(\tau) = \alpha(0)/(1 + b\alpha(0)\tau), \tag{8.5.18}$$

the general form appropriate for QED and QCD, then the differential equation (8.5.17) is easily integrated to

$$\begin{aligned}
\log\left(\frac{\Delta_n(\tau)}{\Delta_n(0)}\right) &= \frac{\alpha(0)A_n}{2\pi} \int_0^\tau \frac{d\tau}{1 + b\alpha(0)\tau} \\[2mm]
&= \frac{A_n}{2\pi b} \log(1 + b\alpha(0)\tau) \\[2mm]
&= \frac{A_n}{2\pi b} \log\left(\frac{\alpha(0)}{\alpha(\tau)}\right).
\end{aligned} \tag{8.5.19}$$

We therefore obtain the simple prediction for the evolution of moments that

$$\Delta_n(\tau)/\Delta_n(0) = (\alpha(\tau)/\alpha(0))^{-A_n/2\pi b}, \tag{8.5.20}$$

as well as an especially simple prediction for

$$\frac{\log(\Delta_n(\tau)/\Delta_n(0))}{\log(\Delta_k(\tau)/\Delta_k(0))} = \frac{A_n}{A_k}. \tag{8.5.21}$$

These specific forms are valid in leading order in renormalization group improved perturbation theory, although it is possible to incorporate higher-order corrections by iteration. The evolution of the moments is completely specified by the exponents A_n, which may be calculated without reference to the electron and photon distribution functions. To describe the evolution of individual moments, rather than moment-by-moment ratios, it is necessary to know or determine the coupling constant $\alpha(0)$.

In fact, nothing of the procedure we have followed is specific to QED. The same method can be adapted to quantum chromodynamics, as was done by Altarelli and Parisi,[27] by identifying the electron, positron, and photon distributions as quark, antiquark, and gluon distributions, and allowing for the possibility of a gluon fluctuating into two gluons.

The splitting functions $P_{B \leftarrow A}(z)$, which are to be computed in perturbation theory, satisfy some obvious sum rules. Fermion number conservation,

$$\int dx \left(\frac{dq^i}{d\tau}(x, \tau) - \frac{d\bar{q}^i}{d\tau}(x, \tau) \right) = 0, \tag{8.5.22}$$

implies that

$$\int_0^1 dz\, P_{q \leftarrow q}(z) = 0. \tag{8.5.23}$$

Here the superscript i has been introduced as a flavor index for the quarks. Momentum conservation,

$$\int_0^1 dx\, x \left(\sum_i \frac{dq^i}{d\tau}(x, \tau) + \sum_i \frac{d\bar{q}^i}{d\tau}(x, \tau) + \frac{dG}{d\tau}(x, \tau) \right) = 0 \tag{8.5.24}$$

imposes two constraints:

$$\int_0^1 dz\, z(P_{q \leftarrow q}(z) + P_{g \leftarrow q}(z)) = 0, \tag{8.5.25}$$

and

$$\int_0^1 dz\, z(2n_f P_{q \leftarrow g}(z) + P_{g \leftarrow g}(z)) = 0, \tag{8.5.26}$$

where n_f denotes the number of quark flavors. In addition, momentum conservation at the elementary vertices requires a number of symmetry properties to hold for $z \neq 1$:

$$P_{q \leftarrow q}(z) = P_{g \leftarrow q}(1 - z), \tag{8.5.27}$$
$$P_{q \leftarrow g}(z) = P_{q \leftarrow g}(1 - z), \tag{8.5.28}$$
$$P_{g \leftarrow g}(z) = P_{g \leftarrow g}(1 - z). \tag{8.5.29}$$

Two methods are in use for the actual evaluation of the splitting functions. We shall sketch the "old-fashioned," which is to say time-ordered, perturbation theory approach taken by Altarelli and Parisi. This has the merit of making plain the generality of the results, and making contact with the techniques of Weizsäcker and Williams. Completely equivalent calculations may be carried out by using the covariant techniques of ordinary Feynman graphs,[28] which are perhaps more transparent but have the appearance of being process-dependent. The procedure is in any case straightforward: for $z \neq 1$, the splitting function $P_{B \leftarrow A}$ can be related to the square of a matrix element.

To see how the connection is made, let us evaluate in the infinite momentum frame the probability $\mathscr{P}_{BA}(z)$ of finding a parton of type B with momentum fraction z in a beam of partons of type A. According to our previous definitions, we may write simply

$$d\mathscr{P}_{BA}(z)\, dz = \frac{\alpha}{2\pi} P_{B \leftarrow A}(z)\, dz\, d\tau. \tag{8.5.30}$$

Consider the reaction

$$A + D \rightarrow C + f \qquad (8.5.31)$$

sketched in Fig. 8-27(a), in which ABC is the calculable elementary vertex of interest and particle D merely provides the kinematic crutch of the process

$$B + D \rightarrow f \qquad (8.5.32)$$

indicated in Fig. 8-27(b). The differential cross section for the compound reaction may be written as

$$d\sigma(A + D \rightarrow C + f) = d\mathscr{P}_{BA}(z)\, d\sigma(B + D \rightarrow f) dz. \qquad (8.5.33)$$

It is then a straightforward exercise in kinematics[29] to show that with four-momenta defined as

$$
\begin{aligned}
k_A &= (P; \vec{0}, P) \\
k_B &= \{[(zP)^2 + p_\perp^2]^{1/2}; \vec{p}_\perp, zP\} \simeq (zP + p_\perp^2/2zP; \vec{p}_\perp, zP) \qquad (8.5.34) \\
k_C &= \{[(1-z)^2 P^2 + p_\perp^2]^{1/2}; -\vec{p}_\perp, (1-z)P\} \\
&\simeq [(1-z)P + p_\perp^2/2(1-z)P; -\vec{p}_\perp, (1-z)P]
\end{aligned}
$$

the differential probability is given by

$$d\mathscr{P}_{BA}(z) = \frac{\alpha}{2\pi} \frac{z(1-z)}{2p_\perp^2} \overline{|\mathscr{M}(A \rightarrow BC)|^2}\, d(\ln p_\perp^2), \qquad (8.5.35)$$

where $\overline{|\mathscr{M}|^2}$ is the spin-averaged square of the matrix element. With the (asymptotic) identification $d(\ln p_\perp^2) \rightarrow d\tau$, this implies that the splitting function is

$$P_{B \leftarrow A}(z) = \frac{z(1-z)}{2p_\perp^2} \overline{|\mathscr{M}(A \rightarrow BC)|^2}, \qquad z \neq 1. \qquad (8.5.36)$$

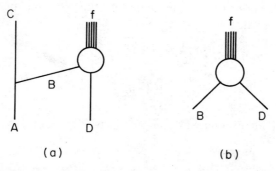

(a)   (b)

FIG. 8-27. (a) The compound process $A + D \rightarrow C + f$ used in the calculation of the splitting function $P_{B \leftarrow A}(z)$. (b) The subprocess $B + D \rightarrow f$.

We therefore find for the quark–quark splitting function

$$P_{q \leftarrow q}(z < 1) = \left(\frac{N^2 - 1}{2N}\right)\left(\frac{1 + z^2}{1 - z}\right), \tag{8.5.37}$$

where the color factor has been written for $SU(N)$. It remains to fix the behavior of $P_{q \leftarrow q}$ at $z = 1$. This is done in two steps. First, we interpret the denominator $1/(1 - z)$ in the sense of a distribution $1/(1 - z)_+$, defined by the property

$$\int_0^1 \frac{dz\, f(z)}{(1 - z)_+} \equiv \int_0^1 dz \frac{f(z) - f(1)}{(1 - z)} = \int_0^1 dz \ln(1 - z)\frac{df}{dz}(z), \tag{8.5.38}$$

for a function f regular at $z = 1$. Second, we add a term proportional to $\delta(1 - z)$ with a coefficient chosen to ensure compliance with the fermion conservation sum rule (8.5.23). This determines the final result for $SU(3)_{\text{color}}$ as

$$P_{q \leftarrow q}(z) = \frac{4}{3}\left[\frac{1 + z^2}{(1 - z)_+} + \frac{3}{2}\delta(1 - z)\right]. \tag{8.5.39}$$

The other splitting functions, which are computed in similar fashion, are

$$P_{g \leftarrow q}(z) = \frac{4}{3}\left[\frac{1 + (1 - z)^2}{z}\right], \tag{8.5.40}$$

$$P_{q \leftarrow g}(z) = \frac{1}{2}\left[z^2 + (1 - z)^2\right], \tag{8.5.41}$$

for each quark flavor, and

$$P_{g \leftarrow g}(z) = 6\left[\frac{z}{(1 - z)_+} + \frac{(1 - z)}{z} + z(1 - z) + \left(\frac{11}{12} - \frac{n_f}{18}\right)\delta(1 - z)\right]. \tag{8.5.42}$$

What is required for the computation of the evolution of the moments of quark distributions is the moments of the splitting functions. As an example, let us evaluate

$$A_n(q \leftarrow q) \equiv \int_0^1 dz\, z^{n-1} P_{q \leftarrow q}(z)$$

$$= \frac{4}{3}\left[\frac{3}{2} + \int_0^1 dz \frac{(z^{n-1} + z^{n+1} - 2)}{1 - z}\right]. \tag{8.5.43}$$

The remaining integral may be computed easily by noting that

$$\int_0^1 dz \frac{z^{n-1} - 1}{1 - z} = -\sum_{j=0}^{n-2} \int_0^1 dz\, z^j = -\sum_{j=1}^{n-1} \frac{1}{j}. \tag{8.5.44}$$

Thus we have

$$A_n(q \leftarrow q) = \frac{4}{3}\left(\frac{3}{2} - \sum_{j=1}^{n-1}\frac{1}{j} - \sum_{k=1}^{n+1}\frac{1}{k}\right)$$

$$= \frac{4}{3}\left[-\frac{1}{2} + \frac{1}{n(n+1)} - 2\sum_{j=2}^{n}\frac{1}{j}\right], \tag{8.5.45}$$

and after similar arithmetic

$$A_n(g \leftarrow q) = \frac{4}{3}\left[\frac{n^2 + n + 2}{n(n^2 - 1)}\right], \tag{8.5.46}$$

$$A_n(q \leftarrow g) = \frac{1}{2}\left[\frac{n^2 + n + 2}{n(n+1)(n+2)}\right], \tag{8.5.47}$$

$$A_n(g \leftarrow g) = 6\left[-\frac{1}{12} + \frac{1}{n(n-1)} + \frac{1}{(n+1)(n+2)} - \sum_{j=2}^{n}\frac{1}{j} - \frac{n_f}{18}\right]. \tag{8.5.48}$$

In any field theory, the splitting functions and hence the exponents are calculable in perturbation theory. A weak-coupling theory such as QED can be expected to give reliable results at low orders in a perturbation expansion. For the strong interactions, only an asymptotically free theory such as QCD presents any hope that low-order perturbation theory should be trustworthy. Even so, because of the uncertainty in the size of α_s, one does not know *a priori* the value of Q^2 at which first-order results become reliable.

The evolution of what is called the nonsinglet moment, which corresponds to the difference of quark and antiquark distribution functions, is particularly simple to evaluate, as the discussion of electromagnetism has shown. It is also accessible experimentally in rather direct fashion because, as noted in Section 7.3, the structure function $\mathscr{F}_3$ measures the difference between quark and antiquark distributions. For an isoscalar target, the difference of neutrino and antineutrino charged-current cross sections is seen from (7.3.71, 86, 87) to yield

$$x\mathscr{F}_3 = u(x) - \bar{u}(x) + d(x) - \bar{d}(x)$$

$$= \left[\frac{d^2\sigma}{dx\,dy}(\nu N \to \mu^- X) - \frac{d^2\sigma}{dx\,dy}(\bar{\nu}N \to \mu^+ X)\right]$$

$$\times \frac{\pi}{G_F^2 ME} \cdot \frac{1}{1 - (1-y)^2}. \tag{8.5.49}$$

Transcribing the results of our QED analysis, we have for the evolution of the nonsinglet moments, defined as

$$\Delta_n(\tau) \equiv \int_0^1 dx\, x^n \mathscr{F}_3(x,\tau) = \int_0^1 dx\, x^{n-1}[q(x,\tau) - \bar{q}(x,\tau)], \quad (8.5.50)$$

the simple and characteristic predictions

$$\Delta_n(\tau)/\Delta_n(0) = (\alpha_s(\tau)/\alpha_s(0))^{-A_n(q \leftarrow q)/2\pi b} \quad (8.5.51)$$

and

$$\frac{\log(\Delta_n(\tau)/\Delta_n(0))}{\log(\Delta_k(\tau)/\Delta_k(0))} = \frac{A_n(q \leftarrow q)}{A_k(q \leftarrow q)}, \quad (8.5.52)$$

where the numerical values of the nonsinglet moments are $A_1(q \leftarrow q) = 0$ (by fermion number conservation), $A_2 = -1.78$, $A_3 = -2.78$, $A_4 = -3.49$, $A_5 = -4.04$, $A_6 = -4.50$, $A_7 = -4.89$.

The logarithmic ratios of the nonsinglet moments have been studied in many experiments. One analysis, based on the high-statistics sample of deeply inelastic neutrino and antineutrino interactions in iron measured in the CERN–Dortmund–Heidelberg–Saclay detector,[30] is represented in Fig. 8-28. The figure shows the slope plot[31] $\log \Delta_5(\tau)$ versus $\log \Delta_3(\tau)$ obtained from their measurements of the structure function $\mathscr{F}_3$. According to the first-order QCD analysis we have just carried out, the slope should be

$$r_{53} = \frac{A_5(q \leftarrow q)}{A_3(q \leftarrow q)} = 1.46, \quad (8.5.53)$$

which is in reasonable agreement with the experimental best fit

$$r_{53} = 1.68 \pm 0.11. \quad (8.5.54)$$

The agreement between the experimental result and the QCD prediction is improved by about one (experimental) standard deviation if second-order corrections are included.

Although the moment-by-moment test is able to provide some discrimination among different theories and has the virtue of giving a single number, it does not confront the predictions of QCD in a particularly differential or incisive fashion. We may probe somewhat deeper by considering the prediction (8.5.51) for the evolution of the moments themselves, which may be rewritten for the special case of QCD as

$$\Delta_n(\tau) = \Delta_n(0)\left(\frac{\alpha_s(0)}{\alpha_s(\tau)}\right)^{6A_n/(33-2n_f)}, \quad (8.5.55)$$

because the evolution of the strong coupling constant is described by

$$b = (33 - 2n_f)/12\pi. \quad (8.5.56)$$

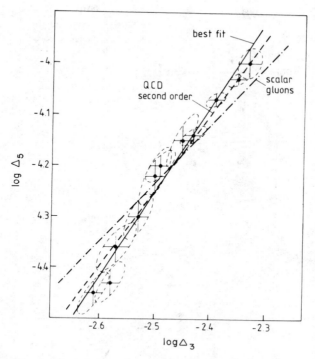

FIG. 8-28. Scatter plot of the fifth versus third nonsinglet moments on a logarithmic scale. The indicated errors include the correlations between the two measurements. (From ref. 30.)

To reparametrize the evolution of α_s given by

$$1/\alpha_s(q^2) = 1/\alpha_s(\mu^2) + b\log(-q^2/\mu^2), \qquad (8.3.28)$$

it has become traditional to introduce the scale parameter Λ by replacing

$$1/\alpha_s(\mu^2) - b\log(\mu^2) \equiv -b\log(\Lambda^2), \qquad (8.5.57)$$

so that

$$1/\alpha_s(q^2) = b\log(-q^2/\Lambda^2). \qquad (8.5.58)$$

It is easy to show, from (8.3.28), that the definition of Λ is independent of the choice of the renormalization point μ^2. Extensive fits to the q^2-dependence of moments and structure functions have been carried out for several processes. A consensus is that the logarithmic evolution of structure functions predicted by QCD is compatible with present data, with a value of the scale parameter that is still uncertain by a factor of 2, but is of order 100 MeV.[32] Representative data for the nonsinglet structure function are shown in Fig. 8-29, together with QCD fits. We may note that if $b > 0$, which is to

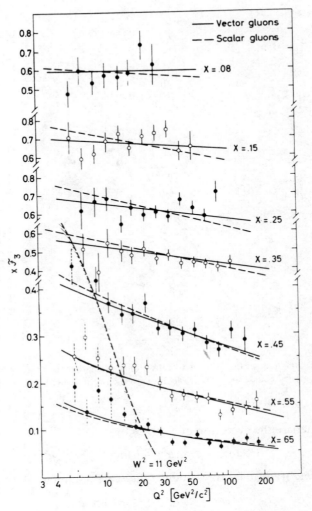

FIG. 8-29. The nonsinglet structure functions as measured by the CDHS experiment (ref. 30). The solid lines are QCD fits to the data.

say that there are fewer than 17 active quark flavors, we have the general expectation that

$$\lim_{Q^2 \to \infty} \Delta_n \to 0, \qquad n > 1, \tag{8.5.59}$$

since $A_n(q \leftarrow q) < 0$. This means in particular that the momentum fraction carried by valence quarks, which is characterized by the nonsinglet moment Δ_2, will vanish as $Q^2 \to \infty$.

The evolution of the remaining moments is somewhat more complicated because of mixing between the gluon distribution $G(x, \tau)$ and the "singlet" quark distribution

$$\Sigma(x, \tau) \equiv \sum_{i=\text{flavors}} (q_i(x, \tau) + \bar{q}_i(x, \tau)). \tag{8.5.60}$$

The evolution of the moments is described by coupled Altarelli–Parisi equations, which may be represented in matrix form as

$$\frac{d}{d\tau}\begin{pmatrix} \Sigma_n \\ G_n \end{pmatrix} = \frac{\alpha}{2\pi}\begin{pmatrix} A_n(q \leftarrow q) & 2n_f A_n(q \leftarrow g) \\ A_n(g \leftarrow q) & A_n(g \leftarrow g) \end{pmatrix}\begin{pmatrix} \Sigma_n \\ G_n \end{pmatrix}. \tag{8.5.61}$$

Of special interest is the behavior of the $n = 2$ moments as $Q^2 \to \infty$, which describe the momentum fractions carried by quarks and gluons in that asymptotic regime. The matrix of moments of the splitting functions for $n = 2$ is

$$A_2 = \begin{pmatrix} -16/9 & n_f/3 \\ 16/9 & -n_f/3 \end{pmatrix}, \tag{8.5.62}$$

which manifestly satisfies the requirements of momentum conservation. To consider the $Q^2 \to \infty$ limit, we diagonalize A_2. It has two eigenvalues,

$$\left.\begin{matrix} \lambda_+ = -(16 + 3n_f)/9 \\ \lambda_- = 0 \end{matrix}\right\}, \tag{8.5.63}$$

the weights of which will evolve with Q^2 as

$$\left(\frac{\alpha_s(0)}{\alpha_s(\tau)}\right)^{6\lambda/(33 - 2n_f)}, \tag{8.5.64}$$

and so the eigenvector corresponding to λ_+ disappears in the limit. The eigenvector corresponding to the zero eigenvalue λ_- is easily found to be

$$\begin{pmatrix} \Sigma_2^\infty \\ G_2^\infty \end{pmatrix} = \frac{1}{16 + 3n_f}\begin{pmatrix} 3n_f \\ 16 \end{pmatrix}. \tag{8.5.65}$$

Thus for (four, six) flavors of quarks, the momentum fraction in gluons is

$$G_2^\infty = (4/7, 8/17), \tag{8.5.66}$$

and each species of quark or antiquark carries

$$\Sigma_2^\infty/2n_f = (3/56, 3/68). \tag{8.5.67}$$

The equilibrium partition reflects both the relative strengths of the quark–antiquark–gluon and three-gluon couplings and the number of available fermion species.

Much has been made of the fact that the evolution of moments of the structure functions in QCD is logarithmic, so that Bjorken scaling may plausibly hold to good approximation over a wide range in Q^2. That this is a special property may be seen simply as follows. Let us imagine a theory of the strong interactions in which α_s is for whatever reason small, so that perturbation theory makes sense, but constant. Such fixed-point theories may indeed be constructed formally. We do not consider asymptotically free theories because only gauge theories would qualify, nor screening theories such as QED in which the coupling constant grows with Q^2. The differential equation

$$\frac{d(\log \Delta_n(\tau))}{d\tau} = \frac{\alpha_s}{2\pi} A_n(q \leftarrow q) \qquad (8.5.17)$$

for the evolution of the nonsinglet moment then has the elementary solution

$$\Delta_n(\tau) = \Delta_n(0)e^{\alpha_s A_n \tau/2\pi}$$
$$= \Delta_n(0)(Q^2/Q_0^2)^{\alpha_s A_n/2\pi} \qquad (8.5.68)$$

which corresponds to power-law violations of Bjorken scaling. Although the existing data do not distinguish adequately between logarithmic deviations from scaling and arbitrary power laws, some specific fixed-point theories have already been ruled out.

We close this brief survey with a remark on the longitudinal cross section in deeply inelastic scattering, which has been noted in Section 7.3 to provide a test of parton spin. It was shown long ago in Problem 1-3 that, for the collinear reaction $\gamma + q \rightarrow q$, the cross section for absorption of longitudinally polarized photons is zero. We remarked, however, that an intrinsic transverse momentum for the partons would give rise to a nonvanishing longitudinal cross section characterized by

$$\sigma_S/\sigma_T \propto \langle p_\perp^2 \rangle/Q^2. \qquad (8.5.69)$$

Evidently quarks and gluons that arise from the virtual dissociation of gluons or quarks need not be collinear with the incident hadron, and thus these processes may also induce $\sigma_S/\sigma_T \neq 0$, even if the partons that carry electroweak charges all are spin-1/2 quarks. A similar effect will be seen in the next section for the photon structure function at the level of the parton model itself. Precision measurements, which would provide an interesting test of QCD, are not at all easy to carry out.

8.6   Two-Photon Processes and the Photon Structure Function

The idea of the equivalent photon (or Weizsäcker–Williams) approximation,[33] which was prominent in our derivation of the parton-splitting functions, leads naturally to the possibility of studying $\gamma\gamma$ collisions in

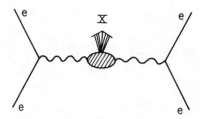

FIG. 8-30. The two-photon process $e^+e^- \rightarrow e^+e^-X$.

electron–positron or electron–electron colliding beams experiments, as indicated in Fig. 8-30. In the double-equivalent photon approximation, the cross section for the reaction

$$e^+e^- \rightarrow e^+e^-X \qquad (8.6.1)$$

at symmetric beam energies E is given by[34]

$$\sigma_{ee \rightarrow eeX}(E) \simeq 2\left(\frac{\alpha}{\pi}\right)^2 \log^2\left(\frac{E}{m_e}\right) \int_0^{4E^2} \frac{ds}{s} f\left(\frac{\sqrt{s}}{2E}\right) \sigma_{\gamma\gamma \rightarrow X}(s), \qquad (8.6.2)$$

where the flux factor is

$$f(x) = (2 + x^2)^2 \log(1/x) - (1 - x^2)(3 + x^2). \qquad (8.6.3)$$

Indeed, QED reactions such as

$$e^+e^- \rightarrow e^+e^-l^+l^- \qquad (8.6.4)$$

were already considered a half-century ago by[35] Landau and Lifshitz and Williams. In a modern notation, their result can be expressed as

$$\sigma \simeq \frac{224\alpha^4}{27\pi m_l^2} \log^2\left(\frac{E}{m_e}\right) \log\left(\frac{E}{m_l}\right). \qquad (8.6.5)$$

This cross section becomes increasingly important at high energies and competes with growing success with the annihilation channel. For example, the ratio

$$\sigma(e^+e^- \rightarrow e^+e^-\mu^+\mu^-)/\sigma(e^+e^- \rightarrow \mu^+\mu^-) \simeq 1 \qquad (8.6.6)$$

for $E \sim 1$ GeV, and grows to $\sim 10^3$ at beam energies of 15–20 GeV. Whatever its physics interest, the two-photon process is an important source of experimental background!

Two-photon production of discrete resonances is also of practical importance. The cross section for the production of a spin-J particle h^0 in the reaction

$$e^+e^- \rightarrow e^+e^-h^0 \qquad (8.6.7)$$

is given by

$$\sigma(E) = 16\alpha^2 \frac{\Gamma(h^0 \to \gamma\gamma)}{M_h^3}(2J + 1)\log^2\left(\frac{E}{m_e}\right)\log\left(\frac{2E}{M_h}\right). \qquad (8.6.8)$$

This implies, for example, a ratio

$$\frac{\sigma(e^+e^- \to e^+e^-\pi^0)}{\sigma(e^+e^- \to \mu^+\mu^-)} \qquad (8.6.9)$$

which is of order unity for $E \simeq 5$ GeV.

From discrete resonances we pass to the inclusive production of hadrons, which is represented in the parton model by the reaction

$$e^+e^- \to e^+e^- q\bar{q} \qquad (8.6.10)$$

depicted in Fig. 8-31. The characteristics of this process are a sizable cross section at high energies,

$$\frac{\sigma(e^+e^- \to e^+e^- + \text{hadrons})}{\sigma(e^+e^- \to e^+e^-\mu^+\mu^-)} \simeq 3 \sum_{\substack{\text{quark} \\ \text{flavors}}} e_q^4 = \frac{34}{27} \qquad (8.6.11)$$

in the usual model of fractionally charged u-, d-, s-, and c-quarks, and an event topology of two large-$p_\perp$ jets of hadrons, which are not emitted back-to-back because of the motion of the $\gamma\gamma$ system with respect to the c.m. frame of the colliding electron beams. As Problem 8-13 will show, the ratio (8.6.11) is specific to the model with fractionally charged quarks and should eventually permit a direct and decisive test of that hypothesis. Recent experiments have established[36] that the predicted class of events exists and that the fractional charge assignment is favored over an integral-charge quark model.

In all these applications, the dominant contribution to the cross section is from two nearly-real photons, because of the enhancement in flux due to the photon propagator. Consider instead the kinematic configuration sketched in Fig. 8-32, in which a large-Q^2 photon is used to probe the structure of the second photon, which we shall take to be nearly real. Such events

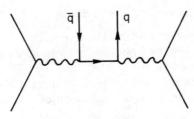

FIG. 8-31. Parton-model mechanism for the two-photon reaction $e^+e^- \to e^+e^- + \text{hadrons}$.

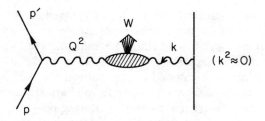

FIG. 8-32. Kinematics for the measurement of the photon structure function in electron–positron collisions.

can be selected by observing the final-state electron and positron. Write the four-momenta of the incident and scattered electrons as usual as

$$p = (E; 0, 0, E)$$
$$p' = (E'; E' \sin \theta, 0, E' \cos \theta),$$

(8.6.12)

so the probe carries momentum $q = p - p'$, and the "target" photon has four-momentum

$$k = (E_\gamma; 0, 0, -E_\gamma).$$

(8.6.13)

As usual for deeply inelastic scattering processes, we define

$$Q^2 = -q^2 = 4EE' \sin^2(\theta/2),$$

(8.6.14)

$$v = k \cdot q = 2E_\gamma[E - E' \cos^2(\theta/2)].$$

(8.6.15)

It is also useful to define the scaling variables

$$x \equiv Q^2/2v = Q^2/(Q^2 + W^2),$$

(8.6.16)

where

$$W^2 = (k + q)^2 = 2v - Q^2,$$

(8.6.17)

and

$$y \equiv v/(p \cdot k) = v/2EE_\gamma = 1 - (E'/E) \cos^2(\theta/2).$$

(8.6.18)

As in the case of lepton–nucleon scattering, we write the target tensor in the form

$$W_{\mu\nu} = W_1 \left(-g_{\mu\nu} + \frac{q_\mu q_\nu}{q^2} \right) + W_2 \left(k_\mu - \frac{k \cdot q}{q^2} q_\mu \right) \left(k_\nu - \frac{k \cdot q}{q^2} q_\nu \right),$$

(8.6.19)

so that

$$\frac{d^2\sigma}{dx \, dy} = \frac{16\pi\alpha^2 EE_\gamma}{Q^4} [(1 - y)F_2(x, Q^2) + xy^2 F_1(x, Q^2)],$$

(8.6.20)

where

$$F_2 = vW_2,$$
$$F_1 = W_1.$$

(8.6.21)

The photon is a particularly interesting target particle for deeply inelastic scattering because of its two-component nature. In many circumstances the photon displays a hadronic character that may be largely understood in the context of the vector meson dominance model. However, as our previous discussion has reminded us, the photon also has a pointlike component because of its elementary coupling to charged particles. This latter component may be observable in hard scattering processes. What is more, as was first recognized by Witten,[37] the contributions of this pointlike component to the structure functions may be calculable a priori even in the presence of the strong interactions.

The structure functions of a photon target may be computed in the parton-model approximation—i.e., neglecting strong-interaction corrections—directly from the box diagrams shown in Fig. 8-33. However, we may also proceed in parallel with our earlier discussion simply by computing to leading order the *electromagnetic* evolution of the quark distribution within the photon. We write, in analogy with (8.5.10),

$$\frac{dq}{d\tau}(x,\tau) = \frac{\alpha(\tau)}{2\pi} \int_x^1 \frac{dy}{y} P_{q \leftarrow \gamma}\left(\frac{x}{y}\right) \gamma(y,\tau), \qquad (8.6.22)$$

where we begin from a monochromatic photon beam characterized by

$$\gamma(y,\tau) = \delta(y-1). \qquad (8.6.23)$$

The essential form of the splitting function has been encountered earlier in (8.5.41). It is nothing but

$$P_{q \leftarrow \gamma}(z) = 3e_q^2[z^2 + (1-z)^2], \qquad (8.6.24)$$

where 3 is a color factor. Consequently, to this approximation, we find

$$\frac{dq}{d\tau}(x,\tau) = \frac{3\alpha e_q^2}{2\pi}[x^2 + (1-x)^2], \qquad (8.6.25)$$

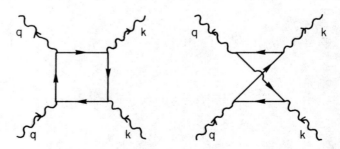

FIG. 8-33. Feynman diagrams for the photon structure function in the parton model.

which can be integrated at once to yield

$$q(x, \tau) = \frac{3\alpha e_q^2}{2\pi} [x^2 + (1 - x)^2] \log\left(\frac{Q^2}{\Lambda^2}\right) + \cdots \tag{8.6.26}$$

plus nonlogarithmic terms that are asymptotically negligible. In anticipation of the later discussion of strong-interaction corrections, we have exercised our freedom to write the scale of Q^2 in the logarithm as Λ^2. The $\log(Q^2)$ dependence, which reflects the intuitive picture of increased resolution of the virtual fluctuations as Q^2 increases, may be traced in the diagram language to the pointlike coupling of $\gamma - q - \bar{q}$ and the integration over transverse momentum in the reaction $\gamma\gamma_{\text{virtual}} \to q\bar{q}$. The same integration over transverse momentum gives rise to a nonvanishing longitudinal cross section, just as happened under the influence of strong-interaction evolution in deeply inelastic lepton scattering.

Now consider the strong-interaction evolution as well, according to

$$\frac{dq_i}{d\tau}(x, \tau) = \frac{1}{2\pi} \int_x^1 \frac{dy}{y} \left\{ \alpha(\tau) P_{q \leftarrow \gamma}\left(\frac{x}{y}\right) \gamma(y, \tau) e_i^2 \right.$$
$$\left. + \alpha_s(\tau) \left[P_{q \leftarrow q}\left(\frac{x}{y}\right) q_i(y, \tau) + P_{q \leftarrow g}\left(\frac{x}{y}\right) G(y, \tau) \right] \right\}, \tag{8.6.27}$$

$$\frac{dG}{d\tau}(x, \tau) = \frac{\alpha_s(\tau)}{2\pi} \int_x^1 \frac{dy}{y} \left[P_{g \leftarrow q}\left(\frac{x}{y}\right) \sum_{\substack{\text{quark} \\ \text{flavors}}} (q_i(y, \tau) + \bar{q}_i(y, \tau)) \right.$$
$$\left. + P_{g \leftarrow g}\left(\frac{x}{y}\right) G(y, \tau) \right], \tag{8.6.28}$$

and for the photon itself an evolution equation analogous to (8.5.12). It is convenient to rewrite these equations in terms of singlet and nonsinglet quark distributions as

$$\Sigma(x, \tau) = \sum_{i = \text{flavors}} [q_i(x, \tau) + \bar{q}_i(x, \tau)] \tag{8.5.60}$$

and

$$\Delta^{(i)}(x, \tau) = q_i(x, \tau) - (1/2n_f)\Sigma(x, \tau), \tag{8.6.29}$$

so that

$$\frac{d\Delta^{(i)}}{d\tau}(x, \tau) = \frac{1}{2\pi} \int_x^1 \frac{dy}{y} \left[\alpha(\tau)(e_i^2 - \langle e^2 \rangle) P_{q \leftarrow \gamma}\left(\frac{x}{y}\right) \gamma(y, \tau) \right.$$
$$\left. + \alpha_s(\tau) P_{q \leftarrow q}\left(\frac{x}{y}\right) \Delta^{(i)}(x, \tau) \right], \tag{8.6.30}$$

where

$$\langle e^2 \rangle \equiv (1/2n_f) \sum_{i=\text{flavors}} e_i^2, \qquad (8.6.31)$$

and

$$\frac{d\Sigma}{d\tau}(x,\tau) = \frac{1}{2\pi} \int_x^1 \frac{dy}{y} \left\{ 2n_f \alpha(\tau) \langle e^2 \rangle P_{q \leftarrow \gamma}\left(\frac{x}{y}\right) \gamma(y,\tau) \right.$$

$$\left. + \alpha_s(\tau) \left[P_{q \leftarrow q}\left(\frac{x}{y}\right) \Sigma(y,\tau) + 2n_f P_{q \leftarrow g}\left(\frac{x}{y}\right) G(y,\tau) \right] \right\}. \quad (8.6.32)$$

Again the solution is straightforward in terms of moments. The important features are illustrated by the computation of the nonsinglet distribution. Taking moments of (8.6.30), we find that

$$\frac{d\Delta_n^{(i)}}{d\tau} = \frac{\alpha_s(\tau)}{2\pi} A_n(q \leftarrow q)\Delta_n^{(i)} + \frac{\alpha}{2\pi}(e_i^2 - \langle e^2 \rangle)A_n(q \leftarrow \gamma), \quad (8.6.33)$$

where

$$A_n(q \leftarrow \gamma) = 3\left[\frac{n^2 + n + 2}{n(n+1)(n+2)} \right]. \qquad (8.6.34)$$

Upon using the expression (8.5.58) for $\alpha_s(\tau)$, this becomes

$$\frac{d\Delta_n^{(i)}}{d\tau} = \frac{A_n(q \leftarrow q)\Delta_n^{(i)}}{2\pi b\tau} + \frac{\alpha}{2\pi}(e_i^2 - \langle e^2 \rangle)A_n(q \leftarrow \gamma), \quad (8.6.35)$$

where

$$b = (33 - 2n_f)/12\pi. \qquad (8.5.56)$$

This inhomogeneous differential equation has the elementary solution

$$\Delta_n^{(i)}(\tau) = \Delta_n^{(i)}(0)[b\tau]^{A_n(q \leftarrow q)/2\pi b} + \frac{\alpha(e_i^2 - \langle e^2 \rangle)A_n(q \leftarrow \gamma)\tau}{2\pi[1 - A_n(q \leftarrow q)/2\pi b]}, \quad (8.6.36)$$

as may readily be verified by direct computation. The second term has the structure of the parton-model result (8.6.26), renormalized by the factor $[1 - A_n(q \leftarrow q)/2\pi b]^{-1}$. This rescaling factor is less than 1 for $n > 1$, because the nonsinglet moments are negative. The first term, which by virtue of the presence of $\Delta_n^{(i)}(0)$ may be said to contain all the details of the hadronic structure of the photon, behaves as a constant (for $n = 1$) or as a negative power of $\tau = \log(Q^2/\Lambda^2)$ (for $n > 1$). Thus it is the pointlike (second) term that dominates as $Q^2 \to \infty$, and the asymptotic behavior of the nonsinglet quark distribution is an absolute prediction, to leading order in the renormalization-group-improved perturbation theory of the strong interactions. The dominant effect of the strong interactions is to soften the quark distribution by gluon radiation, and thus to diminish the structure functions at large x.

The coupled differential equations for the quark singlet and gluon moments may be solved in similar fashion, using the method of Section 8.5, and thus the observable structure functions may be evaluated. The effect of QCD corrections upon F_2^γ is indicated in Fig. 8-34, which shows as well that higher-order corrections tend to diminish further the structure function at large values of x. A first measurement of the photon structure function has been reported by the PLUTO collaboration[38] for $1 \text{ GeV}^2 \lesssim Q^2 \lesssim 15 \text{ GeV}^2$. These early results are consistent with the QCD predictions involving fractionally charged quarks, as shown in Fig. 8-35. Further experimentation promises to test the predictions in detail.

Obviously the prospect of reliable perturbative calculations for the strong interactions is extremely attractive. Therefore there have been, and continue to be, many attempts to apply the methods of perturbative QCD to a large number of processes, including those for which the justification for perturbation theory is not manifest. Among the reactions that have received considerable attention are the entire range of hard-scattering phenomena

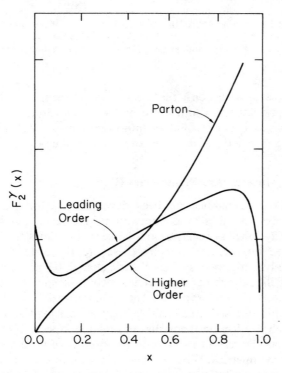

FIG. 8-34. The photon structure function (from W. A. Bardeen, *Proceedings of the 1981 Symposium on Lepton and Photon Interactions at High Energy*, Bonn, edited by W. Pfeil, Physikalisches Institut der Universität Bonn, 1981, p. 432).

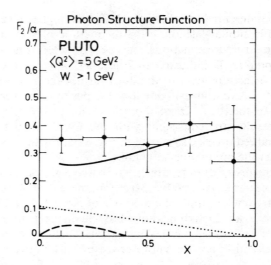

FIG. 8-35. The photon structure function as measured in electron–positron collisions. The curves represent the contributions of the hadronic component of the photon ($\cdots\cdots$) and of the box diagrams of Fig. 8-33 for charmed quarks ($----$), and the sums of the hadronic contribution and the pointlike contribution computed to leading order in QCD with $\Lambda = 0.2$ GeV (———).

in hadron–hadron collisions, the Drell–Yan process, and the production of heavy quark flavors. From perturbative applications of QCD we pass on to brief discussions of two other important topics: symmetries respected by the strong interactions, and the confinement of colored objects.

8.7   Strong-Interaction Symmetries

Quantum chromodynamics has been developed on the basis of an exact color symmetry of the strong interactions. However, as is well known, there are many other symmetries and conservation laws that are respected by the strong interactions. Among these are the exact conservation of additive quantum numbers such as electric charge, baryon number, strangeness, and the like, the flavor symmetries such as isospin and $SU(3)$, and the discrete symmetries under parity, charge conjugation, and time-reversal invariance. There is also a considerable body of work exploiting approximate chiral symmetry, soft-pion theorems, and related aspects of current algebra.

It is of clear interest to investigate the origins of these symmetries and approximate symmetries: Why do they hold, and how do they arise? Before the formulation of QCD, it was common practice to construct effective strong-interaction Lagrangians embodying the important symmetry principles. In such a framework there is no possibility to ask whether symmetries are preserved beyond the lowest order in a perturbative calculation. Moreover, before the development of gauge theories of the weak interactions, it was not possible to contemplate a systematic inquiry into the influence of the (less symmetrical) weak interactions upon the strong. With the availability of supposedly calculable gauge theories of the strong, weak, and electromagnetic interactions, we may seek to penetrate the mystery of the origins of the strong interaction symmetries. We shall mention a few of the important issues, without fully developing either the problems or the solutions. The discussion is abbreviated for several reasons, including the need to develop a considerable phenomenology before presenting a thorough treatment, and the fact that many problems remain open.

Flavor seems to have little to do with quantum chromodynamics. The spectroscopic systematics that were so important for the development of the quark model perhaps emerge from the rule that hadrons be color singlets and the absence of extra quark–antiquark pairs from the static wave functions of hadrons. The unimportance of the sea should be predictable in QCD. It is indeed suggested by the generalization to $SU(N)_{\text{color}}$, with $N \to \infty$. It may then be the case that the different strengths of the strong interactions involving hadrons of various flavors, such as the observation[39] that

$$\sigma_{\text{tot}}(\psi N)/\sigma_{\text{tot}}(\pi N) \simeq 1/10, \qquad (8.7.1)$$

may perhaps have an entirely kinematic origin. An explicit realization of the (transient) importance of kinematics to the shape of Regge trajectories is given[40] by the string model of the hadron spectrum.

Since the notion of flavor seems incidental to QCD, it is natural to ask whether the goodness of isospin symmetry is an accident. We saw in Chapter 7 that the origin of quark masses is apparently in the spontaneous breakdown of the $SU(2)_L \otimes U(1)_Y$ gauge symmetry of the weak and electromagnetic interactions, and specifically in the arbitrary Yukawa interaction term between quarks and scalar bosons. From this perspective, the goodness of isospin symmetry is a consequence of the unexplained fact that

$$m_u \simeq m_d \simeq 0, \qquad (8.7.2)$$

and a small (and also unexplained) quark mass difference

$$m_d > m_u \qquad (8.7.3)$$

may lie at the origin of the nonelectromagnetic (or "explicit") isospin breaking that seems required to explain the sign of the neutron–proton mass difference, among other phenomena.

In contrast, QCD does provide some insight into the successes of current algebra and the long-standing problem of the masses of isoscalar pseudoscalar mesons, which is known as the $U(1)$ problem. In the limit of vanishing quark masses, the QCD Lagrangian will have an exact global $SU(n_f) \otimes SU(n_f)$ chiral symmetry operating independently on the left-handed and right-handed parts of the quark fields for the n_f massless flavors of quarks. That this is approximately so in Nature is evidenced by the success of soft-pion theorems. In the less extreme limit of zero masses for the up, down, and strange quarks, the QCD Lagrangian will generate a flavor octet of exactly conserved axial currents. It is believed that the corresponding chiral symmetry must be spontaneously broken along the lines described by Nambu and Jona-Lasinio[41] in two extremely prescient papers. Accordingly, in the world of three massless quark flavors there should be eight massless Goldstone particles, which we identify with the pseudoscalar octet. Because the up, down, and strange quarks acquire small masses, thanks to electroweak spontaneous symmetry breaking, it follows that the π, K, and η are only approximately massless, although they are presumed to retain some memory of their chiral origin.

In addition to this desirable chiral symmetry, the QCD Lagrangian possesses a vectorial $U(1)$ symmetry associated with baryon number conservation in the strong interactions, and an axial $U(1)$ symmetry that leads to the puzzle of the mass of the flavor singlet pseudoscalar particle. If the corresponding ninth axial current, which corresponds to a flavor singlet, is conserved, the pion should have a light partner η_1 with

$$M(\eta_1) \leq M(\pi)\sqrt{3}. \qquad (8.7.4)$$

The $U(1)$ problem is that the predicted light state does not exist; the obvious candidate $\eta'(957)$ is by no means light. What seems a promising phenomenological explanation is the influence on the spectrum of virtual states composed of glue alone. A formal solution to the $U(1)$ problem was given by 't Hooft,[42] who argued that because of the effects of so-called instanton solutions, the $U(1)$ current has an anomaly that leads to the physical nonconservation of the ninth axial charge. This removes the *raison d'être* for a ninth light pseudoscalar. Other aspects of chiral symmetry breaking have been similarly illuminated.

Let us now turn to the question of the strong-interaction symmetries, which are not respected by the weak and electromagnetic interactions. Parity invariance and strangeness conservation, for example, hold to quite excellent approximation in strong-interaction processes. In particular, experiment shows that they are unbroken at order α^1. However, it is apparent that strong-interaction n-point functions will in general receive corrections from processes involving $SU(2)_L \otimes U(1)_Y$ gauge bosons, as indicated schematically in Fig. 8-36. Such electroweak corrections contain potentially dangerous loop integrals, and, beyond these appearances, one may be put on guard by the phenomenon of nonleptonic enhancement in weak decays, which raises the possibility that symmetry violations may somehow be

FIG. 8-36. Electroweak corrections to strong n-point functions.

amplified. A lengthy analysis by Weinberg[43] shows this not to be the case. Electroweak effects of order α^1 occur only as corrections to the quark mass matrix. As such, they necessarily conserve parity, strangeness, etc., and produce only isovector departures from exact isospin invariance. One facet of the discrete symmetry problem, which may be posed as "Why are the strong interactions not severely polluted by the symmetry violations of the electroweak interactions?" in thus dealt with.

We have not explained, however, why the strong interactions should respect discrete symmetries in the first place, just as we offered no explanation for the parity-violating left-handed form of the weak interactions. In this instance, a problem arises from an unexpected quarter. QCD, as formulated, seems unable to protect itself against self-inflicted *strong* CP violations. Consider adding to the QCD Lagrangian (8.1.2) an additional term of the form

$$\mathscr{L}' = \text{constant} \times \text{tr}(G_{\mu\nu}{}^*G^{\mu\nu}), \tag{8.7.5}$$

where the dual of the field-strength tensor is [compare (3.2.14)] given by

$$^*G^{\mu\nu} = -\tfrac{1}{2}\varepsilon^{\mu\nu\alpha\beta}G_{\alpha\beta}. \tag{8.7.6}$$

In electromagnetism, an analogous term would be of the form

$$\text{tr}(F_{\mu\nu}{}^*F^{\mu\nu}) = 4\mathbf{E}\cdot\mathbf{B}, \tag{8.7.7}$$

which manifestly violates parity, under which

$$\begin{aligned}\mathscr{P}\mathbf{E} &= -\mathbf{E} \\ \mathscr{P}\mathbf{B} &= \mathbf{B}\end{aligned} \tag{8.7.8}$$

but is $\mathscr{C}$-conserving, since

$$\begin{aligned}\mathscr{C}\mathbf{E} &= -\mathbf{E} \\ \mathscr{C}\mathbf{B} &= -\mathbf{B}.\end{aligned} \tag{8.7.9}$$

Consequently the presence of (8.7.5) in the QCD Lagrangian will give rise to violations of $\mathscr{C}\mathscr{P}$ and $\mathscr{P}$ invariance. Because such a term is gauge-invariant and renormalizable, our only basis for excluding it is that it is $\mathscr{C}\mathscr{P}$-violating and hence undesirable. Although not a particularly satisfying course, this would be no worse than what we have done in assuming the left-handed structure of the weak charged-current interaction. However, this procedure is not tenable, at least not without further argumentation. The reason is that, because QCD is a theory of the strong interactions, we are obliged to consider field configurations in which the fields are intense.

The existence of nontrivial topological structures in QCD means that the vacuum cannot be represented simply as a state of vanishing gauge fields, but must be given by a more complex structure that resembles a Bloch wave in condensed-matter physics:

$$|\theta\rangle \equiv \sum_{n=-\infty}^{\infty} e^{in\theta}|n\rangle, \tag{8.7.10}$$

where $|n\rangle$ is the minimum-energy state with topological winding number

$$n = \frac{1}{8\pi^2} \int d^4x \, \mathrm{tr}(G_{\mu\nu} {}^*G^{\mu\nu}). \tag{8.7.11}$$

The parameter θ is a new and arbitrary parameter of QCD. The effect of the θ-vacuum may be represented by a new term in the Lagrangian, of the form

$$\mathscr{L}_\theta = \frac{\theta}{16\pi^2} \mathrm{tr}(G_{\mu\nu} {}^*G^{\mu\nu}). \tag{8.7.12}$$

Although this has the look of a total derivative, or surface term, it cannot be ignored, because the gauge fields do not necessarily vanish at infinity. To comply with observed limits on CP-violating effects such as the upper limit on the neutron's electric dipole moment,[44]

$$|d_n| \lesssim 6 \times 10^{-25} \, e \cdot cm, \tag{8.7.13}$$

the parameter θ must be extraordinarily small,[45]

$$\theta < 10^{-8} - 10^{-9}. \tag{8.7.14}$$

A thoroughly convincing solution to this "strong-CP problem," as it is called, has not yet emerged. The most attractive possibilities at present center on the search for an additional chiral symmetry of the QCD Lagrangian that would permit the angle θ to be rotated to zero.

8.8   Color Confinement

From our abbreviated survey of strong-interaction symmetries we turn to an equally brief treatment of the problem of color confinement. QCD has not been proved to be a confining theory, and thus it is difficult to judge which avenue of approach may ultimately be the most productive. We shall therefore have to be content with a single, and very simplified, argument that makes plausible the possibility of confinement.

We have seen that it is typical in field theories that the coupling constant depend on the distance scale. This dependence can be expressed in terms of a dielectric constant ε. We define

$$\varepsilon(r_0) \equiv 1 \tag{8.8.1}$$

and write

$$g^2(r) = g^2(r_0)/\varepsilon(r). \tag{8.8.2}$$

We assert that the implication of asymptotic freedom is that in QCD the effective color charge decreases at short distances and increases at large distances. In other words, the dielectric "constant" will obey

$$\varepsilon(r) > 1, \qquad \text{for } r < r_0, \tag{8.8.3a}$$
$$\varepsilon(r) < 1, \qquad \text{for } r > r_0. \tag{8.8.3b}$$

Indeed, to second order in the strong coupling we may write

$$\varepsilon(r) = \left[1 + \frac{1}{2\pi}\frac{g^2(r_0)}{4\pi}\left(11 - \frac{2n_f}{3}\right)\ln\left(\frac{r}{r_0}\right) + O(g^4)\right]^{-1} \tag{8.8.4}$$

in QCD, where n_f is the number of active quark flavors.

Let us now consider an idealization based upon electrodynamics. In quantum electrodynamics, we choose

$$\varepsilon_{\text{vacuum}} = 1, \tag{8.8.5}$$

and can show[46] that physical media have $\varepsilon > 1$. The displacement field is

$$\mathbf{D} = \mathbf{E} + \mathbf{P}, \tag{8.8.6}$$

and atoms are polarizable with $\mathbf{P}$ parallel to the applied field $\mathbf{E}$, so that $|\mathbf{D}| > |\mathbf{E}|$. Since the dielectric constant is defined through

$$\mathbf{D} = \varepsilon\mathbf{E} \tag{8.8.7}$$

in these simple circumstances, we conclude that $\varepsilon > 1$.

Now let us consider, in contrast to the familiar situation, the possibility of a dielectric medium with

$$\varepsilon_{\text{medium}} = 0, \tag{8.8.8}$$

a perfect dia-electric medium, or at least

$$\varepsilon_{\text{medium}} \ll 1, \tag{8.8.9}$$

a very effective dia-electric medium. We can easily show that, if a test charge is placed within the medium, a hole will develop around it.

To see this, consider the arrangement depicted in Fig. 8-37(a), a positive charge distribution ρ_+ placed in the medium. Suppose that a hole is formed. Then, because the dielectric constant of the medium is less than unity, the induced charge on the inner surface of the hole will also be positive. The test charge and the induced charge thus repel, and the hole is stable against

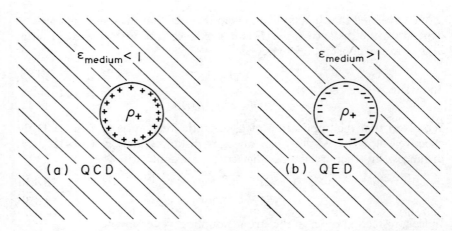

Fig. 8-37. Charge induced by a positive test charge placed at the center of a hole in a dielectric medium. (a) Dia-electric case $\varepsilon_{\text{medium}} < 1$ hoped to resemble QCD. (b) Dielectric case $\varepsilon_{\text{medium}} > 1$ of normal electrodynamics.

collapse. In normal QED, the induced charge will be negative, as indicated in Fig. 8-37(b), and will attract the test charge. The hole is thus unstable against collapse.

The radius of the hole can be estimated on the basis of energetics. Within the hole the electrical energy W_{in} is finite and independent of the dielectric constant of the medium. The displacement field is radial and hence continuous across the spherical boundary. Thus it is given outside the hole by

$$\mathbf{D}_{\text{out}}(r > R) = \hat{\mathbf{r}}Q/r^2, \tag{8.8.10}$$

where Q is the total test charge. The induced charge density on the surface of the hole is

$$\sigma_{\text{induced}} = (1 - \varepsilon)|\mathbf{D}(R)|/4\pi\varepsilon$$
$$= (1 - \varepsilon)Q/4\pi\varepsilon R^2, \tag{8.8.11}$$

which has the same sign as Q, as earlier asserted. Outside the hole, the electric field is determined by the total interior charge

$$Q + (1 - \varepsilon)Q/\varepsilon = Q/\varepsilon, \tag{8.8.12}$$

so that

$$\mathbf{E}_{\text{out}}(r > R) = \hat{\mathbf{r}}Q/\varepsilon r^2. \tag{8.8.13}$$

The energy stored in electric fields outside the hole is then

$$W_{\text{out}} = \frac{1}{8\pi} \int d^3\mathbf{r}\, \mathbf{D}_{\text{out}}(r) \cdot \mathbf{E}_{\text{out}}(r)$$

$$= \frac{1}{2} \int_R^\infty r^2\, dr \frac{Q^2}{\varepsilon r^4} = \frac{Q^2}{2\varepsilon R}. \tag{8.8.14}$$

As the dielectric constant of the medium approaches zero, W_{out} becomes large compared with W_{in}, so that the total electric energy

$$W_{\text{el}} \equiv W_{\text{in}} + W_{\text{out}} \to W_{\text{out}}, \qquad \text{as } \varepsilon \to 0. \qquad (8.8.15)$$

One must consider as well the energy required to hew such a hole out of the medium. For a hole of macroscopic size, it is reasonable to suppose that

$$W_{\text{hole}} = \frac{4\pi R^3}{3} v + 4\pi R^2 s + \cdots, \qquad (8.8.16)$$

where v and s are nonnegative constants. The total energy of the system,

$$W = W_{\text{el}} + W_{\text{hole}}, \qquad (8.8.17)$$

can now be minimized with respect to R. In the regime where the volume term dominates W_{hole}, the minimum occurs at

$$R = \left(\frac{Q^2}{2\varepsilon} \times \frac{1}{4\pi v} \right)^{1/4} \neq 0, \qquad (8.8.18)$$

for which

$$W_{\text{el}} \simeq \left(\frac{Q^2}{2\varepsilon} \right)^{3/4} (4\pi v)^{1/4} \qquad (8.8.19)$$

and

$$W_{\text{hole}} \simeq \frac{1}{3} \left(\frac{Q^2}{2\varepsilon} \right)^{3/4} (4\pi v)^{1/4}, \qquad (8.8.20)$$

so that

$$W \simeq \frac{4}{3} \left(\frac{Q^2}{2\varepsilon} \right)^{3/4} (4\pi v)^{1/4}. \qquad (8.8.21)$$

Thus, in a very effective dia-electric medium, a test charge will induce a bubble or hole of finite radius. Notice, however, that in the limit of a perfect dia-electric medium

$$W \to \infty \qquad \text{as } \varepsilon \to 0. \qquad (8.8.22)$$

An isolated charge in a perfect dia-electric thus has infinite energy. This is the promised analog of the argument used in Section 8.1 to wish away isolated colored objects.

If instead of an isolated charge we place a test dipole within the putative hole in the medium, we can again show that the minimum-energy con-figuration occurs for a hole of finite radius about the test dipole. In this case, however, the field lines need not extend to infinity, so the hole radius remains finite as $\varepsilon \to 0$, and so does the total energy of the system. The analogy between the exclusion of chromoelectric flux from the QCD vacuum and the exclusion of magnetic flux from a superconductor is now suggestive. To separate the dipole charges to $\pm \infty$ requires an infinite amount of work, as shown in the previous example. This is the would-be analog of quark confinement.

Two issues arise in this line of reasoning. One is the question of color confinement. The other is what form does the vacuum of sourceless QCD take if it may be regarded as a perfect, or very effective, dia-electric medium? Is the QCD vacuum unstable against the formation of domains containing dipole pairs in the electrostatic model, corresponding to gluons in color-singlet, spin-singlet configurations?

Quantum chromodynamics incorporates many of the observed systematics of the strong interactions in an elegant way that is in accord with current theoretical prejudices. It promises calculability for the strong interactions in an unspecified, asymptotically free regime. Some observables, such as the total cross section in electron–positron annihilation into hadrons and the structure functions in deeply inelastic lepton–nucleon scattering, hint that the domain of computability is not far away, and may already be accessible to experiment. In the nonperturbative regime, several broad goals of a theory of strong interactions remain unattained, but no longer appear unattainable. Included among unfulfilled aspirations are these: to compute the properties of hadrons, explain the absence of unseen species, and predict the existence of new varieties of hadrons; to explain why quarks and gluons are not observed; and to derive the interactions among hadrons as a collective effect of the interactions among constituents. Experimentally, it is most important to test the confinement hypothesis by searching for free quarks, or for signatures of unconfined color. Sensitive negative searches for quarks continue to be interesting, and the definitive observation of free quarks would be revolutionary.

Problems

8-1.   If the quark colors are designated as red (R), blue (B), and green (G), the gluons may be represented conveniently as $\bar{R}B$, $\bar{R}G$, $\bar{B}R$, $\bar{G}R$, $\bar{B}G$, $\bar{G}B$, $(\bar{R}R - \bar{B}B)/\sqrt{2}$, $(\bar{R}R + \bar{B}B - 2\bar{G}G)/\sqrt{6}$. The last two are color-preserving forms that are orthogonal to the color-singlet combination $(\bar{R}R + \bar{B}B + \bar{G}G)/\sqrt{3}$. The elementary quark–gluon interactions will be of the form

$$\text{Red quark} + \bar{R}B \text{ gluon} \rightarrow \text{Blue quark},$$

etc. Repeat in this language the maximally attractive channel analysis of Section 8.1, and show that the color singlet $q\bar{q}$ and qqq configurations are energetically favored. (References: R. P. Feynman, in *Weak and Electromagnetic Interactions at High Energy*, 1976 Les Houches Lectures, edited by R. Balian and C. H. Llewellyn Smith, North-Holland, Amsterdam, 1977, p. 120; C. Quigg, in *Techniques and Concepts of High-Energy Physics*, edited by T. Ferbel, Plenum, New York, 1981, p. 143.)

8-2. Consider the representations of the group $SU(N)$ that correspond to quark–quark and quark–antiquark states. For the symmetric and anti-symmetric representations of two quarks, and for the singlet and adjoint quark–antiquark representations,
(a) calculate the dimension of the representation;
(b) decompose each representation with respect to $SU(2)_{isospin}$ and $SU(N - 2)$;
(c) evaluate the quadratic Casimir operator $\langle \mathbf{T}^2 \rangle$, where the T_i are the normalized generators of $SU(N)$.

8-3. For quark–antiquark bound states [for a color symmetry group $SU(N)$], calculate the quantity $\langle \mathbf{T}^{(1)} \cdot \mathbf{T}^{(2)} \rangle$ that characterizes the interaction energy to lowest order in perturbation theory. What is the state of lowest energy?

8-4. Consider the Schrödinger equation for s-wave bound states of a $1/r$ potential in N space dimensions:

$$[\nabla^2 + 2\mu(E + \alpha/r)]\,\psi(r) = 0.$$

(a) Show that the radial equation is

$$\left[\frac{d^2}{dr^2} + \frac{(N - 1)}{r}\frac{d}{dr} + 2\mu\left(E + \frac{\alpha}{r} \right) \right]\psi(r) = 0. \qquad (*)$$

(b) Now take the limit of large N, so that $(N - 1) \to N$. Introduce a reduced radial wave function

$$u = r^{N/2}\psi$$

and a scaled radial coordinate

$$R = r/N^2.$$

Show that the Schrödinger equation becomes

$$\left[\frac{1}{N^2}\frac{d^2u}{dR^2} - \frac{u}{4R^2} + 2\mu\left(N^2E + \frac{\alpha}{R} \right)u \right] = 0.$$

(c) Apart from the factor N^2, which sets the scale of E, this equation describes a particle with effective mass μN^2 moving in an effective potential

$$V_{\rm eff} = \frac{1}{8\mu R^2} - \frac{\alpha}{R}.$$

Find the energy of the ground state in the limit as $N \to \infty$, for which the kinetic energy vanishes. Show that it is given by the absolute minimum of $V_{\rm eff}$, so that

$$E_{N\to\infty} = -2\mu\alpha^2/N^2.$$

Corrections to this result may be computed by expanding V_{eff} about the minimum and treating the additional terms as perturbations.

(d) The exact solution to the exact eigenvalue problem (*) is easily verified to be

$$E_{\text{exact}} = -2\mu\alpha^2/(N-1)^2.$$

Show that the exact eigenvalue can be recast in the form of an expansion in powers of $1/N$ as

$$E_{\text{exact}} = -\frac{2\mu\alpha^2}{N^2} \sum_{j=1}^{\infty} jN^{1-j}$$

$$= E_{N\to\infty}\left(1 + \sum_{j=2}^{\infty} jN^{1-j}\right),$$

so that the $N \to \infty$ result may form the basis for a systematic approximation scheme. How many terms must be retained to obtain a 1% approximation for $N = 3$? [Reference: L. D. Mlodinow and N. Papanicolaou, *Ann. Phys. (NY)* **128**, 314 (1980).]

8-5.   Modify the calculation (Problem 6-7) of the process $e^+e^- \to \gamma\gamma$ to describe the reaction $q\bar{q} \to gg$ in quantum chromodynamics. In this case, the two diagrams shown in Problem 6-7 are not by themselves gauge-invariant.

(a) Show that in QCD the quantities $k_{1\nu}(A^{\mu\nu} + \tilde{A}^{\mu\nu})$ and $k_{2\mu}(A^{\mu\nu} + \tilde{A}^{\mu\nu})$ are proportional to $[\lambda^a, \lambda^b]$, where a and b are the $SU(3)$ color indices of the two gluons.

(b) What is the resolution of this noninvariance?

(c) For the full gauge invariant amplitude described by $\varepsilon_{1\nu}^*\varepsilon_{2\mu}^*T^{\mu\nu}$, under what conditions are the requirements $k_{1\nu}T^{\mu\nu} = 0 = k_{2\mu}T^{\mu\nu}$ fulfilled?

8-6.   Repeat the one-loop calculation of the charge renormalization in scalar electrodynamics using the Pauli–Villars regularization. Relate the relative magnitudes of the charge renormalization in scalar and spinor electrodynamics to the cross-section ratio $\sigma(e^+e^- \to \sigma^+\sigma^-)/\sigma(e^+e^- \to \mu^+\mu^-)$ developed in Problems 1-4 and 1-5.

8-7.   Compute the one-loop charge renormalization in (spinor) QED using the method of dimensional regularization.

8-8.   Consider the one-loop modifications to Coulomb scattering in the limit of low momentum transfer $-q^2 \ll m^2$. Beginning from equation (8.2.31), show that the amplitude is modified by a factor

$$\left[1 - \frac{\alpha_R}{15\pi}\frac{q^2}{m^2} + O(\alpha_R^2)\right].$$

Show that this corresponds, in position space, to an additional interaction
of the form

$$\frac{4}{15}\frac{\alpha_R^2}{m^2}\delta^3(\mathbf{x}),$$

and estimate the first-order shift in the energy levels of the hydrogen atom.
(Reference: Uehling, ref. 13.)

8-9. Using the Feynman rules for the Faddeev–Popov ghost given in
Fig. 8-15 and Section 8.3, verify that the modification to the gluon propagator
due to one ghost loop as shown in Fig. 8-16 is given by (8.3.14).

8-10. Consider a gauge theory of the strong interactions based on the
color symmetry group $SO(3)$, in which both quarks and gluons are assigned
to the adjoint representation. By appropriately modifying the color factors
entering the expressions leading to (8.3.27), evaluate the running coupling
constant in one-loop order. What is the condition for asymptotic freedom
in this theory?

8-11. Consider the dissociation of an electron of momentum p into a
photon with momentum $\mathbf{k} = (\vec{k}_\perp, zp)$ and an electron in QED. Calculate
the square of the matrix element for dissociation, and thus the probability
to find such a photon associated with the electron beam. (Reference:
Weizsäcker and Williams, ref. 33.)

8-12. Following the method of Altarelli and Parisi, compute the splitting
function $P_{q\leftarrow q}(z)$ for a theory of colored quarks interacting by means of
scalar gluons for the color group $SU(N)$. Assume that the theory has a
fixed coupling constant α^*. Calculate the Q^2-evolution of the nonsinglet
moments, and predict the slope of the logarithmic ratio

$$\frac{\log(\Delta_n(\tau)/\Delta_n(0))}{\log(\Delta_k(\tau)/\Delta_k(0))}$$

for $(n,k) = (5,3)$ and $(6,4)$. [References: M. Glück and E. Reya, *Phys. Rev.*
D16, 3242 (1977); D. Bailin and A. Love, *Nucl. Phys.* **B75**, 159 (1974).]

8-13. The model of M.-Y. Han and Y. Nambu [*Phys. Rev.* **149B**, 1006
(1965)] is an integer-charge alternative to the fractional-charge quark model,
with charges assigned as

color \ flavor	u	d	s
R	0	-1	-1
G	1	0	0
B	1	0	0

(a) Show that below the threshold for color liberation, the ratio

$$R \equiv \sigma(e^+e^- \to \text{hadrons})/\sigma(e^+e^- \to \mu\mu)$$

is $R = 2$, as in the fractional-charge model, and that $R = 4$ if color can be liberated.

(b) Consider the reaction

$$\gamma\gamma \to \text{hadrons},$$

viewed as $\gamma\gamma \to q\bar{q}$. Show that with fractionally charged quarks

$$\sigma(\gamma\gamma \to \text{hadrons}) \propto \sum_i e_i^4 = \frac{2}{3},$$

and that in the Han–Nambu model

$$\sigma(\gamma\gamma \to \text{hadrons}) \propto \begin{cases} 2 & \text{below color threshold} \\ 4 & \text{above color threshold} \end{cases}$$

[References: F. E. Close, *An Introduction to Quarks and Partons*, Academic, New York, 1979, Chapter 8; M. Chanowitz, in *Particles and Fields-1975*, edited by H. J. Lubatti and P. M. Mockett, University of Washington, Seattle, 1975, p. 448; H. J. Lipkin, *Nucl. Phys.* **B155**, 104 (1979).]

For Further Reading

ORIGINS OF QCD. The idea of a vector gluon theory may be found in
Y. Nambu, in *Preludes in Theoretical Physics in Honor of V. F. Weisskopf*, edited by A. De-Shalit, H. Feshbach, and L. Van Hove, North-Holland, Amsterdam, 1966, p. 133.
Motivation for a color gauge theory is given in
M. Gell-Mann, *Acta Phys. Austriaca Suppl.* **IV**, 733 (1972).
W. A. Bardeen, H. Fritzsch, and M. Gell-Mann, in *Scale and Conformal Symmetry in Hadron Physics*, edited by R. Gatto, Wiley, New York, 1973, p. 139.
Clear formulations of the theory, after the recognition of asymptotic freedom, appear in
D. J. Gross and F. Wilczek, *Phys. Rev.* **D8**, 3633 (1973).
S. Weinberg, *Phys. Rev. Lett.* **31**, 494 (1973).
For general reviews of the theory, consult
W. J. Marciano and H. Pagels, *Phys. Rep.* **36C**, 137 (1978).
F. Wilczek, *Ann. Rev. Nucl. Part. Sci.* **32**, 177 (1982).
QUANTIZATION OF GAUGE THEORIES. The technical aspects and questions of consistency are treated by
E. S. Abers and B. W. Lee, *Phys. Rep.* **9C**, 1 (1973).
R. P. Feynman, in *Weak and Electromagnetic Interactions at High Energy*, 1976 Les Houches Lectures, edited by R. Balian and C. H. Llewellyn Smith, North-Holland, Amsterdam, 1977, p. 120.
L. D. Faddeev and A. A. Slavnov, *Gauge Fields, Introduction to Quantum Theory*, Benjamin, Reading, Massachusetts, 1980.
C. Itzykson and J.-B. Zuber, *Quantum Field Theory*, McGraw-Hill, New York, 1980.
T. D. Lee, *Particle Physics and Introduction to Field Theory*, Harwood Academic, Chur, London, New York, 1981.

CHARGE SCREENING IN QED. An early reference is the paper by
V. F. Weisskopf, *Phys. Rev.* **56**, 72 (1939).
A classic commentary on the evolution of the coupling constant is given by
L. D. Landau, in *Niels Bohr and the Development of Physics*, edited by W. Pauli, Pergamon, London, 1955, p. 52.
The sign of the dielectric constant is the subject of the review article by
O. V. Dolgov, D. A. Kirzhnits, and E. G. Maksimov, *Rev. Mod. Phys.* **53**, 81 (1981).
Experimental consequences of the Uehling term in the atomic potential may be seen in the review of muonic atoms by
E. F. Borie and G. A. Rinker, *Rev. Mod. Phys.* **54**, 67 (1982).
ASYMPTOTIC FREEDOM. For derivations and discussion of the asymptotic freedom of non-Abelian gauge theories, see
V. B. Berestetskii, *Usp. Fiz. Nauk* **120**, 439 (1976) [English translation: *Sov. Phys.-Uspekhi* **19**, 934 (1976)].
T.-P. Cheng and L.-F. Li, *Gauge Theory of Elementary Particle Physics*, Oxford University Press, Oxford, 1983.
C. Itzykson and J.-B. Zuber, *op. cit.*
H. D. Politzer, *Phys. Rep.* **14C**, 129 (1974).
E. Reya, *Phys. Rep.* **69**, 195 (1981).
The proof that the only renormalizable theories that are asymptotically free in four dimensions are non-Abelian gauge theories is due to
S. Coleman and D. J. Gross, *Phys. Rev. Lett.* **31**, 851 (1973).
Some insight into the physical origin of antiscreening is provided by calculations in (non-covariant) Coulomb and axial gauge, including those by
J. Frenkel and J. C. Taylor, *Nucl. Phys.* **B109**, 439 (1976); **B117**, 546E (1976).
J. D. Bjorken, in *Quantum Chromodynamics*, Proceedings of the 7th SLAC Summer Institute on Particle Physics, 1979, edited by A. Mosher, SLAC, Stanford, California, 1980, p. 219.
S. D. Drell, in *A Festschrift for Maurice Goldhaber*, edited by G. Feinberg, A. W. Sunyar, and J. Weneser, *Trans. NY Acad. Sci.* Series II, **40**, 76 (1980).
V. N. Gribov, "Instability of Non-Abelian Gauge Theories and Impossibility of Choice of Coulomb Gauge," lectures at the 12th Winter School of the Leningrad Nuclear Physics Institute, available in English as SLAC-Trans-176 (1977).
A. Duncan, *Phys. Rev.* **D13**, 2866 (1976).
A magnetic moment interpretation of the antiscreening has been given by
N. K. Nielsen, *Am. J. Phys.* **49**, 1171 (1981).
R. J. Hughes, *Phys. Lett.* **97B**, 246 (1980); *Nucl. Phys.* **B186**, 376 (1981).
and nicely reviewed by
K. Johnson, "The Physics of Asymptotic Freedom," in *Asymptotic Realms of Physics*, edited by A. Guth, K. Huang, and R. L. Jaffe, MIT Press, Cambridge, Mass., 1983.
These discussions focus on asymptotic freedom as a paramagnetic, rather than dielectric, effect.
GROUP THEORY FOR GAUGE THEORY CALCULATIONS. Easily applied graphical methods have been devised by
P. Cvitanovic, *Phys. Rev.* **D14**, 1536 (1976).
RENORMALIZATION GROUP METHODS. The renormalization group as a technique for summing to all orders in perturbation theory in electrodynamics was invented by
E. C. G. Stueckelberg and A. Peterman, *Helv. Phys. Acta* **26**, 499 (1953).
M. Gell-Mann and F. E. Low, *Phys. Rev.* **95**, 1300 (1954).
A thorough review of early applications appears in the textbook by
N. N. Bogoliubov and D. V. Shirkov, *Introduction to the Theory of Quantized Fields*, Wiley-Interscience, New York, 1959, Chapter 8.

The modern formulation of the renormalization group equations is due to
   C. Callan, *Phys. Rev.* **D2**, 1541 (1970).
   K. Symanzik, *Comm. Math. Phys.* **18**, 227 (1970).
The power of renormalization group methods for a wide range of physical problems was recognized by
   K. Wilson, *Phys. Rev.* **D3**, 1818 (1971).
For a general introduction see
   K. Wilson, *Sci. Am.* **241**, 140 (August, 1979).
Newcomers to the subject would be well advised to begin their studies with
   S. Coleman, "Dilatations," in *Properties of the Fundamental Interactions*, Vol. 9A, 1971 International School of Physics 《E. Majorana》, Erice, edited by A. Zichichi, Editrice Compositore, Bologna, 1973, p. 358.
Accessible later reviews include
   K. Wilson and J. Kogut, *Phys. Rep.* **12C**, 75 (1974).
   K. Wilson, *Rev. Mod. Phys.* **47**, 773 (1975).
   S.-K. Ma, *Rev. Mod. Phys.* **45**, 589 (1973).
   S.-K. Ma, *Modern Theory of Critical Phenomena*, Benjamin, Reading, Massachusetts, 1976.
Applications of the renormalization group to QCD are stressed by
   D. J. Gross, in *Methods in Field Theory*, 1975 Les Houches Lectures, edited by R. Balian and J. Zinn-Justin, North-Holland, Amsterdam, 1976, p. 140.
   A. Peterman, *Phys. Rep.* **53C**, 157 (1979).
THE $1/N$ EXPANSION. The use of the $N \to \infty$ limit of $SU(N)_{color}$ as a strategy for deriving the consequences of QCD was pioneered by
   G. 't Hooft, *Nucl. Phys.* **B72**, 461 (1974); **B75**, 461 (1974).
Clear introductions to the method, with allusions to different physical situations, are given by
   S. Coleman, "$1/N$," in *Pointlike Structures Inside and Outside Hadrons*, 1979 Erice School, edited by A. Zichichi, Plenum, New York, 1982, p. 11.
   E. Witten, *Nucl. Phys.* **B160**, 57 (1979); *Phys. Today* **33**, 38 (July, 1980); and in *Recent Developments in Gauge Theories*, 1979 Cargèse Lectures, edited by G. 't Hooft *et al.*, Plenum, New York, 1980, p. 403.
For additional applications in atomic physics, see
   L. D. Mlodinow and N. Papanicolaou, *Ann. Phys.* (*NY*) **131**, 1 (1981).
INFRARED–FINITE OBSERVABLES IN e^+e^- ANNIHILATIONS. The Sterman–Weinberg predictions for the energy dependence of the size of a quark jet were extended to the evolution of gluon jets by
   K. Shizuya and S.-H. H. Tye, *Phys. Rev. Lett.* **41**, 787 (1978).
   M. B. Einhorn and B. G. Weeks, *Nucl. Phys.* **B146**, 445 (1978).
Recent assessments of this general topic are given by
   S. D. Ellis, in *Perturbative Quantum Chromodynamics*, Tallahassee, 1981, edited by D. W. Duke and J. F. Owens, American Institute of Physics, New York, 1981, p. 1.
   G. Sterman, *ibid.* p. 22.
CANCELLATION OF INFRARED DIVERGENCES. The classic treatment of the "infrared catastrophe" is due to
   F. Bloch and A. Nordsieck, *Phys. Rev.* **52**, 54 (1937).
For modern treatments with extensive references to the literature, see
   D. R. Yennie, S. C. Frautschi, and H. Suura, *Ann. Phys.* (*NY*) **13**, 379 (1961).
   G. Grammer and D. R. Yennie, *Phys. Rev.* **D8**, 4332 (1973).
The generalization to the case of massless fermions was made by
   T. Kinoshita, *J. Math. Phys.* **3**, 650 (1962).
   T. D. Lee and M. Nauenberg, *Phys. Rev.* **133B**, 1549 (1964).
The infrared finiteness of Yang–Mills theories was shown by
   T. Appelquist, J. Carazzone, H. Kluberg-Stern, and M. Roth, *Phys. Rev. Lett.* **36**, 768 (1976).

DEEPLY INELASTIC SCATTERING. The approach based on evolving parton distributions has been thoroughly reviewed by
  G. Altarelli, *Phys. Rep.* **81**, 1 (1982).
Early applications of asymptotic freedom within the context of the operator product expansion were made by
  H. Georgi and H. D. Politzer, *Phys. Rev.* **D9**, 416 (1974).
  A. DeRujula, H. Georgi, and H. D. Politzer, *Ann. Phys. (NY)* **103**, 315 (1977).
  D. J. Gross and F. Wilczek, *Phys. Rev.* **D9**, 980 (1974).
Higher-order QCD corrections are discussed systematically in
  A. J. Buras, *Rev. Mod. Phys.* **52**, 199 (1980).
A "physical gauge" approach is emphasized by
  Yu. L. Dokshitzer, D. I. Dyakonov, and S. I. Troyan, *Phys. Rep.* **58**, 269 (1980).
Other facets of perturbative calculations are treated by
  A. V. Efremov and A. V. Radyushkin, *Riv. del Nuovo Cim.* **3**, No. 2 (1980).
  A. H. Mueller, *Phys. Rep.* **73**, 237 (1981).
HADRONIC COMPONENT OF THE PHOTON. Many manifestations are treated in the comprehensive article by
  T. H. Bauer, R. D. Spital, D. R. Yennie, and F. M. Pipkin, *Rev. Mod. Phys.* **50**, 260(1978).
The vector meson dominance philosophy is expounded in
  J. J. Sakurai, *Currents and Mesons*, University of Chicago Press, Chicago, 1969.
TWO-PHOTON PROCESSES. Detailed reviews of theoretical prospects have been given by
  H. Terazawa, *Rev. Mod. Phys.* **45**, 615 (1973).
  V. M. Budnev, I. F. Ginzburg, G. V. Meledin, and V. G. Serbo, *Phys. Rep.* **15C**, 181 (1975).
PHOTON STRUCTURE FUNCTION. The proposal to measure the photon structure function is due to
  S. J. Brodsky, T. Kinoshita, and H. Terazawa, *Phys. Rev. Lett.* **27**, 280 (1971).
  T. F. Walsh, *Phys. Lett.* **36B**, 121 (1971).
The parton-model calculation was carried out by
  R. L. Kingsley, *Nucl. Phys.* **B60**, 45 (1973).
  T. F. Walsh and P. Zerwas, *Phys. Lett.* **44B**, 195 (1973).
The QCD calculation in the Altarelli–Parisi framework is reported in
  R. J. DeWitt, L. M. Jones, J. D. Sullivan, D. E. Willen, and H. W. Wyld, Jr., *Phys. Rev.* **D19**, 2046 (1979).
  C. Peterson, T. F. Walsh, and P. Zerwas, *Nucl. Phys.* **B174**, 424 (1980).
  A. Nicolaidis, *Nucl. Phys.* **B163**, 156 (1980).
Higher-order corrections were evaluated by
  W. A. Bardeen and A. J. Buras, *Phys. Rev.* **D20**, 166 (1979).
CURRENT ALGEBRA AND CHIRAL SYMMETRY. There are a number of excellent books:
  S. L. Adler and R. Dashen, *Current Algebras*, Benjamin, New York, 1968.
  J. Bernstein, *Elementary Particles and Their Currents*, W. H. Freeman, San Francisco, 1968.
  B. W. Lee, *Chiral Dynamics*, Gordon and Breach, New York, 1972.
  B. Renner, *Current Algebras and Their Applications*, Pergamon, London, 1968.
  S. B. Treiman, R. Jackiw, and D. J. Gross, *Current Algebra and Its Applications*, Princeton University Press, Princeton, 1971.
THE U(1) PROBLEM. Clear statements of the problem of the ninth pseudoscalar meson were formulated by
  M. Gell-Mann, R. J. Oakes, and B. Renner, *Phys. Rev.* **175**, 2195 (1968).
  S. Weinberg, *Phys. Rev.* **D11**, 3583 (1975).
The mixing with gluonic intermediate states is discussed at a phenomenological level in
  A. DeRujula, H. Georgi, and S. L. Glashow, *Phys. Rev.* **D12**, 147 (1975).
  N. Isgur, *Phys. Rev.* **D13**, 122 (1976).

The connection between mixing and the instanton-induced breaking of the $U(1)$ symmetry is made in the context of the $1/N$ expansion by

E. Witten, *Nucl. Phys.* **B156**, 269 (1979); *Ann. Phys. (NY)* **128**, 363 (1980).

G. Veneziano, *Nucl. Phys.* **B159**, 213 (1979).

P. DiVecchia, *Phys. Lett.* **85B**, 357 (1979).

Other aspects of chiral symmetry breaking in QCD are treated by

G. 't Hooft, in *Recent Developments in Gauge Theories*, 1979 Cargèse Lectures, edited by G. 't Hooft *et al.*, Plenum, New York, 1980, p. 135.

S. Coleman and E. Witten, *Phys. Rev. Lett.* **45**, 100 (1980).

H. Pagels, *Phys. Rev.* **D19**, 3080 (1979).

G. G. Ross, *Rep. Prog. Phys.* **44**, 655 (1981).

INSTANTONS. The topological structure of the gauge potentials and the possibility of tunneling between distinct vacua was identified by

A. M. Polyakov, *Phys. Lett.* **59B**, 82 (1975).

A. A. Belavin, A. M. Polyakov, A. S. Schwartz, and Yu. S. Tyupkin, *Phys. Lett.* **59B**, 85 (1975).

There are several accessible reviews, including

S. Coleman, "The Uses of Instantons," in *The Whys of Subnuclear Physics*, 1977 Erice School, edited by A. Zichichi, Plenum, New York and London, 1979, p. 805.

J. D. Bjorken, *op. cit.*

R. Jackiw, *Rev. Mod. Phys.* **49**, 681 (1977).

A. Actor, *Rev. Mod. Phys.* **51**, 461 (1979).

D. Olive, S. Sciuto, and R. J. Crewther, *Riv. del Nuovo Cim.* **2**, No. 2 (1979).

THE STRONG CP PROBLEM. The danger to CP-invariance posed by the θ-vacuum was developed in

G. 't Hooft, *Phys. Rev. Lett.* **37**, 8 (1976); *Phys. Rev.* **D14**, 3432 (1976).

R. Jackiw and C. Rebbi, *Phys. Rev. Lett.* **37**, 172 (1976).

C. Callan, R. Dashen, and D. Gross, *Phys. Lett.* **63B**, 334 (1976).

The possibility of resolving the problem by imposing a new chiral symmetry was raised by

R. D. Peccei and H. R. Quinn, *Phys. Rev.* **D16**, 1791 (1977).

That the spontaneous breakdown of such a symmetry should give rise to a new light particle, the axion, was remarked by

S. Weinberg, *Phys. Rev. Lett.* **40**, 223 (1978).

F. Wilczek, *Phys. Rev. Lett.* **40**, 279 (1978).

DIELECTRIC ANALOGY TO CONFINEMENT. The picture was formulated by

J. B. Kogut and L. Susskind, *Phys. Rev.* **D9**, 3501 (1974).

A variation appears in

T. D. Lee, *op. cit.*, Chapter 17.

For attempts to deduce an effective dia-electric theory from QCD, see

S. L. Adler, *Phys. Rev.* **D23**, 2905 (1981); **D24**, 1063E (1981).

H. B. Nielsen and A. Patkós, *Nucl. Phys.* **B195**, 137 (1982).

THE MIT BAG MODEL. This phenomenological formulation of a model of confined quarks has been extremely influential in hadron spectroscopy. Good reviews appear in

K. Johnson, *Acta Phys. Polon.* **B6**, 865 (1975).

K. Johnson, in *Fundamentals of Quark Models*, Scottish Universities Summer School in Physics, edited by I. M. Barbour and A. T. Davies, SUSSP, Edinburgh, 1977, p. 245.

R. L. Jaffe, "The Bag," in *Pointlike Structures inside and outside Hadrons*, 1979 Erice School, edited by A. Zichichi, Plenum, New York, 1982, p. 99.

P. Hasenfratz and J. Kuti, *Phys. Rep.* **40C**, 75 (1978).

CONFINEMENT AND NONCONFINEMENT. A general reference on theories of quark confinement is

M. Bander, *Phys. Rep.* **75**, 205 (1981).

If QCD is indeed a confining theory, it is of interest to ask whether spontaneous breaking of the color symmetry can lead to liberation of color. This question is investigated in

A. DeRujula, R. Giles, and R. L. Jaffe, *Phys. Rev.* **D17**, 285 (1978); **D22**, 227 (1980).

H. Georgi, *Phys. Rev.* **D22**, 225 (1980).

L. B. Okun and M. Shifman, *Z. Phys.* **C8**, 17 (1981).

R. Slansky, T. Goldman, and G. L. Shaw, *Phys. Rev. Lett.* **47**, 887 (1981).

COLORED PARTICLES. Some properties of explicitly colored particles built of integrally charged quarks are presented by

M. Y. Han and Y. Nambu, *Phys. Rev.* **D10**, 674 (1974).

F. E. Close, *Acta Phys. Polon.* **B6**, 785 (1975).

References

[1] C. N. Yang and R. L. Mills, *Phys. Rev.* **96**, 191 (1954).

[2] J. J. Sakurai, *Ann. Phys. (NY)* **11**, 1 (1960).

[3] Y. Ne'eman, *Nucl. Phys.* **26**, 222 (1961).

[4] F. Englert and R. Brout, *Phys. Rev. Lett.* **13**, 321 (1964).

[5] G. 't Hooft, *Nucl. Phys.* **B35**, 167 (1971).

[6] G. Altarelli and G. Parisi, *Nucl. Phys.* **B126**, 298 (1977).

[7] Y. Nambu, in *Preludes in Theoretical Physics in Honor of V. F. Weisskopf*, edited by A. De-Shalit, H. Feshbach and L. Van Hove, North-Holland, Amsterdam; Wiley, New York, 1966, p. 133.

[8] O. W. Greenberg and D. Zwanziger, *Phys. Rev.* **150**, 1177 (1966); H. J. Lipkin, *Phys. Lett.* **45B**, 267 (1973); H. Fritzsch, M. Gell-Mann, and H. Leutwyler, *ibid.* **47B**, 365 (1973).

[9] M. Gell-Mann, California Institute of Technology Synchrotron Laboratory Report CTSL-20, 1961; reprinted in M. Gell-Mann and Y. Ne'eman, *The Eightfold Way*, Benjamin, New York, 1964, p. 11.

[10] Useful references are R. L. Jaffe, *Phys. Rev.* **D15**, 281 (1977) (in which the normalizations differ from those adopted here), and J. L. Rosner, in *Techniques and Concepts of High Energy Physics*, St. Croix, 1980, edited by T. Ferbel, Plenum, New York, 1981, p. 1.

[11] G. S. LaRue, J. D. Phillips, and W. M. Fairbank, *Phys. Rev. Lett.* **46**, 967 (1981).

[12] W. Pauli and F. Villars, *Rev. Mod. Phys.* **21**, 434 (1949).

[13] E. A. Uehling, *Phys. Rev.* **48**, 55 (1935). For an early application, see R. Serber, *ibid.* p. 49.

[14] G. 't Hooft and M. Veltman, *Nucl. Phys.* **B44**, 189 (1972).

[15] The factors of i by which these rules appear to differ from those given in Fig. 6-14 for the Weinberg–Salam model are due to the use of a cartesian basis for the gluon color indices, and a spherical basis for the intermediate boson charges.

[16] That the result is indeed gauge-independent may be verified from the expressions given above.

[17] D. J. Gross and F. Wilczek, *Phys. Rev. Lett.* **30**, 1343 (1973); H. D. Politzer, *ibid.* p. 1346. See also the work of G. 't Hooft, *Phys. Lett.* **61B**, 455 (1973); **62B**, 444 (1973), and the interesting calculation by I. B. Khriplovich, *Yad. Fiz.* **10**, 409 (1969) [English translation: *Sov. J. Nucl. Phys.* **10**, 235 (1970)], in which the possibility of an antiscreening effect in non-Abelian gauge theories was noted.

[18] E. de Rafael and J. L. Rosner, *Ann. Phys. (NY)* **82**, 369 (1974), and earlier references cited therein.

[19] W. Caswell, *Phys. Rev. Lett.* **33**, 244 (1974); D. R. T. Jones, *Nucl. Phys.* **B75**, 531 (1974).

[20] R. Jost and J. M. Luttinger, *Helv. Phys. Acta* **23**, 201 (1950).

[21] T. Appelquist and H. Georgi, *Phys. Rev.* **D8**, 4000 (1973); A. Zee, *ibid.* p. 4038. Continuation of the running coupling constant into the timelike regime is discussed in R. P. Feynman, *Phys. Rev.* **76**, 769 (1949).

[22] K. G. Chetyrkin, A. L. Kataev, and F. V. Tkachov, *Phys. Lett.* **85B**, 277 (1979); M. Dine and J. Sapirstein, *Phys. Rev. Lett.* **43**, 668 (1979); W. Celmaster and R. J. Gonsalves, *ibid.* **44**, 560 (1980).

[23] R. Brandelik *et al.*, TASSO Collaboration, *Phys. Lett.* **113B**, 499 (1982).

[24] J. Ellis, M. K. Gaillard, and G. G. Ross, *Nucl. Phys.* **B111**, 253 (1976).

[25] G. Sterman and S. Weinberg, *Phys. Rev. Lett.* **39**, 1436 (1977).

[26] B. Rossi, *High-Energy Particles*, Prentice-Hall, Englewood Cliffs, New Jersey, 1952, Chapter 5.

[27] G. Altarelli and G. Parisi, *Nucl. Phys.* **B126**, 298 (1977).

[28] See, for example, E. Reya, *Phys. Rep.* **69**, 195 (1981).

[29] For a general reference on kinematics, consult E. Byckling and K. Kajantie, *Particle Kinematics*, Wiley, New York, 1973.

[30] H. Abramowicz *et al.*, CDHS Collaboration, *Z. Phys.* **C13**, 199 (1982).

[31] Target-mass corrections are made by using the definition of moments due to O. Nachtmann, *Nucl. Phys.* **B117**, 50 (1976).

[32] For a survey, see A. J. Buras, *Proceedings of the 1981 Symposium on Lepton and Photon Interactions at High Energy*, Bonn, edited by W. Pfeil, Physikalisches Institut der Universität Bonn, 1981, p. 636.

[33] The spirit of this method may be traced to E. Fermi, *Z. Phys.* **29**, 315 (1924), who employed it to calculate the ionization of atoms by α-particles. The development for radiation theory is due to C. F. von Weizsäcker, *ibid.* **88**, 612 (1934) and E. J. Williams, *Phys. Rev.* **45**, 729 (1934).

[34] F. E. Low, *Phys. Rev.* **120**, 582 (1960); a misprint occurs in the expression for $f(x)$.

[35] L. D. Landau and E. M. Lifshitz, *Phys. Z. Sowjetunion* **6**, 244 (1934); E. J. Williams, *Det Kgl. Danske Videnskab. Selskab. Mat.-Fys. Med.* **XIII**, No. 4 (1935).

[36] W. Bartel *et al.*, JADE Collaboration, *Phys. Lett.* **107B**, 163 (1981); R. Brandelik *et al.*, TASSO Collaboration, *ibid.* p. 290.

[37] E. Witten, *Nucl. Phys.* **B120**, 189 (1977).

[38] Ch. Berger *et al.*, PLUTO Collaboration, *Phys. Lett.* **107B**, 168 (1981).

[39] M. Binkley *et al.*, *Phys. Rev. Lett.* **48**, 73 (1982); J. J. Aubert *et al.*, *Phys. Lett.* **89B**, 267 (1980); A. R. Clark *et al.*, *Phys. Rev. Lett.* **43**, 187 (1979).

[40] Y. Nambu, *Phys. Rev.* **D10**, 4262 (1974).

[41] Y. Nambu and J. Jona-Lasinio, *Phys. Rev.* **122**, 345 (1961); **124**, 246 (1961). See also Y. Nambu, *Phys. Rev. Lett.* **4**, 380 (1960) and *The Last Decade in Particle Theory*, edited by E. C. G. Sudarshan and Y. Ne'eman, Gordon and Breach, New York, 1973, p. 33.

[42] G. 't Hooft, *Phys. Rev. Lett.* **37**, 8 (1976).

[43] S. Weinberg, *Phys. Rev.* **D8**, 605, 4482 (1973).

[44] I. S. Altarev *et al.*, *Phys. Lett.* **102B**, 13 (1981); see also N. F. Ramsey, *Phys. Rep.* **43**, 401 (1978), and W. B. Dress *et al.*, *Phys. Rev.* **D15**, 9 (1977).

[45] V. Baluni, *Phys. Rev.* **D19**, 2227 (1979); R. Crewther, P. diVecchia, G. Veneziano, and E. Witten, *Phys. Lett.* **88B**, 123 (1979).

[46] For an elementary argument, see L. D. Landau and E. M. Lifshitz, *Electrodynamics of Continuous Media*, Addison-Wesley, Reading, Massachusetts, 1960, Section 14.

CHAPTER 9

UNIFIED THEORIES

In the early chapters of this book we stressed the economy and elegance of the gauge principle as a guide to the construction of theories of the fundamental interactions among the elementary constituents of matter. Subsequently these ideas were put into practice, and there emerged a satisfying picture of the weak and electromagnetic interactions as well as a promising description of the strong interactions among quarks and gluons. In a systematic if somewhat descriptive fashion, the Weinberg–Salam model and quantum chromodynamics account for many of the prominent experimental observations in subnuclear physics and provide a large measure of understanding of the relationships among different phenomena. Both of these gauge theories face many more experimental tests—the most sharply posed being the predictions for the properties of the intermediate bosons—and many consequences of the theories remain to be worked out, especially in the strong-coupling regime of QCD. Nevertheless, the $SU(3)_{\text{color}} \otimes SU(2)_L \otimes U(1)_Y$ structure serves as a theoretical paradigm and provides the framework in which experimental prospects are considered and experimental results are analyzed.

In earlier discussions we have emphasized the logic, the consequences, and the successes of these theories, while remarking from time to time on the incompleteness or arbitrariness of the descriptions they provide. An important lesson to be drawn from these considerations—both theoretical and phenomenological—is that no obstacles have arisen to the general program of constructing interactions from local gauge symmetries, and that nothing seems to prevent extending the program. In this chapter, we shall not continue to celebrate the successes of the $SU(3)_c \otimes SU(2)_L \otimes U(1)_Y$ picture but shall instead begin by concentrating upon its shortcomings.

269

Some of these will be seen to invite a further unification of the strong, weak, and electromagnetic interactions. Most of the chapter will therefore be devoted to a brief exposition of the simplest theory that unifies these three forces: the theory of Georgi and Glashow[1] based on the gauge group $SU(5)$.

The $SU(5)$ model is by no means the only imaginable "grand unified" theory, nor will it be the answer to all our prayers, but it will nicely illustrate the general strategy and analytical techniques of unification without introducing encumbering complications. Furthermore, it makes a number of predictions that are interesting both as prototypes and as specific targets for experiment. These include the fixing of the weak mixing parameter $\sin^2 \theta_W$ and the expectation that the nucleon be unstable, with its implications for the baryon number of the universe.

9.1   Why Unify?

With quantum chromodynamics and the standard model of weak and electromagnetic interactions in hand, at least as theoretical constructions, what remains to be understood? If both theories are correct, can they also be complete? Actually, there are many observations that are explained only in part, or not at all, by the separate gauge theories of the strong and the electroweak interactions. It is instructive to list the most prominent among these.

- The weak mixing parameter $x_W = \sin^2 \theta_W$ is arbitrary, and there are three distinct coupling constants, which may be characterized by α_s, α_{EM}, and $\sin^2 \theta_W$. This reflects the fact that the gauge group is not simple, but is given by the direct product $SU(3)_c \otimes SU(2)_L \otimes U(1)_Y$. Could the number of independent parameters be reduced to two or one?

- Both quarks and leptons are spin-1/2 particles that are structureless at present resolution. Are they related in any way?

- The leptonic and hadronic charged weak currents are identical in form, being characterized by

$$\begin{pmatrix} v_e \\ e \end{pmatrix}_L \quad \text{and} \quad \begin{pmatrix} u \\ d_\theta \end{pmatrix}_L, \tag{9.1.1}$$

and so forth. This both requires and makes possible the anomaly cancellation that is a prerequisite for the renormalizability of the electroweak theory and suggests the grouping of quark and lepton doublets into fermion "generations." Why should this pattern hold? How many fermion generations are there?

- Why is electric charge quantized? Why is[2] $Q(e) + Q(p) \equiv 0$? Why is $Q(v) - Q(e) \equiv Q(u) - Q(d)$? Why is $Q(d) = (1/3)Q(e)$? Why is $Q(v) + Q(e) + 3Q(u) + 3Q(d) \equiv 0$?

- Fermion masses and mixings are arbitrary. (Why) is the neutrino massless? Higgs boson interactions seem entirely arbitrary. Leaving aside parameters due to the nonperturbative vacuum, the theories entail the *3* coupling parameters noted above, *6* mass parameters for the 6 quarks plus *3* generalized Cabibbo angles, *1* CP-violating phase, *2* parameters for the Higgs potential, and either *3* or *10* mass, mixing, and phase parameters for the leptons (corresponding to massless or massive neutrinos), for a total of *18* or *25* independent parameters.

- Gravitation is absent.

In addition, we may note that in the standard electroweak model the strengths of weak and electromagnetic interactions become comparable for s (or Q^2) $\gg M_W^2$. For example, the cross section for the reaction

$$\bar{v}_e e \to \bar{v}_\mu \mu \tag{9.1.2}$$

is asymptotically

$$\sigma(\bar{v}_e e \to \bar{v}_\mu \mu) \xrightarrow[s \to \infty]{} \frac{G_F^2 M_W^4}{3\pi s} = \frac{\pi \alpha^2}{6 x_W^2 s}, \tag{9.1.3}$$

which is of the same size as the s-channel photon contribution to electron–positron annihilation into muons,

$$\sigma_\gamma(e^+ e^- \to \mu^+ \mu^-) \xrightarrow[s \to \infty]{} \frac{4\pi \alpha^2}{3s}. \tag{9.1.4}$$

Moreover, in view of the evolution of the strong coupling constant given by (8.3.27), it is conceivable that $\alpha_s(Q^2)$ itself approaches α_{EM} for very large values of Q^2.

These questions and observations fall into several categories. Some argue for a qualitative quark–lepton connection. Others inspire a more complete unification of weak and electromagnetic interactions, perhaps in the form of a larger gauge symmetry group

$$G \supset SU(2)_L \otimes U(1)_Y \tag{9.1.5}$$

which would fix x_W. Still others suggest a grand unification of the strong, weak, and electromagnetic interactions, which would automatically complete the electroweak unification. Finally, it is possible to envisage a "super-unification" that would include gravitation as well, perhaps through the medium of a supersymmetric field theory. We shall not discuss this last, and most ambitious, program any further in this volume.

One cannot fail to notice that both QCD and the standard model of the weak and electromagnetic interactions are gauge theories. It is therefore natural to base their unification upon a simple group

$$G \supset SU(3)_{\text{color}} \otimes SU(2)_L \otimes U(1)_Y, \qquad (9.1.6)$$

in order that interactions be determined by one single coupling constant. The scale at which the larger symmetry is attained and the unification is realized is set by the energy at which the running coupling constants coincide, which is approximately 10^{15} GeV. The enlarged gauge group G will imply extra gauge bosons beyond the photon, the intermediate bosons, and the gluons. These new bosons, which will carry both flavor and color properties, are presumably extremely massive, because their effects are unfamiliar to us. One may speculate that all colored gauge bosons will be confined, as the quarks and gluons are presumed to be.

Once unification is undertaken, there is no reason not to assign quarks and leptons to the same representation of the symmetry group. In the absence of symmetry principles that require the exact conservation of lepton number and baryon number, quark–lepton transitions may be expected to occur. What consequences these will have will depend on the details of the unification, but proton decay is a natural, and extremely interesting, outcome.

Why unify? Why not?

9.2   The $SU(5)$ Model

The choice of a unifying gauge group G is to be guided by requirements implicit in our discussion of the motivation for unification. It is worthwhile to formulate these requirements somewhat more precisely. It will be necessary to assume, as indicated in (9.1.6), that G contains $SU(3) \otimes SU(2) \otimes U(1)$. Next, because of the fermion content we wish to build into the theory, G must admit complex representations. To see why this is so, let us examine the $SU(3)_{\text{color}} \otimes SU(2)_L \otimes U(1)_Y$ content of the "first generation" of fermions u, d, v_e, and e. (There is no need to introduce the complication of Cabibbo mixing at this stage.)

It will be convenient to express all the fermions in terms of left-handed fields only. We have frequently used the chiral decomposition of a Dirac field,

$$\psi = \tfrac{1}{2}(1 - \gamma_5)\psi + \tfrac{1}{2}(1 + \gamma_5)\psi = \psi_L + \psi_R, \qquad (9.2.1)$$

which is particularly useful because the gauge-field couplings preserve chirality. If the field ψ is understood to annihilate a particle, the charge-conjugate field

$$\psi^c \equiv C\bar{\psi}^T, \qquad (9.2.2)$$

where T designates transpose, annihilates an antiparticle. In the (standard) representation of Dirac matrices that we have adopted, the charge-conjugation matrix is given by

$$C = i\gamma^2\gamma^0 = -C^{-1} = -C^\dagger = -C^T, \tag{9.2.3}$$

for which

$$C\gamma_\mu C^{-1} = -\gamma_\mu^T. \tag{9.2.4}$$

It is then easy to see that the charge conjugate of a right-handed field is left-handed, for

$$\psi_L^c = \tfrac{1}{2}(1 - \gamma_5)\psi^c = \tfrac{1}{2}(1 - \gamma_5)C\bar\psi^T$$
$$= C\tfrac{1}{2}(1 - \gamma_5)\bar\psi^T = C[\bar\psi\tfrac{1}{2}(1 - \gamma_5)]^T = C(\bar\psi_R)^T, \tag{9.2.5}$$

and similarly

$$\bar\psi_L^c = \psi_R^T C = -\psi_R^T C^{-1}. \tag{9.2.6}$$

Thus we enumerate the fermion fields of the first generation in terms of their $SU(3)_{color} \otimes SU(2)_L \otimes U(1)_Y$ quantum numbers as

$$\begin{array}{rl}
u_L, d_L: & (\mathbf{3}, \mathbf{2})_{Y=1/3} \\
d_L^c: & (\mathbf{3^*}, \mathbf{1})_{2/3} \\
u_L^c: & (\mathbf{3^*}, \mathbf{1})_{-4/3} \\
v_L, e_L: & (\mathbf{1}, \mathbf{2})_{-1} \\
e_L^c: & (\mathbf{1}, \mathbf{1})_2
\end{array} \right\} \tag{9.2.7}$$

If the neutrino is massive, we may wish to add

$$v_L^c: \quad (\mathbf{1}, \mathbf{1})_0. \tag{9.2.8}$$

Evidently this set of representations is not equivalent to its complex conjugate, so we must require that G possess complex representations to accommodate these fermions. As a final requirement, we demand that G have only a single coupling constant. For the purpose of this discussion, that will be what is meant by unification.

One may search systematically for candidate unifying groups that meet these (or indeed, other) criteria. The smallest such group is $SU(5)$. Let us now see how it meets the requirements we have set out. To analyze the structure of the model, it is helpful to refer to the $SU(3) \otimes SU(2) \otimes U(1)$ decomposition of some of the low-dimensioned representations of $SU(5)$ given in Table 9.1. The **15**-dimensional representation, which might be the first hope to accommodate the 15 fermions of the first generation, contains color-sextext quarks, and thus is not acceptable. Similarly, the **45**-dimensional representation, which could be hoped to contain the three known fermion generations, contains color octet and color antisextet elements.

TABLE 9.1: BRANCHING RULES FOR $SU(5) \to SU(3) \otimes SU(2) \otimes U(1)$

Young Tableau	Dimension	$(SU(3), SU(2))_Y$
□	5	$(3,1)_{-2/3} \oplus (1,2)_1$
(tableau)	10	$(3,2)_{1/3} \oplus (3^*,1)_{-4/3} \oplus (1,1)_2$
(tableau)	15	$(6,1)_{-4/3} \oplus (3,2)_{1/3} \oplus (1,3)_2$
(tableau)	$24 = 24^*$	$(8,1)_0 \oplus (3,2)_{-5/3} \oplus (3^*,2)_{5/3} \oplus (1,3)_0 \oplus (1,1)_0$
(tableau)	45	$(8,2)_1 \oplus (6^*,1)_{-2/3} \oplus (3,3)_{-2/3} \oplus (3^*,2)_{-7/3} \oplus (3,1)_{-2/3}$ $\oplus (3^*,1)_{8/3} \oplus (1,2)_1$
(tableau)	50	$(8,2)_1 \oplus (6,1)_{8/3} \oplus (6^*,3)_{-2/3} \oplus (3^*,2)_{-7/3}$ $\oplus (3,1)_{-2/3} \oplus (1,1)_{-4}$

However, the first-generation fermions may be assigned to the $\mathbf{5}^*$ and $\mathbf{10}$ representations as

$$\mathbf{5}^*: \quad \psi_{jL} = \begin{bmatrix} d_1^c \\ d_2^c \\ d_3^c \\ \hline e \\ -v_e \end{bmatrix}_L \tag{9.2.9}$$

and

$$\mathbf{10}: \quad \psi_L^{jk} = \frac{1}{\sqrt{2}} \left(\begin{array}{ccc|cc} 0 & u_3^c & -u_2^c & -u_1 & -d_1 \\ -u_3^c & 0 & u_1^c & -u_2 & -d_2 \\ u_2^c & -u_1^c & 0 & -u_3 & -d_3 \\ \hline u_1 & u_2 & u_3 & 0 & -e^c \\ d_1 & d_2 & d_3 & e^c & 0 \end{array} \right)_L \tag{9.2.10}$$

where the quark colors (Red, Blue, Green) have been denoted $(1, 2, 3)$. The factor $1/\sqrt{2}$ in (9.2.10) is a convenient normalization, and many of the signs are matters of convention. The identification of $(e, -v_e)^T$ as a $\mathbf{2}^*$ of weak isospin follows from the assignment of (v_e, e) as a $\mathbf{2}$. An additional neutrino state (9.2.8) could be assigned as an $SU(5)$ singlet. Although it would have been pleasing to assign all the particles of the first generation to a single irreducible representation, as can be done in the closely analogous group $SO(10)$ (see Problem 9-1), there is nothing objectionable about this assignment.

We see that the $SU(5)$ model can accommodate, if not predict, the known elementary fermions in terms of 15-member generations assigned to the (reducible) $\mathbf{5^*} \oplus \mathbf{10}$ representations. These assignments bring with them some agreeable features. First, since the electric charge operator Q is to be a generator of $SU(5)$, the sum of electric charges over any representation must be zero. (Indeed, this requirement influenced the assignments presented above.) This means in particular that

$$Q(d^c) = (-1/3)Q(e), \qquad (9.2.11)$$

which explains the quantization of electric charge. In fact, although it appears here almost as an arithmetic triviality, charge quantization is deeply related to the existence of magnetic monopoles[3] in unified theories.

We know that the $SU(3)_c \otimes SU(2)_L \otimes U(1)_Y$ subgroup upon which the nonunified gauge theories of strong, weak, and electromagnetic interactions are founded is anomaly-free for the representation we have chosen. This was in fact a part of the motivation for identifying quark and lepton doublets as belonging to the same fermion generation. It remains to verify that the unified theory is also anomaly-free. This may be done either by direct computation, or by using the fact[4] that the anomaly for the representation R may be characterized by

$$\text{Tr}(\{T^a, T^b\}T^c) = A(R)d^{abc}, \qquad (9.2.12)$$

where the T^a are the normalized generators of the representation R and the d^{abc} are the symmetric structure constants of the Lie algebra. For a representation that is given by the completely antisymmetric product of p fundamental representations of $SU(N)$, the coefficient $A(R)$ is given by

$$A(R) = \frac{(N-3)!(N-2p)}{(N-p-1)!(p-1)!}. \qquad (9.2.13)$$

Thus for the $\mathbf{5^*}$ ($p = 4$) we find

$$A(\mathbf{5^*}) = -1, \qquad (9.2.14)$$

and for the $\mathbf{10}$ ($p = 2$) we have

$$A(\mathbf{10}) = 1. \qquad (9.2.15)$$

Thus, remarkably, the $\mathbf{5^*}$ and $\mathbf{10}$ have equal and opposite anomalies, and the unified theory is entirely anomaly-free.

Let us now verify that the gauge boson content is also what is required by the known phenomenology. Constructing a gauge theory by standard methods, we shall encounter 24 gauge bosons corresponding to the 24 elements of the adjoint representation of $SU(5)$. Using the decomposition of Table 9.1, we may identify the gauge bosons as follows:

$$\left.\begin{array}{l} (\mathbf{8,1})_0 \leftrightarrow \text{gluons} \\ (\mathbf{1,3})_0 \leftrightarrow W^+, W^-, W_3 \\ (\mathbf{1,1})_0 \leftrightarrow \mathscr{A} \\ (\mathbf{3,2})_{-5/3} \leftrightarrow X^{-4/3}, Y^{-1/3} \\ (\mathbf{3^*,2})_{5/3} \leftrightarrow X^{4/3}, Y^{1/3} \end{array}\right\} \qquad (9.2.16)$$

The first twelve of these are precisely the known gauge bosons of the $SU(3)_c \otimes SU(2)_L \otimes U(1)_Y$ theory: the eight gluons and the four electroweak bosons that are to become, upon spontaneous symmetry breaking, the physical W^+, W^-, Z^0, and photon. The last dozen objects are new in the $SU(5)$ theory. They are often called leptoquark bosons because they mediate transitions between quarks and leptons. The $SU(5)$ gauge bosons may be displayed in matrix form as

$$V\sqrt{2} = \left(\begin{array}{c|cc} & X_1 & Y_1 \\ \text{gluons} & X_2 & Y_2 \\ & X_3 & Y_3 \\ \hline X_{\bar{1}} X_{\bar{2}} X_{\bar{3}} & W_3/\sqrt{2} & W^+ \\ Y_{\bar{1}} \, Y_{\bar{2}} \, Y_{\bar{3}} & W^- & -W_3/\sqrt{2} \end{array}\right)$$

$$+ \frac{\mathscr{A}}{\sqrt{30}} \begin{pmatrix} -2 & & & & \\ & -2 & & \mathbf{0} & \\ & & -2 & & \\ \mathbf{0} & & & 3 & \\ & & & & 3 \end{pmatrix}, \qquad (9.2.17)$$

where the matrix V is written in terms of the gauge bosons and the normalized generators of the fundamental representation t^a as

$$V \equiv \sum_{a=1}^{24} t^a V^a. \qquad (9.2.18)$$

We have separated the $U(1)_Y$ gauge field to emphasize a crucial difference between the $SU(5)$ theory and the $SU(2)_L \otimes U(1)_Y$ model, which is that the hypercharge coupling is no longer a free parameter. Its strength, as we shall see explicitly below, is fixed by its position in the $SU(5)$ gauge group.

Until now, we have only constructed theories in which the fermions lie in the fundamental representation of the gauge group. (A modest exception to this restriction occurred in Problem 8-10.) The fermion assignments we have chosen for the $SU(5)$ theory necessitate an extension of our methods to include fermions in the **5*** and **10** representations. The transformation properties of these representations and the construction of minimal interaction terms are, however, easily worked out. For each fermion generation, the interaction term in the $SU(5)$ Lagrangian may be written

$$\mathscr{L}_{\text{int}} = -\frac{g_5}{2} G_\mu^a (\bar{u}\gamma^\mu \lambda^a u + \bar{d}\gamma^\mu \lambda^a d)$$

$$-\frac{g_5}{2} W_\mu^i (\overline{L}_u \gamma^\mu \tau^i L_u + \overline{L}_e \gamma^\mu \tau^i L_e)$$

$$-\frac{g_5}{2}\frac{3}{5} \mathscr{A}_\mu \sum_{\substack{\text{fermion} \\ \text{species}}} \bar{f}\gamma^\mu Y f$$

$$-\frac{g_5}{\sqrt{2}} [X_{\mu;\alpha}^-(\bar{d}_R^\alpha \gamma^\mu e_R^c + \bar{d}_L^\alpha \gamma^\mu e_L^c + \varepsilon_{\alpha\beta\gamma}\bar{u}_L^{c\gamma}\gamma^\mu u_L^\beta) + \text{h.c.}]$$

$$+\frac{g_5}{\sqrt{2}} [Y_{\mu;\alpha}^-(\bar{d}_R^\alpha \gamma^\mu v_R^c + \bar{u}_L^\alpha \gamma^\mu e_L^c + \varepsilon_{\alpha\beta\gamma}\bar{u}_L^{c\beta}\gamma^\mu d_L^\gamma) + \text{h.c.}], \quad (9.2.19)$$

where the notation λ^a has been used for the $SU(3)$ matrices, gluons are denoted by G_μ^a, where the color index a runs from 1 to 8, and the color indices α, β, γ run over 1, 2, 3, or Red, Blue, Green. The $SU(2)_L$ doublets are

$$L_u = \begin{pmatrix} u \\ d_\theta \end{pmatrix}_L, \qquad L_e = \begin{pmatrix} v_e \\ e \end{pmatrix}_L. \qquad (9.2.20)$$

The first three terms of (9.2.19) are the standard interactions of the unbroken $SU(3)_c \otimes SU(2)_L \otimes U(1)_Y$ model, provided we identify the coupling constants g_{strong} and $g_{SU(2)_L}$ with g_5. In the weak-hypercharge term we must identify

$$g' = g_5 \sqrt{3/5}, \qquad (9.2.21)$$

where g' corresponds to the canonical definition in the Weinberg–Salam model, as expressed in (6.3.12). The numerical factor arose, as earlier remarked, from the requirement that the field $\mathscr{A}_\mu$ couple to the current of a normalized generator of $SU(5)$. This result may be retrieved in more pedestrian fashion as follows. Because the electric charge Q is a generator of the group, it may be written in the form

$$Q = T_3 + \kappa T_0, \qquad (9.2.22)$$

where T_3 is a generator of $SU(2)_L$, and T_0 is a weak-isosinglet generator of $SU(5)$. For $SU(N)$, we then have that

$$\sum_{\text{representation}} Q^2 = (1 + \kappa^2) \sum_{\text{representation}} T_3^2. \qquad (9.2.23)$$

An explicit computation for the **5*** then yields

$$\tfrac{4}{3} = (1 + \kappa^2)\tfrac{1}{2}, \qquad (9.2.24)$$

so that

$$\kappa^2 = \tfrac{5}{3}. \qquad (9.2.25)$$

This shows that the normalization of T_0 differs by a factor of $\sqrt{3/5}$ from the conventional $U(1)$ hypercharge operator Y.

The important consequence of this information is that the weak-hypercharge coupling is no longer an independent parameter, but is given by

$$g'^2 = \tfrac{3}{5} g^2_{SU(2)}. \qquad (9.2.26)$$

Recalling the definition (6.3.49) of the weak mixing angle, we find that

$$x_{\mathrm{W}} = \sin^2 \theta_{\mathrm{W}} = \frac{g'^2}{g^2_{SU(2)} + g'^2} = \frac{3}{8} \qquad (9.2.27)$$

in the unbroken $SU(5)$ theory.

The fourth and fifth terms in the interaction Lagrangian correspond to new transitions that occur in the unified theory. The new vertices, summarized in Fig. 9-1, mediate processes such as proton decay, which proceeds by the three elementary transitions shown in Fig. 9-2. Each of these changes both baryon number and lepton number by -1 unit and changes the number of elementary fermions by -4.

The intermediate bosons W^+, W^-, and Z^0 are expected to acquire masses according to something resembling the usual mechanism of spontaneous symmetry breaking. The leptoquark bosons X and Y must *a fortiori* be endowed with enormous masses ($\sim 10^{15}$ GeV/c^2) by means of a similar scheme, for the theory to survive the existing bounds[5]

$$\tau(p) > 3 \times 10^{31} \text{ years} \cdot \frac{\Gamma(p \to \mu + \text{anything})}{\Gamma(p \to \text{all})} \qquad (9.2.28)$$

on the proton lifetime.

The necessary symmetry breaking can be achieved in two steps. First, a **24** of real scalar fields is introduced to break

$$SU(5) \xrightarrow[\text{24}]{} SU(3)_c \otimes SU(2)_L \otimes U(1)_Y. \qquad (9.2.29)$$

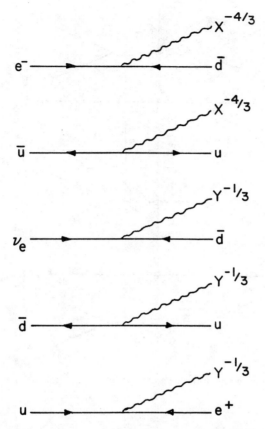

FIG. 9-1. New fermion–fermion transitions that appear in the $SU(5)$ unified theory.

At this point the X and Y leptoquark bosons acquire mass. Next, a **5**-dimensional representation of complex scalar fields is employed to accomplish the breaking

$$SU(3)_c \otimes SU(2)_L \otimes U(1)_Y \underset{\mathbf{5}}{\rightarrow} SU(3)_c \otimes U(1)_{\text{EM}}. \qquad (9.2.30)$$

This is the straightforward extension of the spontaneous symmetry breaking in the Weinberg–Salam model described in Section 6.3, in which a complex scalar $SU(2)_L$ doublet breaks $SU(2)_L \otimes U(1)_Y$ down to $U(1)_{\text{EM}}$. The spontaneous symmetry breaking gives rise to many physical Higgs bosons. It is instructive to consider the logic of the breakdown in some detail.

To accomplish the symmetry breaking at the leptoquark scale, we assume that the adjoint **24** representation of real scalar bosons acquires a vacuum

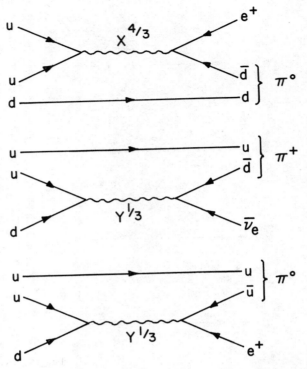

FIG. 9-2. Some mechanisms for proton decay in the $SU(5)$ model of unification.

expectation value

$$\langle \Phi_{24} \rangle_0 = \begin{pmatrix} 1 & & & & \\ & 1 & & & \\ & & 1 & & \\ \hline & & & -\frac{3}{2} & \\ & & & & -\frac{3}{2} \end{pmatrix} \cdot v_{24}, \qquad (9.2.31)$$

where v_{24} is the parameter that will set the scale of masses of the X and Y bosons. It is obvious that this preserves separately the $SU(3)$ symmetry of the first three indices and the $SU(2)$ symmetry of the last two indices. However, the symmetry under the $SU(5)$ generators that couple the color indices to the $SU(2)_L$ indices will be broken, and the corresponding gauge bosons will acquire mass. Referring to the $SU(3) \otimes SU(2) \otimes U(1)$ decomposition of the **24** given in Table 9.1, we see that the X and Y bosons corresponding to the $(\mathbf{3},\mathbf{2})_{-5/3}$ and $(\mathbf{3^*},\mathbf{2})_{5/3}$ representations of the subgroup acquire masses, whereas the gluons $(\mathbf{8},\mathbf{1})_0$, incipient W bosons $(\mathbf{1},\mathbf{3})_0$, and hypercharge boson $(\mathbf{1},\mathbf{1})_0$ remain massless. These are accompanied by twelve physical scalar particles, or Higgs bosons, corresponding to the $(\mathbf{8},\mathbf{1})_0$, $(\mathbf{1},\mathbf{3})_0$, and $(\mathbf{1},\mathbf{1})_0$ representations. The masses of these Higgs bosons will be comparable to those acquired by the leptoquark bosons. The **24** will not endow the fermions with similar masses because it does not occur in the $\bar{\mathrm{L}}\mathrm{R}$ products [cf. (7.1.34) and (6.3.18)]

$$\mathbf{5^*} \otimes \mathbf{10} = \mathbf{5} \oplus \mathbf{45}$$
$$\mathbf{10} \otimes \mathbf{10} = \mathbf{5^*} \oplus \mathbf{45^*} \oplus \mathbf{50^*} \tag{9.2.32}$$

that generate fermion mass terms.

The complex **5** contains the complex $(\mathbf{1},\mathbf{2})_1$ representation used for spontaneous symmetry breaking in the standard model of the weak and electromagnetic interactions. It is therefore appropriate to choose the vacuum expectation value

$$\langle\phi_5\rangle_0 = \begin{pmatrix} 0 \\ 0 \\ 0 \\ 0 \\ 1 \end{pmatrix} v_5/\sqrt{2}, \tag{9.2.33}$$

where $v_5 \simeq 246$ GeV [cf. (6.3.45)]. Of the ten real fields of the complex **5**, only three are absorbed into the longitudinal components of the now-massive $W^\pm$ and Z^0. In addition to the normal Higgs scalar H^0, still with unknown mass, there are six additional physical scalar particles, the color triplet $h^{\pm 1/3}$, corresponding to the

$$(\mathbf{3},\mathbf{1})_{-2/3} \quad \text{and} \quad (\mathbf{3^*},\mathbf{1})_{2/3} \tag{9.2.34}$$

representations. These are potentially troublesome objects because they can mediate nucleon decay through processes such as

$$u + d \to h^{1/3} \to \begin{cases} e^+ + \bar{u} \\ \bar{v}_e + \bar{d} \end{cases} \tag{9.2.35}$$

The Yukawa couplings are as usual closely tied to the fermion masses. Unless these vanish, the only way to suppress the rate for $h^{\pm 1/3}$-mediated proton decay is to arrange that the color-triplet scalar particles receive masses comparable to those of the X and Y bosons. This may be achieved[6] by introducing locally gauge-invariant interactions among Φ_{24} and ϕ_5, which would in any event arise from quantum corrections to the lowest-order interaction terms. However, maintaining a ratio of thirteen orders of magnitude between the electroweak scale characterized by v_5 and the leptoquark scale characterized by v_{24} requires an exceedingly delicate tuning of parameters in the Higgs potentials. This balancing act is neither natural nor likely to survive radiative corrections. Its precarious nature may well be a symptom of the incompleteness of this minimal example of a unified theory.

We noted earlier, in our discussion of the symmetry-breaking effects of the **24**, that scalars belonging to the fundamental **5** representation could generate fermion masses in the manner familiar from the Weinberg–Salam model. The necessary Yukawa term in the Lagrangian is

$$\mathscr{L}_{\text{Yukawa}} = \{G_d \overline{\psi}^c_{Rj} \psi^{jk}_L \phi^\dagger_k + G_u \varepsilon_{jklmn} \overline{\psi}^{cjk}_L \psi^{lm}_L \phi^n\} + \text{h.c.} \qquad (9.2.36)$$

for a single generation. If we write the vacuum expectation value of the scalar field as

$$\langle \phi_k \rangle_0 = \delta^5_k v_5 / \sqrt{2}, \qquad (9.2.37)$$

we see that the first of the Yukawa terms simply becomes

$$\mathscr{L}_d = \frac{G_d v_5}{\sqrt{2}} (\overline{\psi}^c_{Rj} \psi^{j5}_L + \text{h.c.})$$

$$= \frac{-G_d v_5}{\sqrt{2}} (\overline{d}^\alpha_R d^\alpha_L + \overline{e}^c_R e^c_L + \text{h.c.})$$

$$= \frac{-G_d v_5}{\sqrt{2}} (\overline{d}d + \overline{e}e), \qquad (9.2.38)$$

where α is a color index for the quarks. The masses of the electron and the down quark are therefore given by

$$m_d = m_e = G_d v_5 / \sqrt{2}. \qquad (9.2.39)$$

That is, the $SU(5)$ symmetry requires the down quark and electron mass matrices to be the same. In similar fashion, for several generations we obtain in addition the predictions

$$m_s = m_\mu, \qquad (9.2.40)$$

$$m_b = m_\tau. \qquad (9.2.41)$$

The mass of the up quark, generated by the second term in (9.2.36), continues to be a free parameter of the theory. Although the relations (9.2.39–41) support in a qualitative fashion our identification of the quarks and leptons that make up each fermion generation, they are not quantitatively successful. However, as we have seen earlier for coupling constants, masses in field theory are to be interpreted as parameters that depend on the momenta at which they are measured. The equalities (9.2.39–41) should then be interpreted as predictions that apply for $Q^2 \gtrsim M^2_X$. At lower values of Q^2, the equalities are then broken by the differing rates of evolution of the fermion mass operators, with results that are at least in schematic agreement with experiment.[6] We shall not pursue this matter any further but shall return instead to the subject of the coupling constants and their evolution.

In motivating the unification of the strong, weak, and electromagnetic interactions, we remarked on the possibility that the running coupling constants of the $SU(3)_{color}$, $SU(2)_L$, and $U(1)_Y$ gauge groups might evolve to a common value at some large value of Q^2. We have now seen that in the unbroken $SU(5)$ theory the equality of coupling constants indeed emerges when the $U(1)_Y$ coupling is suitably normalized. This was exhibited in (9.2.19) and reflected in the prediction (9.2.27) for the weak mixing parameter $\sin^2 \theta_W$. It is now appropriate to ask, at what value of Q^2 does this equality obtain, or, in other terms, what have the predictions of the unified theory to do with the low-energy world in which we live? The general analysis of these questions was given by Georgi, Quinn, and Weinberg.[7]

The evolution of running coupling constants has been investigated in Sections 8.2 and 8.3 for Abelian and non-Abelian gauge theories, respectively. In both cases, the coupling-constant evolution is influenced by the number of particles that can appear in bubble corrections to the gauge boson propagator. For QCD, for example, the number of quark flavors influences the rate of change of the coupling constant. Before transcribing the appropriate results from Chapter 8, we need only cite an important technical result from field theory:[8] At each momentum or mass scale μ, particles that have masses $M \gg \mu$ effectively decouple from matrix elements involving ordinary external particles, with masses $\lesssim \mu$. This means, for present purposes, that we may ignore the contribution of a particle to the coupling-constant evolution when the particle's mass exceeds the momentum scale of interest. The practical effect of this decoupling theorem is to excuse us from worrying about the effects of superheavy particles.[9]

Let us now consider the evolution of the three coupling constants in turn. Writing as usual $\alpha \equiv g^2/4\pi$, we have from (8.3.27) that the strong $SU(3)_{color}$ coupling constant behaves as

$$1/\alpha_3(Q^2) = 1/\alpha_3(\mu^2) + b_3 \log(q^2/\mu^2), \qquad (9.2.42)$$

with

$$4\pi b_3 = 11 - 2n_f/3, \qquad (9.2.43)$$

where n_f is the number of flavors of quarks with masses less than $\sqrt{Q^2}$. It will be more apt for present purposes to count the number of fermion generations n_g, rather than the number of individual flavors, so that

$$4\pi b_3 = 11 - 4n_g/3. \qquad (9.2.44)$$

Now referring to (8.3.26) we may write the evolution of the $SU(2)_L$ weak-isospin coupling constant as

$$1/\alpha_2(Q^2) = 1/\alpha_2(\mu^2) + b_2 \log(Q^2/\mu^2), \qquad (9.2.45)$$

with

$$4\pi b_2 = (22 - 4n_g)/3, \tag{9.2.46}$$

neglecting a small contribution from Higgs bosons. The first term in b_2 is the standard gauge boson contribution for an $SU(2)$ theory. The second is the contribution of fermion loops, which is composed as follows. In each generation there are four left-handed fermion doublets: one lepton doublet and the quark doublet in three colors. Each of these doublets contributes the standard QED value $(-4/3)$ times $(1/2)$ from the group factor

$$\text{tr}\left(\frac{\tau^a}{2} \cdot \frac{\tau^b}{2}\right) = \frac{\delta^{ab}}{2} \tag{9.2.47}$$

analogous to (8.3.9), times a spin factor $(1/2)$ because only left-handed fermions appear in the loop (see Problem 9-4). Thus we have a total fermion contribution of

$$4 \cdot n_g \cdot \left(-\frac{4}{3}\right) \cdot \left(\frac{1}{2}\right) \cdot \left(\frac{1}{2}\right) = -\frac{4n_g}{3}. \tag{9.2.48}$$

Finally, for the $U(1)_Y$ weak hypercharge coupling, which evolves as

$$(5/3) \cdot 1/\alpha_1(Q^2) = 1/\alpha_Y(Q^2) = 1/\alpha_Y(\mu^2) + b_Y \log(Q^2/\mu^2), \tag{9.2.49}$$

we simply have the QED contribution of the fermion loops weighted by the square of the hypercharge rather than the square of the electric charge for each species:

$$4\pi b_Y = \left(-\frac{4}{3}\right)\left[\frac{1}{2}\text{tr}(Y_L^2) + \frac{1}{2}\text{tr}(Y_R^2)\right]n_g$$

$$= \left(-\frac{4}{3}\right) \cdot \frac{5}{3} \cdot n_g = -\frac{20n_g}{9}. \tag{9.2.50}$$

The factor 5/3 recovered here simply reflects the now-familiar ratio (9.2.21) between the normalizations of α_Y and α_1.

If we now impose the equality of the $SU(3)_c$, $SU(2)_L$, and $U(1)_Y$ coupling constants at the unification scale u, we have that

$$1/\alpha_3(Q^2) - 1/\alpha_2(Q^2) = (11/12\pi)\log(Q^2/u^2). \tag{9.2.51}$$

Then with the definition [compare (6.3.48)] of the electromagnetic coupling constant

$$1/\alpha \equiv 1/\alpha_Y + 1/\alpha_2, \tag{9.2.52}$$

which has but a formal meaning for momenta above the scale of electroweak symmetry breaking, we may write

$$1/\alpha(Q^2) = 1/\alpha_Y(u^2) + 1/\alpha_2(u^2) + (b_Y + b_2)\log(Q^2/u^2)$$

$$= (8/3) \cdot 1/\alpha_3(u^2) + (b_Y + b_2)\log(Q^2/u^2). \tag{9.2.53}$$

We may then form the combination

$$(8/3) \cdot 1/\alpha_3(Q^2) - 1/\alpha(Q^2) = (8b_3/3 - b_Y - b_2)\log(Q^2/u^2)$$
$$= (11/2\pi)\log(Q^2/u^2), \qquad (9.2.54)$$

which does not depend upon the number of (light) fermion generations. This may be used, together with experimentally determined values of the strong and electromagnetic coupling constants, to estimate the scale at which unification—in the sense of equal values of the coupling constants—is achieved.

Clearly, if we choose to evaluate the coupling constants in the low-energy regime, the expression for $\log(Q^2/u^2)$ will be dominated by $1/\alpha$. It is worth noting that the unification scale u is exponentially sensitive to the value of α, and so it is important to take into account the Q^2-evolution of α itself. If we choose to work at the electroweak scale of about $M_W \simeq 100$ GeV/c^2, we find that

$$1/\alpha(M_W^2) = 1/\alpha(m_e^2) - (1/3\pi)\sum_{\substack{\text{light}\\\text{fermions}}} Q_f^2\log(M_W^2/m_e^2)$$

$$\simeq 128. \qquad (9.2.55)$$

Then with

$$\alpha_3(M_W^2) = 0.11, \qquad (9.2.56)$$

a plausible but by no means well-established value, we deduce that the unification scale is

$$u \simeq 10^{15} \text{ GeV/c}^2. \qquad (9.2.57)$$

Although this is enormous on the scale of laboratory energies (a fact that underlines the danger of insouciant extrapolations), it is small compared with the Planck mass

$$M_P \equiv (\hbar c/G_{\text{Newton}})^{1/2} = 1.22 \times 10^{19} \text{ GeV/c}^2, \qquad (9.2.58)$$

at which the strength of gravitational interactions approaches unity. This may be seen to justify *a posteriori* the neglect of gravity in the construction of the unified theory. How two mass scales as different as u and M_W may arise and be preserved remains mysterious.

Having roughly estimated the unification mass, we are now in a position to calculate the values of the $SU(3)_c \otimes SU(2)_L \otimes U(1)_Y$ coupling constants at low energy. These are shown, in one-loop approximation, for three fermion generations in Fig. 9-3. There we show the evolution of the $SU(3)_c$, $SU(2)_L$, and $U(1)$ couplings α_3, α_2, and α_1, which attain a common value at the unification scale u, as well as the weak-hypercharge coupling α_Y and, as a derived quantity, the "electromagnetic" coupling constant defined by (9.2.52).

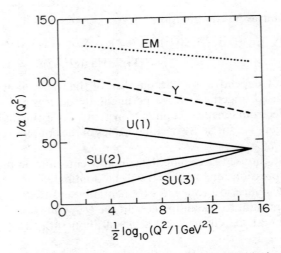

FIG. 9-3. Evolution of running coupling constants in leading logarithmic approximation in the $SU(5)$ model. Three fermion generations are assumed.

Of special interest is the weak mixing parameter

$$x_W = \sin^2 \theta_W = \alpha/\alpha_2, \qquad (9.2.59)$$

which is plotted in Fig. 9-4. A convenient expression for this quantity may by obtained as follows. Subtract from the definition (9.2.52) of the electromagnetic coupling an amount

$$(8/3) \cdot 1/\alpha_2 = (8/3) \cdot x_W/\alpha \qquad (9.2.60)$$

to obtain

$$[1 - 8x_W(Q^2)/3] \cdot 1/\alpha(Q^2) = 1/\alpha_Y(Q^2) - (5/3) \cdot 1/\alpha_2(Q^2)$$
$$= (b_Y - 5b_2/3)\log(Q^2/u^2). \qquad (9.2.61)$$

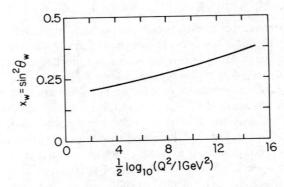

FIG. 9-4. Evolution of the $\gamma - Z^0$ mixing parameter in the $SU(5)$ model (same assumptions as for Fig. 9-3).

This may be solved at once to yield

$$x_W(Q^2) = \frac{3}{8} - \alpha(Q^2)\frac{(3b_Y - 5b_2)}{8}\log\left(\frac{Q^2}{u^2}\right)$$

$$= \frac{3}{8} + \frac{55\alpha(Q^2)}{48\pi}\log\left(\frac{Q^2}{u^2}\right). \qquad (9.2.62)$$

The implied value of the weak mixing parameter at the electroweak scale,

$$x_W(M_W^2) \simeq 0.21 \qquad (9.2.63)$$

is consistent with the value[10]

$$x_W = 0.233 \pm 0.009\,(\pm 0.005) \qquad (9.2.64)$$

extracted from experiments. This would seem to be a remarkable triumph of the minimal unification hypothesis. It remains to be seen whether refined calculations and experiments will sustain this apparent agreement. In this instance, giving a precise definition of the theoretical uncertainties is no less important than reducing experimental errors.

9.3 Nucleon Decay

We have seen, in a minimal unified theory, the possibility of sharpening the low-energy predictions of the $SU(2)_L \otimes U(1)_Y$ electroweak theory by predicting the value of the weak mixing parameter. This has definite implications for the properties of the intermediate vector bosons of the weak interactions. Beyond the esthetic arguments for unification, this represents an important advance in predictive power, although—as we shall discuss at greater length in Section 9.5—the theory is not yet free of arbitrariness. However, the most dramatic consequence of unification, in a qualitative sense, is the prediction that the nucleon be unstable.

It was already remarked in connection with Problem 3-7 that the exact conservation of baryon number is not implied by any of the known symmetry principles. The unified theory goes beyond merely reminding us of this fact by providing specific mechanisms by which nucleons may decay. Without serious calculation we may note from the resemblance of the muon decay process to the diagrams relevant to proton decay shown in Figs. 9-1 and 9-2 that the proton lifetime must be given approximately by

$$\tau_p \simeq \tau_\mu \left(\frac{m_\mu}{m_p}\right)^5 \left(\frac{u}{M_W}\right)^4, \qquad (9.3.1)$$

where all coupling constants have been assumed equal and the unification scale u has been taken to represent the mass of the leptoquark bosons. Using the measured muon and proton masses in the somewhat schematic phase-space factor together with the observed muon lifetime

$$\tau_\mu = 7 \times 10^{-14} \text{ year} \qquad (9.3.2)$$

and the extremely crude conclusion (9.2.57) that

$$u/M_W \simeq 10^{13}, \qquad (9.3.3)$$

we estimate that

$$\tau_p \simeq 10^{34} \text{ years}. \qquad (9.3.4)$$

Clearly, this estimate is both very uncertain and exquisitely (but not exponentially!) sensitive to the unification scale.

Before looking into the elements of a more serious calculation, however, let us note that this rough prediction is comfortably larger than the experimental lower bound quoted in (9.2.28), so the theory is not trivially ruled out. Moreover, the predicted rate is sufficiently near to the existing limits that further experimentation is invited. Recalling that 1 cm^3 of water contains Avogadro's number (6×10^{23}) of nucleons, we readily find that a year's careful scrutiny of a cube of water 10 meters on a side (6×10^{32} nucleons) is the magnitude of effort required to reach the decay rate suggested by the $SU(5)$ theory. Although not a table-top experiment, the required apparatus is not at all unthinkable. Indeed, a number of new searches for nucleon instability have been mounted, with theoretical sensitivities reaching to 10^{33} years. Cosmic-ray backgrounds may well prove to be the ultimate limits to sensitivity.

The detailed computation of nucleon decay rates is a rather involved undertaking, which requires the development of more technology than is appropriate to this introduction. Apart from the desirability of estimating the unification scale with greater care, there is the need to compute transition matrix elements within physical hadrons, which requires understanding the strong-interaction or bound-state corrections to the elementary vertices of Fig. 9-2. This necessitates the intervention of models for hadronic structure as well as a more systematic application of the renormalization group so that elementary vertices computed at the unification scale may be related to transition amplitudes relevant on the scale of hadronic dimensions. Some specialized reviews are included in the bibliography to this chapter.

In contrast, it is quite straightforward to enumerate the allowed channels for nucleon decay. The leptoquark terms in the interaction Lagrangian (9.2.19) generate an effective four-fermion Lagrangian in the local (large unification scale) limit, which may be written as

$$\mathscr{L}_{\text{eff}} = \frac{4G_u}{\sqrt{2}} \varepsilon_{\alpha\beta\gamma} [\overline{u}^{c\gamma}\gamma^\mu u_L^\beta(\overline{e}_L^c\gamma_\mu d_L^\alpha + \overline{e}_R^c\gamma_\mu \overline{d}_R^\alpha)$$

$$+ \overline{u}_L^{c\beta}\gamma^\mu d_L^\gamma(\overline{v}_R^c\gamma_\mu d_R^\alpha + \overline{e}_L^c\gamma_\mu u_L^\alpha)] + \text{h.c.}, \qquad (9.3.5)$$

where, in analogy with the Fermi constant defined in (6.3.44) by

$$G_F/\sqrt{2} = g_2^2/8M_W^2 \qquad (9.3.6)$$

we have introduced

$$G_u/\sqrt{2} = g_5^2/8u^2. \qquad (9.3.7)$$

As usual, the indices α, β, γ run over the three quark colors. After a Fierz rearrangement of the final term in (9.3.5), the effective Lagrangian takes the form

$$\mathscr{L}_{\text{eff}} = \frac{4G_u}{\sqrt{2}} \varepsilon_{\alpha\beta\gamma} [\overline{u}_L^{c\gamma} \gamma^\mu u_L^\beta (2\overline{e}_L^c \gamma_\mu d_L^\alpha + \overline{e}_R^c \gamma_\mu d_R^\alpha) + \overline{u}_L^{c\beta} \gamma^\mu d_L^\gamma \overline{v}_R^c \gamma_\mu d_R^\alpha] + \text{h.c.} \qquad (9.3.8)$$

for a single generation, ignoring the effects of mixing between generations.

The possible semifinal states for proton decay are therefore

$$p \to d\overline{d}e^+, \qquad u\overline{u}e^+, \qquad u\overline{d}\,\overline{v}_e, \qquad (9.3.9)$$

and, generalizing to a second generation,

$$p \to u\overline{s}\,\overline{v}_\mu, \qquad d\overline{s}\mu^+. \qquad (9.3.10)$$

Similarly for neutron decay we have

$$n \to d\overline{u}e^+, \qquad d\overline{d}\,\overline{v}_e, \qquad d\overline{s}\,\overline{v}_\mu. \qquad (9.3.11)$$

Nucleon decay always leads, in the $SU(5)$ model, to antileptons and non-negative strangeness in the final state, but the conclusion that $B - L$, the difference of baryon number and lepton number, is conserved is more generally valid in unified theories.[11] In specific calculations, the most important decays are

$$p \to \pi^0 e^+, \qquad \rho^0 e^+, \qquad \text{and } \omega^0 e^+ \qquad (9.3.12)$$

and

$$n \to \pi^- e^+, \qquad \rho^- e^+, \qquad \text{and } \overline{v} + \text{anything.} \qquad (9.3.13)$$

Those that lead to photons and charged particles in the final state should be detectable with high efficiency.

Similarly, detailed calculations show that within the $SU(5)$ model the proton lifetime is likely to lie in the range

$$\tau_p = 3 \times 10^{30 \pm 2} \text{ years}, \qquad (9.3.14)$$

assuming the existence of three generations of quarks and leptons and a single electroweak Higgs multiplet. Even within the simplified discussion we have given, it is apparent that there is a close relation in this model between many experimental observables: the strong coupling constant, the masses and widths of the intermediate bosons, and the spectrum of elementary fermions. Refinements in our knowledge of any of these sharpens the predictions for the others. With the experimental initiatives now under way, incisive tests of specific unified models and perhaps of the idea of unification will soon be at hand.

9.4   The Baryon Number of the Universe

While the specific, if not yet highly quantitative, prediction of baryon decay is a dramatic consequence of unification, the difficulty of studying leptoquark transitions in the laboratory is clear. We live in a world in which energies are low compared with the unification scale. However, the discovery[12] of the cosmic microwave background radiation together with many supporting pieces of evidence makes it extremely likely that the universe began in a hot big bang of extraordinarily high energy density. Many aspects of the observed universe find natural explanations in terms of this standard cosmological model.[13] What is particularly striking is how much can be understood on the basis of laboratory experience and the hypothesis that the universe has remained in thermal equilibrium since its temperature was about 1 MeV.

Other features of the universe are not so easily understood in terms of common experience and innocuous hypotheses. Prominent among these is the net baryon number of the universe. These more difficult questions are of special interest for the window they may provide on the early thermal history of the universe and the interactions among fundamental particles at exceedingly high energies. This realization, catalyzed by the prediction of baryon-number violation in unified theories, has stimulated a new symbiosis between elementary particle physics and cosmology. Our current ignorance in both domains precludes a complete and credible explanation of the baryon asymmetry of the universe. Nevertheless, the issue is of such interest— it has to do with our very existence—that it is appropriate to conclude our brief study of unified theories with an outline of the principal ingredients of a calculation.

We wish to understand why matter dominates over antimatter in the universe, or at least in our region of the universe. To be slightly more specific, observations indicate that the density of antibaryons is negligible, whereas the average density of matter, characterized by the baryon-to-photon ratio

$$n_B/n_\gamma \simeq 10^{-9.9 \pm 1} \tag{9.4.1}$$

is small but nonzero.[14] It is attractive to assume that the universe began from a symmetric state of zero baryon number. But then how did the present asymmetric universe evolve?

Three necessary elements for the evolution to a state with net baryon number are

(1) the existence of fundamental processes that violate baryon number;

(2) microscopic CP violation; and

(3) a departure from thermal equilibrium during the epoch in which baryon-number-violating processes were important.

Unified theories entail transitions that violate baryon number and must incorporate CP violation to explain the behavior of the neutral kaon system. The processes that do not conserve baryon number may be forced out of thermal equilibrium by the expansion of the universe. Let us see why these three conditions are required, and then examine their relevance to unified theories.

The need for B-violating interactions requires no comment. That CP violation must be present in the B-nonconserving processes may be argued as follows. Suppose that a heavy boson X decays into two channels characterized by baryon numbers B_1 and B_2, with branching ratios

$$\frac{\Gamma(X \to B_1)}{\Gamma(X \to \text{all})} \equiv f,$$

$$\frac{\Gamma(X \to B_2)}{\Gamma(X \to \text{all})} = 1 - f. \tag{9.4.2}$$

The antiparticle $\bar{X}$ decays into channels with baryon numbers $-B_1$ and $-B_2$, with branching ratios

$$\frac{\Gamma(\bar{X} \to -B_1)}{\Gamma(\bar{X} \to \text{all})} \equiv \bar{f},$$

$$\frac{\Gamma(\bar{X} \to -B_2)}{\Gamma(\bar{X} \to \text{all})} = 1 - \bar{f}. \tag{9.4.3}$$

The equality of the total decay rates,

$$\Gamma(X \to \text{all}) = \Gamma(\bar{X} \to \text{all}), \tag{9.4.4}$$

is a consequence of CPT invariance. Decays from an initial state with equal numbers of bosons and antibosons will then lead to a net baryon number

$$\Delta B = (f - \bar{f})B_1 + [(1 - f) - (1 - \bar{f})]B_2$$
$$= (f - \bar{f})(B_1 - B_2). \tag{9.4.5}$$

For this to be nonzero requires $B_1 \neq B_2$, which is to say baryon-number nonconservation in the decays, as was already apparent, and also

$$f \neq \bar{f}. \tag{9.4.6}$$

However, CP invariance requires the equality of the S-matrix elements

$$S(X \to B_1) = S(\bar{X} \to \bar{B}_1), \tag{9.4.7}$$

which would imply the equality of f and $\bar{f}$. Hence CP violation is seen to be a prerequisite to the generation of net baryon number in the decay process.

To see that equilibrium of the environment prevents the development of a baryon–antibaryon asymmetry, let us consider the implications of unitarity and CPT invariance. The unitarity of the S-matrix may be expressed as

$$SS^\dagger = S^\dagger S = I. \tag{9.4.8}$$

Writing

$$S_{ab} = S(a \to b), \tag{9.4.9}$$

we then have

$$(SS^\dagger)_{aa} = \sum_i S(a \to i)S^*(a \to i)$$

$$= \sum_i |S(a \to i)|^2 = \sum_i |S(i \to a)|^2, \tag{9.4.10}$$

where the summation over intermediate states i includes both particle and antiparticle states. Invariance under CPT implies that

$$S(a \to b) = S(\bar{b} \to \bar{a}), \tag{9.4.11}$$

whereupon

$$\sum_i |S(a \to i)|^2 = \sum_i |S(i \to a)|^2 = \sum_{\bar{i}} |S(\bar{i} \to \bar{a})|^2. \tag{9.4.12}$$

Since the summation runs over both particle and antiparticle states, we may relabel $i \to \bar{i}$ in the last term, whereupon the equality

$$\sum_i |S(i \to a)|^2 = \sum_i |S(i \to \bar{a})|^2 \tag{9.4.13}$$

is obtained as a consequence of CPT and unitarity alone, though it would also be implied by exact CP invariance. In thermal equilibrium, all the states i corresponding to a given energy are equally populated, so the total rates leading to particle and antiparticle final states are equal. Accordingly, no baryon number asymmetry can develop, unless the system is out of thermal equilibrium. This can occur if the expansion rate of the universe, and thus its cooling rate, is large compared with the reaction rates involved in baryon-number-violating processes.

We have now identified the basic elements required to generate a net baryon number in the universe. Specific calculations are quite sensitive to the early thermal history of the universe, as well as the details of baryon number nonconservation and CP violation in the unified theory. For example, we have seen already from very general considerations that departures from CP invariance must persist up to the unification scale if this sort of mechanism is to succeed. This kind of guidance may be crucial in fashioning an ultimately satisfactory theory. Although the explanation of the matter–antimatter asymmetry from the perspective of unified theories is only at a qualitative stage, the continuation of this program is an exciting prospect.

9.5   An Assessment

The appeal that a unified theory of the strong, weak, and electromagnetic interactions seems to hold was reviewed in Section 9.1. Having now looked at the basic elements of the simplest example of a unified theory, we are in a position to take stock of what has been accomplished and what remains to be done.

The minimal $SU(5)$ theory has numerous desirable attributes and has successfully dealt with a number of the issues that motivate the search for a unified theory.

- It contains the standard $SU(3)_{color} \otimes SU(2)_L \otimes U(1)_Y$ gauge group and thus brings with it the attractive features of quantum chromodynamics and the Weinberg–Salam model.

- The number of independent gauge coupling constants is reduced from three to one.

- The predicted value of the weak mixing angle at the electroweak scale, $\sin^2 \theta_W \simeq 0.21$, which governs the relative strength of the vector and axial vector couplings of the weak neutral current, is in good agreement with experiment.

- The charged-current weak interactions are correctly described.

- Electric charge is quantized.

- The mass of the b-quark is qualitatively predicted.

- The masses of the predicted leptoquark bosons, which mediate new sorts of interactions, can be made very large, so the exotic processes are feeble effects at low energies. However, the leptoquark masses are small compared with the Planck mass at which the neglect of gravitation would be unjustified.

- Proton decay is possible and may lead to an understanding of the baryon excess in the universe.

- The neutrino is automatically massless in the simplest model. In more complicated versions of the theory, a small mass may be acquired. Pending the outcome of future experiments, this appears to be an advantage.

Whether or not it is correct or complete, the $SU(5)$ model provides an existence proof for unified theories. It appears to show that a satisfying unification of the strong, weak, and electromagnetic interactions can meaningfully be achieved without gravitation. There are, however, several areas in which accomplishments fall short of the announced aspirations, and there are also a number of specific problems to be faced.

- No particular insight has been gained into the nature of fermion masses or mixing angles. CP violation in the weak interactions does not arise gracefully.

- The number of parameters of the theory has not been materially reduced. The loss of two gauge coupling parameters and three mass parameters (corresponding to the charged leptons) is approximately compensated by the increased number of parameters in the Higgs sector of the theory.

- Although the idea that fermion generations are meaningful gains support, no understanding of the pattern has emerged. We still do not know why generations repeat or how many generations there are.

- It seems inelegant that each generation is assigned to a reducible representation of the gauge group.

- Gravitation is omitted, although the unification scale is only four orders of magnitude from the Planck mass.

- The most serious structural problem is the requirement that there be a dozen orders of magnitude between the electroweak and leptoquark mass scales. Is it possible to maintain the result $M_W/M_X \ll 1$ beyond low orders of perturbation theory?

So far as experimental implications are concerned, unified theories provide the important reminder that we do not understand the basis of baryon-number and lepton-number conservation. Searches for nucleon decay, neutrino masses, and lepton-number oscillations make up an exciting and significant complement to the standard fare of accelerator experiments.

Problems

9-1.   Consider a unified theory based on the gauge group $SO(10)$.
(a) By referring to the $SU(5) \otimes U(1)$ decomposition of the representations of $SO(10)$, show that each fermion generation can be accommodated in an irreducible **16**-dimensional representation, which also has a place for a left-handed antineutrino.
(b) Show that the adjoint **45** representation contains the gauge bosons of the $SU(5)$ theory.

(c) Now examine the branching of $SO(10)$ into $SU(4) \otimes SU(2) \otimes SU(2)$, and the subsequent branching of $SU(4)$ into $SU(3) \otimes U(1)$. Use the $SU(3) \otimes SU(2) \otimes SU(2)$ decomposition of the fermion representation to show that $SO(10)$ contains the left–right symmetric electroweak group as a subgroup.

(d) Give the transformation properties of the 45 gauge bosons under $SU(3) \otimes SU(2) \otimes SU(2)$, and identify the $SU(5)$ gauge bosons among them. [Reference: H. Fritzsch and P. Minkowski, *Ann. Phys. (NY)* **93**, 193 (1975).]

9-2. The groups $SO(N)$ are generated by antisymmetric tensors $T_{ij} = - T_{ji}$ $(i, j = 1 \ldots N)$, which satisfy the commutation relations $[T_{ij}, T_{kl}] = i(\delta_{ik} T_{jl} - \delta_{il} T_{jk} - \delta_{jk} T_{il} + \delta_{jl} T_{ik})$. Show that for $N > 6$ the quantity $\mathrm{tr}(\{T_{ij}, T_{kl}\} T_{mn})$ must vanish, so that all representations of $SO(N \geq 7)$ are anomaly-free. This helps to explain the apparently miraculous anomaly cancellation in the $SU(5)$ model. [Reference: H. Georgi and S. L. Glashow, *Phys. Rev.* **D6**, 429 (1972).]

9-3. Consider the first stage of spontaneous symmetry breaking in the $SU(5)$ model, namely $SU(5) \xrightarrow[24]{} SU(3)_c \otimes SU(2)_L \otimes U(1)_Y$. Suppose that the effective potential for the scalars is given by

$$V(\Phi^2) = \frac{\mu^2}{2} \mathrm{tr}(\Phi^2) + \frac{a}{4}[\mathrm{tr}(\Phi^2)]^2 + \frac{b}{2} \mathrm{tr}(\Phi^4),$$

where $\Phi \in$ **24**.

(a) Show that for $\mu^2 < 0$ the asymmetric vacuum state characterized by (9.2.31) corresponds to the absolute minimum of the classical potential, provided that $15a + 7b > 0$ and $b > 0$.

(b) Express the parameter v_{24} of the vacuum expectation value in terms of μ, a and b.

(c) Compute the masses of the superheavy gauge bosons and Higgs scalars that occur at this stage in the symmetry breaking. [Reference: A. J. Buras, J. Ellis, M. K. Gaillard, and D. V. Nanopoulos, *Nucl. Phys.* **B135**, 66 (1978).]

9-4. Calculate separately the contributions of loops containing left-handed and right-handed electrons to the renormalization of the photon propagator.

For Further Reading

MOTIVATION FOR UNIFICATION. The case for unified gauge theories is summarized cogently by

H. Harari, *Phys. Rep.* **42C**, 235 (1978).

J. Iliopoulos, *Proceedings of the XVII International Conference on High Energy Physics*, London, 1974, edited by J. R. Smith, Rutherford Laboratory, Chilton, U. K., 1974, p. III-79.

296   Unified Theories

F. Wilczek, *Proceedings of the* 1979 *International Conference on Lepton and Photon Interactions at High Energies*, edited by T. B. W. Kirk and H. D. I. Abarbanel, Fermilab, Batavia, Illinois, 1979, p. 437.

M. K. Gaillard, *Comments Nucl. Part. Phys.* **9**, 39 (1980).

H. Georgi, *Nature (London)* **288**, 649 (1980).

Issues of unification and proton stability were raised in the context of gauge theories with somewhat different structure than we have discussed by

J. C. Pati and A. Salam, *Phys. Rev.* **D8**, 1240 (1973); *Phys. Rev. Lett.* **31**, 661 (1973); *Phys. Rev.* **D10**, 275 (1974).

PARTIAL UNIFICATION. Various scenarios for reducing the three coupling constants of the $SU(3)_c \otimes SU(2)_L \otimes U(1)_Y$ theory to two independent parameters by completing the electroweak unification have been studied in

P. Q. Hung, J. D. Bjorken, and A. J. Buras, *Phys. Rev.* **D25**, 805 (1982).

INCORPORATION OF GRAVITATION. An ambitious example is provided by the $N = 8$ supergravity model of

E. Cremmer and B. Julia, *Phys. Lett.* **80B**, 48 (1978); *Nucl. Phys.* **B159**, 141 (1979).

An accessible introduction to supersymmetry, which relates fermionic and bosonic degrees of freedom, is given by

J. Wess and J. Bagger, *Supersymmetry and Supergravity*, Princeton University Press, Princeton, New Jersey, 1983.

UNIFIED THEORIES OF THE STRONG, WEAK, AND ELECTROMAGNETIC INTERACTIONS. Among reviews of specific models and their predictions, see

J. Ellis, in *Gauge Theories and Experiments at High Energies*, 1980 Scottish Universities Summer School, edited by K. C. Bowler and D. G. Sutherland, SUSSP, Edinburgh, 1981, p. 201; and in *Gauge Theories in High Energy Physics*, 1981 Les Houches Lectures, edited by M. K. Gaillard and R. Stora, North-Holland, Amsterdam, 1983.

P. Langacker, *Phys. Rep.* **72**, 185 (1981).

S.-H. H. Tye, "Introduction to the $SU(5)$ Grand Unified Theory and Related Topics," Cornell preprint CLNS-82/527, to appear in the Proceedings of the 1981 Summer School on Particle Physics, Hefei, China.

GROUP THEORY FOR UNIFIED MODEL BUILDING. Convenient sources for the properties of groups and their representations are

J. Patera and D. Sankoff, *Tables of Branching Rules for Representations of Simple Lie Algebras*, Université de Montréal, Montréal, 1973.

W. McKay and J. Patera, *Tables of Dimensions, Indices, and Branching Rules for Representations of Simple Algebras*, Dekker, New York, 1981.

Specific discussions of unification are given by

R. Slansky, *Phys. Rep.* **79**, 1 (1981).

H. Georgi, *Lie Groups in Particle Physics*, Benjamin, Reading, Massachusetts, 1982.

A very useful group-theoretical analysis of spontaneous symmetry breaking appears in

L.-F. Li, *Phys. Rev.* **D9**, 1723 (1974).

GAUGE HIERARCHIES. The problem of sustaining widely different mass scales in spontaneously broken theories is studied in

E. Gildener, *Phys. Rev.* **D14**, 1667 (1976).

S. Weinberg, *Phys. Lett.* **82B**, 387 (1979).

NUCLEON INSTABILITY. Recent summaries of the status of baryon number and lepton number conservation include

F. Reines and J. Schultz, *Surveys in HEP* **1**, 89 (1980).

M. Goldhaber, P. Langacker, and R. Slansky, *Science* **210**, 851 (1980).

H. Primakoff and S. P. Rosen, *Ann. Rev. Nucl. Part. Sci.* **31**, 145 (1981).

S. Weinberg, *Sci. Am.* **244**, 52 (June, 1981).

SU(5) PREDICTIONS OF THE $\gamma - Z^0$ MIXING PARAMETER. Representative of the current generation of careful evaluations are

W. J. Marciano and A. Sirlin, *Phys. Rev. Lett.* **46**, 163 (1981).

C. H. Llewellyn Smith and J. F. Wheater, *Phys. Lett.* **105B**, 486 (1981).

COSMOLOGY. Valuable introductions to this subject are given by

P. J. E. Peebles, *Physical Cosmology*, Princeton University Press, Princeton, New Jersey, 1971.

G. Steigman, *Ann. Rev. Nucl. Part. Sci.* **29**, 313 (1979).

S. Weinberg, *Gravitation and Cosmology*, Wiley, New York, 1972.

F. Wilczek, 1981 Erice School, Santa Barbara preprint NSF-ITP-81-91.

BARYON NUMBER OF THE UNIVERSE. The link with baryon- and lepton-number nonconservation has been understood in general terms for some time. See, for example, the remark on p. 482 of

S. Weinberg, in *Lectures in Particles and Field Theory*, 1964 Brandeis lectures, edited by S. Deser and K. Ford, Prentice-Hall, Englewood Cliffs, New Jersey, 1965, p. 405.

A specific scenario for baryogenesis was given by

A. D. Sakharov, *ZhETF Pis'ma* **5**, 32 (1967) [English translation: *Sov. Phys. - JETP Lett.* **5**, 24 (1967)].

In the context of gauge theories, the issue was reopened by

M. Yoshimura, *Phys. Rev. Lett.* **41**, 281 (1978); **42**, 746E (1979).

The importance of the nonequilibrium condition was emphasized in

D. Toussaint, S. B. Treiman, F. Wilczek, and A. Zee, *Phys. Rev.* **D19**, 1036 (1979).

Other influential works include the papers by

S. Dimopoulos and L. Susskind, *Phys. Rev.* **D18**, 4500 (1978).

S. Weinberg, *Phys. Rev. Lett.* **42**, 850 (1979).

For an interesting detailed investigation, see

E. W. Kolb and S. Wolfram, *Nucl. Phys.* **B172**, 224 (1980).

A popular account is given by

F. Wilczek, *Sci. Am.* **243**, 60 (December, 1980).

MAGNETIC MONOPOLES. An excellent account of the Dirac monopole is given by

E. Amaldi and N. Cabibbo, *Aspects of Quantum Theory*, edited by A. Salam and E. P. Wigner, Cambridge University Press, Cambridge, 1972, p. 183.

The monopoles that appear in unified theories are reviewed in

P. Goddard and D. Olive, *Rep. Prog. Phys.* **41**, 1357 (1978).

References

[1] H. Georgi and S. L. Glashow, *Phys. Rev. Lett.* **32**, 438 (1974).

[2] On the assumption that the neutron charge is equal to the sum of the proton and electron charges, H. F. Dylla and J. G. King, *Phys. Rev.* **A7**, 1224 (1973) have demonstrated the neutrality of matter at the level of $|Q(p) + Q(e)| < 10^{-21}\,e$. Their article contains a review of earlier experimental work.

[3] This is explained in S. Coleman, "Classical Lumps and Their Quantum Descendants," in *New Phenomena in Subnuclear Physics*, 1975 Erice School, edited by A. Zichichi, Plenum, New York, 1977, p. 297. See in particular Section 3.

[4] J. Banks and H. Georgi, *Phys. Rev.* **D14**, 1159 (1976). See also S. Okubo, *Phys. Rev.* **D16**, 3528 (1977). Very general and convenient methods for evaluating Casimir operators are due to A. M. Perelomov and V. S. Popov, *Yad. Fiz.* **3**, 24, 1127 (1966) [English translation: *Sov. J. Nucl. Phys.* **3**, 676, 819 (1966)]; V. S. Popov and A. M. Perelomov, *Yad. Fiz.* **5**, 693 (1967); **7**, 460 (1968) English translation: *Sov. J. Nucl. Phys.* **5**, 489 (1967); **7**, 290 (1968)].

[5] This limit is due to M. L. Cherry *et al.*, *Phys. Rev. Lett.* **47**, 1507 (1981). See also the review articles cited at the end of this chapter.

[6] A. J. Buras, J. Ellis, M. K. Gaillard, and D. V. Nanopoulos, *Nucl. Phys.* **B135**, 66 (1978).

[7] H. Georgi, H. R. Quinn, and S. Weinberg, *Phys. Rev. Lett.* **33**, 451 (1974).

[8] T. Appelquist and J. Carazzone, *Phys. Rev.* **D11**, 2856 (1975).

[9] A less brutal treatment of thresholds is clearly desirable. For one precise formulation of the evolution problem, see I. Antoniadis, C. Bouchiat, and J. Iliopoulos, *Phys. Lett.* **97B**, 367 (1980).

[10] This is taken from a global fit by J. E. Kim, P. Langacker, M. Levine, and H. H. Williams, *Rev. Mod. Phys.* **53**, 211 (1981).

[11] See, for example, the operator analyses by S. Weinberg, *Phys. Rev. Lett.* **43**, 1566 (1979), and by F. Wilczek and A. Zee, *ibid.* p. 1571.

[12] A. A. Penzias and R. W. Wilson, *Astrophys. J.* **142**, 419 (1965); the interpretation is due to R. H. Dicke, P. J. E. Peebles, P. G. Roll, and D. T. Wilkinson, *ibid.* p. 414. See also the Nobel lectures by A. A. Penzias, *Rev. Mod. Phys.* **51**, 425 (1979) and R. W. Wilson, *ibid.* p. 433.

[13] For accessible introductions see J. Silk, *The Big Bang*, Freeman, San Francisco, 1980, and S. Weinberg, *The First Three Minutes*, Basic Books, New York, 1977.

[14] The observational case for the absence of antimatter is reviewed by G. Steigman, *Ann. Rev. Astron. Astrophys.* **14**, 339 (1976) and by J. D. Barrow, *Surveys in HEP* **1**, 183 (1980). The estimate of the baryon-to-photon ratio is due to K. A. Olive, D. N. Schramm, G. Steigman, M. S. Turner, and J. Yang, *Astrophys. J.* **246**, 557 (1981).

Epilogue

I began this book by invoking the widespread belief that a grand synthesis of physical law may be at hand. The subsequent discussions have exhibited the power and promise of gauge principles as the origins of the fundamental interactions. Gauge theories are renormalizable and, under the right circumstances, may be asymptotically free. Particular models have been seen to agree with experiment with respect to both systematic patterns and specific observables, and to provide inspiration for new kinds of measurements. Whether or not they will bring an ultimate understanding of the laws of Nature, gauge theories unquestionably provide us with an extraordinarily unified and unifying language for the description of natural phenomena.

Beyond the technical questions and unsolved problems that we have encountered in this introduction to gauge theories, there are some important general issues to be acknowledged. One of these has to do with the choice of a gauge group or, at a more fundamental level, with the origin of local gauge symmetries. A second concerns the asymmetry between the gauge bosons and all the other particles in the theories we have discussed. The spectrum of spin-one bosons is prescribed by the local gauge symmetry, but the fermions and scalars are added with considerable arbitrariness. A completely unified model, it would seem, might relate all the fundamental fields in the theory and, as a result, reduce the number of independent mass parameters. Connections of this general type exist in so-called super-symmetric models, which link fermionic and bosonic degrees of freedom, but a model that gives a complete and straightforward description of this world has not yet emerged. A third problem is posed by the proliferation of the fundamental fermions and the puzzle of quark–lepton generations, which many physicists take as a strong hint for a new level of elementary

299

particles, not yet attained by experiment. These are the kinds of issues that the future must face.

I hope in these chapters not only to have communicated a few facts and conveyed a point of view, but also to have evoked an awareness that there is much to be done that is significant and exciting. In theory and experiment alike, there are many opportunities to contribute to the numinous intellectual adventure in which we are privileged to share.

APPENDIX A: NOTATIONS AND CONVENTIONS

This book conforms closely to the conventions of Bjorken and Drell,[1] except for the normalization of Dirac spinors. It will generally be advantageous to adopt natural units in which the quantities $\hbar$ and c are set equal to 1. Numerical values and conversion factors are given in Appendix C.

A.1 Four-Vectors and Scalar Product

Space–time coördinates $(t; x, y, z) = (t; \mathbf{x})$ are denoted by the contravariant four-vector

$$x^\mu \equiv (t; x, y, z) \equiv (x^0; x^1, x^2, x^3). \tag{A.1.1}$$

The metric tensor

$$g_{\mu\nu} = \begin{pmatrix} 1 & 0 & 0 & 0 \\ 0 & -1 & 0 & 0 \\ 0 & 0 & -1 & 0 \\ 0 & 0 & 0 & -1 \end{pmatrix} \tag{A.1.2}$$

generates the covariant four-vector

$$x_\mu \equiv (x_0; x_1, x_2, x_3) \equiv g_{\mu\nu}x^\nu = (t; -x, -y, -z). \tag{A.1.3}$$

Unless specifically indicated to the contrary, repeated indices are summed. The scalar product is

$$x^2 \equiv x^\mu x_\mu = t^2 - \mathbf{x}^2. \tag{A.1.4}$$

Four-vectors are printed in italic type: a; (spatial) three-vectors are in boldface: $\mathbf{a}$. Thus a general scalar product is

$$a \cdot b = a^\mu b_\mu = a^0 b^0 - \mathbf{a} \cdot \mathbf{b}. \tag{A.1.5}$$

301

Momentum vectors are written as

$$p^\mu = (E; p_x, p_y, p_z).\tag{A.1.6}$$

A convenient notation for the four-gradient is

$$\partial^\mu = \partial/\partial x_\mu \qquad \text{or} \qquad \partial_\mu = \partial/\partial x^\mu,\tag{A.1.7}$$

in terms of which the position space momentum operator is

$$\mathsf{p}^\mu = i\partial^\mu = (i\partial/\partial t; -i\nabla).\tag{A.1.8}$$

Thus

$$\mathsf{p}^\mu \mathsf{p}_\mu = -\partial^\mu \partial_\mu = -[(\partial^2/\partial t^2) - \nabla^2] \equiv -\square,\tag{A.1.9}$$

where ∇^2 is the Laplacian and $\square$ is the d'Alembertian operator.

A.2   Dirac Matrices

The Dirac γ-matrices satisfy the anticommutation relations

$$\{\gamma^\mu, \gamma^\nu\} \equiv \gamma^\mu \gamma^\nu + \gamma^\nu \gamma^\mu = 2g^{\mu\nu},\tag{A.2.1}$$

so that

$$\gamma^\mu \gamma_\mu = 4 \cdot I,\tag{A.2.2}$$

where I is the 4×4 identity matrix. In contexts in which the matrix character of I is obvious, it will be convenient simply to write 1. Other useful identities follow at once from the anticommutation relations:

$$[\gamma^\mu \gamma^\nu, \gamma^\rho] \equiv \gamma^\mu \gamma^\nu \gamma^\rho - \gamma^\rho \gamma^\mu \gamma^\nu = 2(\gamma^\mu g^{\nu\rho} - \gamma^\nu g^{\mu\rho});\tag{A.2.3}$$

$$\gamma^\mu \gamma_\nu \gamma_\mu = -2\gamma_\nu;\tag{A.2.4}$$

$$\gamma^\mu \gamma_\nu \gamma_\rho \gamma_\mu = 4g_{\nu\rho};\tag{A.2.5}$$

$$\gamma^\mu \gamma_\nu \gamma_\rho \gamma_\sigma \gamma_\mu = -2\gamma_\sigma \gamma_\rho \gamma_\nu;\tag{A.2.6}$$

$$\gamma^\mu \gamma_\nu \gamma_\rho \gamma_\sigma \gamma_\tau \gamma_\mu = 2(\gamma_\tau \gamma_\nu \gamma_\rho \gamma_\sigma + \gamma_\sigma \gamma_\rho \gamma_\nu \gamma_\tau).\tag{A.2.7}$$

When an explicit representation is required, we shall adopt

$$\gamma^0 = \begin{pmatrix} I & 0 \\ 0 & -I \end{pmatrix}, \qquad \gamma = \begin{pmatrix} 0 & \sigma \\ -\sigma & 0 \end{pmatrix},\tag{A.2.8}$$

where I is the 2×2 identity matrix and the 2×2 Pauli matrices are given by

$$\sigma_x = \begin{pmatrix} 0 & 1 \\ 1 & 0 \end{pmatrix}; \qquad \sigma_y = \begin{pmatrix} 0 & -i \\ i & 0 \end{pmatrix}; \qquad \sigma_z = \begin{pmatrix} 1 & 0 \\ 0 & -1 \end{pmatrix}.\tag{A.2.9}$$

The spin tensor is

$$\sigma^{\mu\nu} = (i/2)[\gamma^\mu, \gamma^\nu] = i(\gamma^\mu \gamma^\nu - g^{\mu\nu}),\tag{A.2.10}$$

for which

$$[\sigma^{\mu\nu}, \gamma^\rho] = 2i(\gamma^\mu g^{\nu\rho} - \gamma^\nu g^{\mu\rho}).\tag{A.2.11}$$

In the standard representation the spin tensor has components (Latin indices run over 1, 2, 3)

$$\sigma^{ij} = \varepsilon^{ijk}\begin{pmatrix} \sigma^k & 0 \\ 0 & \sigma^k \end{pmatrix}, \tag{A.2.12}$$

where the antisymmetric three-index symbol takes the values

$$\varepsilon^{ijk} = \begin{cases} +1, & \text{for even permutations of 123} \\ -1, & \text{for odd permutations} \\ 0, & \text{otherwise,} \end{cases} \tag{A.2.13}$$

and

$$\sigma^{0j} = \begin{pmatrix} 0 & i\sigma^j \\ i\sigma^j & 0 \end{pmatrix}. \tag{A.2.14}$$

The remaining important combination is

$$\begin{aligned} \gamma^5 &\equiv i\gamma^0\gamma^1\gamma^2\gamma^3 = -i\gamma_0\gamma_1\gamma_2\gamma_3 \equiv \gamma_5 \\ &= (i/4!)\varepsilon_{\mu\nu\rho\sigma}\gamma^\mu\gamma^\nu\gamma^\rho\gamma^\sigma \\ &= (i/4!)\varepsilon^{\mu\nu\rho\sigma}\gamma_\mu\gamma_\nu\gamma_\rho\gamma_\sigma, \end{aligned} \tag{A.2.15}$$

where the Levi–Città tensor is defined as

$$\varepsilon_{\mu\nu\rho\sigma} = \begin{cases} +1, & \text{for even permutations of 0123} \\ -1, & \text{for odd permutations} \\ 0, & \text{otherwise,} \end{cases} \tag{A.2.16}$$

and $\varepsilon^{\mu\nu\rho\sigma} = -\varepsilon_{\mu\nu\rho\sigma}$. Evidently

$$(\gamma_5)^2 = I, \tag{A.2.17}$$

and

$$\{\gamma^5, \gamma^\mu\} = 0. \tag{A.2.18}$$

It follows from the definition (A.2.15) that

$$\gamma^5\gamma^\sigma = (i/3!)\varepsilon^{\mu\nu\rho\sigma}\gamma_\mu\gamma_\nu\gamma_\rho. \tag{A.2.19}$$

In the standard representation,

$$\gamma_5 = \begin{pmatrix} 0 & I \\ I & 0 \end{pmatrix}. \tag{A.2.20}$$

The frequently encountered scalar product of a four-vector and a γ-matrix is denoted by

$$\gamma \cdot a \equiv \gamma_\mu a^\mu \equiv \not{a} = \gamma^0 a^0 - \boldsymbol{\gamma} \cdot \mathbf{a} \tag{A.2.21}$$

so that, in particular,

$$\gamma \cdot p \equiv i\gamma \cdot \partial = i\gamma_\mu \partial^\mu = i\not{\partial}. \tag{A.2.22}$$

A.3   Trace Theorems and Tensor Contractions

In evaluating the traces of products of γ-matrices that occur in the computation of transition matrix elements, the following theorems are useful:

$$\text{tr}(I) = 4 \tag{A.3.1}$$

$$\text{tr}(AB) = \text{tr}(BA) \tag{A.3.2}$$

$$\text{tr}(\gamma_\mu) = 0 \tag{A.3.3}$$

$$\text{tr(odd number of } \gamma\text{'s)} = 0 \tag{A.3.4}$$

$$\text{tr}(\gamma_\mu \gamma_\nu) = 4g_{\mu\nu} \tag{A.3.5a}$$

$$\text{tr}(\not a \not b) = 4a \cdot b \tag{A.3.5b}$$

$$\text{tr}(\gamma_\mu \gamma_\nu \gamma_\rho \gamma_\sigma) = 4[g_{\mu\nu}g_{\rho\sigma} - g_{\mu\rho}g_{\nu\sigma} + g_{\mu\sigma}g_{\nu\rho}] \tag{A.3.6a}$$

$$\text{tr}(\not a \not b \not c \not d) = 4[(a \cdot b)(c \cdot d) - (a \cdot c)(b \cdot d) + (a \cdot d)(b \cdot c)] \tag{A.3.6b}$$

$$\text{tr}(\gamma_5) = 0 \tag{A.3.7}$$

$$\text{tr}(\gamma_5 \gamma_\mu) = 0 \tag{A.3.8}$$

$$\text{tr}(\gamma_5 \gamma_\mu \gamma_\nu) = 0 \tag{A.3.9}$$

$$\text{tr}(\gamma_5 \gamma_\mu \gamma_\nu \gamma_\rho) = 0 \tag{A.3.10}$$

$$\text{tr}(\gamma_5 \gamma_\mu \gamma_\nu \gamma_\rho \gamma_\sigma) = 4i\varepsilon_{\mu\nu\rho\sigma} \tag{A.3.11a}$$

$$\text{tr}(\gamma_5 \not a \not b \not c \not d) = 4i\varepsilon_{\mu\nu\rho\sigma}a^\mu b^\nu c^\rho d^\sigma \tag{A.3.11b}$$

Derivations and extensions of these results may be found in many places, including Section 7.2 of Bjorken and Drell[1] and Section 28 of Berestetskii et al.[2]

Some useful results from tensor calculus are these:

$$g^{\lambda\mu}g_{\mu\nu} = \delta_\nu^\lambda = \begin{cases} 1, & \lambda = \nu \\ 0, & \text{otherwise} \end{cases} \tag{A.3.12}$$

$$-\varepsilon^{\alpha\lambda\mu\nu}\varepsilon_{\alpha\rho\sigma\tau} = \delta_\rho^\lambda(\delta_\sigma^\mu\delta_\tau^\nu - \delta_\tau^\mu\delta_\sigma^\nu) - \delta_\sigma^\lambda(\delta_\rho^\mu\delta_\tau^\nu - \delta_\tau^\mu\delta_\rho^\nu) + \delta_\tau^\lambda(\delta_\rho^\mu\delta_\sigma^\nu - \delta_\sigma^\mu\delta_\rho^\nu); \tag{A.3.13}$$

$$-\varepsilon^{\alpha\beta\mu\nu}\varepsilon_{\alpha\beta\sigma\tau} = 2(\delta_\sigma^\mu\delta_\tau^\nu - \delta_\tau^\mu\delta_\sigma^\nu); \tag{A.3.14}$$

$$-\varepsilon^{\alpha\beta\gamma\nu}\varepsilon_{\alpha\beta\gamma\tau} = 6\delta_\tau^\nu; \tag{A.3.15}$$

$$-\varepsilon^{\alpha\beta\gamma\delta}\varepsilon_{\alpha\beta\gamma\delta} = 24. \tag{A.3.16}$$

A.4   Dirac Equation and Dirac Spinors

A free spin-1/2 particle of mass m with four-momentum $p = (\sqrt{\mathbf{p}^2 + m^2}; \mathbf{p})$ and spin s is described by the positive energy spinor $u(p, s)$. The four-vector s satisfies $s \cdot p = 0$, $s^2 = -1$. In the rest frame of the particle it is the polarization vector

$$s^\mu = (0; \hat{\mathbf{s}}), \qquad \hat{\mathbf{s}} \cdot \hat{\mathbf{s}} = 1. \tag{A.4.1}$$

The positive energy spinor satisfies the Dirac equation

$$(\not{p} - m)u(p, s) = 0,$$ (A.4.2)

while the adjoint spinor

$$\bar{u}(p, s) \equiv u^\dagger(p, s)\gamma^0$$ (A.4.3)

satisfies

$$\bar{u}(p, s)(\not{p} - m) = 0.$$ (A.4.4)

The negative energy solutions $v(p, s)$ and $\bar{v}(p, s) = v^\dagger(p, s)\gamma^0$ correspond to antiparticles. They satisfy the Dirac equations

$$(\not{p} + m)v(p, s) = 0,$$ (A.4.5)

$$\bar{v}(p, s)(\not{p} + m) = 0.$$ (A.4.6)

It is frequently a convenience to work in a helicity basis in which the spinors $u_\lambda(p)$ are eigenstates of the operator $\gamma_5\not{s}$ (defined for convenience as twice the helicity), with eigenvalues $\lambda = \pm 1$ for spin aligned parallel or antiparallel to the direction of motion. The spinors are normalized such that

$$\bar{u}_\lambda(p)u_\mu(p) = 2m\delta_{\lambda\mu},$$ (A.4.7)

$$\bar{v}_\lambda(p)v_\mu(p) = -2m\delta_{\lambda\mu},$$ (A.4.8)

$$\bar{u}_\lambda(p)v_\mu(p) = 0 = \bar{v}_\lambda(p)u_\mu(p).$$ (A.4.9)

The projection operators Λ that occur in the evaluation of matrix elements are then

$$2m\Lambda(p) \equiv \sum_\lambda u_\lambda(p)\bar{u}_\lambda(p) = m + \not{p},$$ (A.4.10)

$$2m\Lambda(-p) \equiv -\sum_\lambda v_\lambda(p)\bar{v}_\lambda(p) = m - \not{p},$$ (A.4.11)

$$2m\Lambda_\lambda(p) \equiv u_\lambda(p)\bar{u}_\lambda(p) \qquad \text{not summed}$$

$$= (m + \not{p})\left(\frac{1 + \lambda\gamma_5\not{s}}{2}\right),$$ (A.4.12)

$$2m\Lambda_\lambda(-p) \equiv -v_\lambda(p)\bar{v}_\lambda(p) \qquad \text{not summed}$$

$$= (m - \not{p})\left(\frac{1 - \lambda\gamma_5\not{s}}{2}\right).$$ (A.4.13)

These imply the completeness relation

$$\Lambda(p) + \Lambda(-p) = I.$$ (A.4.14)

It is sometimes essential to have an explicit form for the spinor. The positive energy spinor with momentum $|\mathbf{p}|$ along the positive z-axis and helicity $\lambda/2$ is

$$u_\lambda(p) = \sqrt{E+m}\begin{pmatrix} \chi_\lambda \\ \dfrac{2\lambda|\mathbf{p}|}{E+m}\chi_\lambda \end{pmatrix}, \tag{A.4.15}$$

where

$$\chi_+ = \begin{pmatrix} 1 \\ 0 \end{pmatrix}, \qquad \chi_- = \begin{pmatrix} 0 \\ 1 \end{pmatrix}. \tag{A.4.16}$$

For the corresponding antiparticle spinor it is convenient to substitute

$$v_\lambda(p) = -\lambda\gamma_5 u_{-\lambda}(p). \tag{A.4.17}$$

The spinor appropriate to a particle with momentum $\mathbf{p}$ such that $\hat{\mathbf{p}}\cdot\hat{\mathbf{z}} = \cos\theta$ is obtained by a rotation about the y-axis:

$$u_+(p) = \sqrt{E+m}\begin{bmatrix} \cos(\theta/2) \\ \sin(\theta/2) \\ \cos(\theta/2)|\mathbf{p}|/(E+m) \\ \sin(\theta/2)|\mathbf{p}|/(E+m) \end{bmatrix}; \tag{A.4.18}$$

$$u_-(p) = \sqrt{E+m}\begin{bmatrix} -\sin(\theta/2) \\ \cos(\theta/2) \\ \sin(\theta/2)|\mathbf{p}|/(E+m) \\ -\cos(\theta/2)|\mathbf{p}|/(E+m) \end{bmatrix}; \tag{A.4.19}$$

The operators $(1\pm\gamma_5)/2$ are spin projection operators in the limit $m\to 0$. Thus, the equations

$$\frac{1}{2}(1\pm\gamma_5)u_\lambda(p) = \frac{1}{2}\left(1\pm\frac{\lambda|\mathbf{p}|}{E+m}\right)\begin{pmatrix} \chi_\lambda \\ \pm\chi_\lambda \end{pmatrix} \tag{A.4.20}$$

become, for massless particles,

$$\tfrac{1}{2}(1+\gamma_5)u_\pm(p) = \begin{cases} u_+(p) \\ 0 \end{cases}; \tag{A.4.21a}$$

$$\tfrac{1}{2}(1-\gamma_5)u_\pm(p) = \begin{cases} 0 \\ u_-(p) \end{cases}. \tag{A.4.21b}$$

Hermitian conjugates of matrix elements are encountered routinely in calculations. They may be reexpressed as

$$[\bar{u}(p')\Lambda u(p)]^\dagger = \bar{u}(p)\bar{\Lambda}u(p'), \tag{A.4.22}$$

where $\bar{\Lambda} \equiv \gamma^0\Lambda^\dagger\gamma^0$ is the Dirac conjugate of the operator Λ. Simple and commonly occurring examples are

$$\bar{1} = \gamma^0 1^\dagger \gamma^0 = 1; \tag{A.4.23}$$

$$\bar{\gamma}^\mu = \gamma^0\gamma^{\mu\dagger}\gamma^0 = \gamma^\mu, \tag{A.4.24}$$

from which

$$\overline{\not{q}_1\not{q}_2\cdots\not{q}_n} = \not{q}_n\not{q}_{n-1}\cdots\not{q}_1; \tag{A.4.25}$$

$$\overline{\sigma^{\mu\nu}} = \gamma^0\sigma^{\mu\nu\dagger}\gamma^0 = \sigma^{\mu\nu}; \tag{A.4.26}$$

$$\overline{\gamma^5} = \gamma^0\gamma^{5\dagger}\gamma^0 = -\gamma^5; \tag{A.4.27}$$

and

$$\overline{\gamma^\mu\gamma^5} = \gamma^0\gamma^{5\dagger}\gamma^{\mu\dagger}\gamma^0 = \gamma^\mu\gamma^5. \tag{A.4.28}$$

The Fierz reordering transformation[3] is of value in computing amplitudes represented by a sum of Feynman graphs. In the operator basis $\Lambda_i = (1, \gamma_\mu, \sigma_{\mu\nu}, \gamma_\mu\gamma_5, \gamma_5)$, a transition among four arbitrary spinors u_i may be expressed as

$$\bar{u}_3\Lambda_i u_2\bar{u}_1\Lambda_i u_4 = \sum_{j=1}^{5} \lambda_{ij}\bar{u}_1\Lambda_j u_2\bar{u}_3\Lambda_j u_4, \tag{A.4.29}$$

where

$$\lambda_{ij} = \frac{1}{4}\begin{pmatrix} 1 & 1 & 1 & 1 & 1 \\ 4 & -2 & 0 & 2 & -4 \\ 6 & 0 & -2 & 0 & 6 \\ 4 & 2 & 0 & -2 & -4 \\ 1 & -1 & 1 & -1 & 1 \end{pmatrix}. \tag{A.4.30}$$

Because the spins are arbitrary, one may replace $u_4 \to \gamma_5 u_4$ and recover the same result for matrix elements of the form $\bar{u}_1\Lambda_i u_2\bar{u}_3\Lambda_i\gamma_5 u_4$, etc. Of particular utility is the result

$$\bar{u}_3\gamma_\mu(1 - \gamma_5)u_2\bar{u}_1\gamma^\mu(1 - \gamma_5)u_4 = -\bar{u}_1\gamma_\mu(1 - \gamma_5)u_2\bar{u}_3\gamma^\mu(1 - \gamma_5)u_4. \tag{A.4.31}$$

References

[1] J. D. Bjorken and S. D. Drell, *Relativistic Quantum Mechanics*, McGraw-Hill, New York, 1964; *Relativistic Quantum Fields*, McGraw-Hill, New York, 1965.

[2] V. B. Berestetskii, E. M. Lifshitz, and L. P. Pitaevski, *Relativistic Quantum Theory*, Part 1, translated by J. B. Sykes and J. S. Bell, Pergamon, Oxford, 1971, 1979.

[3] M. Fierz, *Z. Phys.* **88**, 161 (1934); see also ref. 2, Sections 22–28.

APPENDIX B: OBSERVABLES AND FEYNMAN RULES

Feynman rules will be given for the evaluation of the invariant amplitude $\mathcal{M}$, the specific form of which depends on the dynamics of a process. The connection between invariant amplitude and observables is then entirely kinematic. This appendix summarizes the rules for writing down the invariant amplitude, some methods for evaluating integrals over undetermined loop momenta, and the computation of the simplest measurable quantities.

B.1 Phase-Space Formulas: Decay Rates and Cross Sections

It is straightforward to derive the following useful relations between amplitudes and observables.

For a general decay process $\alpha \to (1, 2, \ldots, n) \equiv \beta$, the differential decay rate is

$$d\Gamma_{\beta\alpha} = \frac{(2\pi)^4 \delta^{(4)}(p_\beta - p_\alpha)|\mathcal{M}_{\beta\alpha}|^2}{2p_\alpha^0} \prod_{i=1}^{n} \frac{\tilde{d}p_i}{(2\pi)^3} \cdot S, \qquad \text{(B.1.1)}$$

where

$$\tilde{d}p_i = d^3\mathbf{p}_i / [2(\mathbf{p}_i^2 + m^2)^{1/2}] \qquad \text{(B.1.2)}$$

is the Lorentz-invariant phase-space volume element and the statistical weight S contains a factor $1/n!$ if there are n identical particles in the final state:

$$S = \prod_{k = \text{species}} 1/n_k! \qquad \text{(B.1.3)}$$

308

For the important special case $\alpha \rightarrow (1, 2) \equiv \beta$ of the two-body decay of a particle at rest, the differential decay rate is

$$\frac{d\Gamma_{\beta\alpha}}{d\Omega_{\text{c.m.}}} = \frac{|\mathcal{M}_{\beta\alpha}|^2}{64\pi^2} \frac{\mathcal{S}_{12}}{m_\alpha^3} \cdot S \qquad\qquad (\text{B.1.4})$$

Here the quantity

$$\mathcal{S}_{12} = [s - (m_1 + m_2)^2]^{1/2}[s - (m_1 - m_2)^2]^{1/2}, \qquad (\text{B.1.5})$$

with $s = m_\alpha^2$ the square of the c.m. energy, is equal to $2m_\alpha$ times the three-momentum of the products.

A general two-body collision cross section $\alpha \equiv (1, 2) \rightarrow (3, 4, \ldots, n) \equiv \beta$ is written as

$$d\sigma_{\beta\alpha} = \frac{(2\pi)^4 \delta^{(4)}(p_\beta - p_\alpha)|\mathcal{M}_{\beta\alpha}|^2}{2\,\mathcal{S}_{12}} \prod_{i=3}^{n} \frac{d\boldsymbol{p}_i}{(2\pi)^3} \cdot S. \qquad (\text{B.1.6})$$

Again it is the case of a two-body final state $\alpha \equiv (1, 2) \rightarrow (3, 4) \equiv \beta$ that is of particular interest. The differential cross section is

$$\frac{d\sigma}{d\Omega_{\text{c.m.}}} = \frac{|\mathcal{M}_{\beta\alpha}|^2}{64\pi^2 s} \left(\frac{\mathcal{S}_{34}}{\mathcal{S}_{12}}\right) \cdot S \qquad\qquad (\text{B.1.7})$$

or, in terms of the invariant momentum transfer $t = (p_1 - p_3)^2 = (p_2 - p_4)^2$,

$$\frac{d\sigma}{dt} = \frac{|\mathcal{M}_{\beta\alpha}|^2}{16\pi\mathcal{S}_{12}^2} \cdot S. \qquad\qquad (\text{B.1.8})$$

In all the above expressions, an optional summation over final spins and average over initial spins is implicit.

B.2   Feynman Rules: Generalities

The invariant amplitude that corresponds to a Feynman diagram is given as a product of factors associated with external lines (particles in the initial or final state), internal lines, and interactions. For lowest-order perturbation calculations, the factors that do not depend on details of the interaction are these.

External Lines:

(*i*) For each external spinless boson, a factor 1.

(*ii*) For absorption (emission) of a spin-one boson, a factor $\varepsilon_\mu(k, \lambda)(\varepsilon_\mu(k, \lambda)^*)$, where ε is the polarization four-vector for a boson with momentum k and helicity λ. Only the transverse polarization states of massless vector bosons are populated. In the case of propagation along the z-direction the polarization vectors are thus $\varepsilon_\mu(k, \pm 1) = (0; 1, \pm i, 0)/\sqrt{2}$. A massive vector at rest is also permitted the longitudinal state of polarization specified by $\varepsilon_\mu(k, 0) = (0; 0, 0, -1)$. All these polarization vectors satisfy $\varepsilon_\mu(k, \lambda)^* \varepsilon^\mu(k, \lambda') = -\delta_{\lambda\lambda'}$. For the massless case, $k \cdot \varepsilon = 0$ as well.

(*iii*) For a spin-1/2 fermion with momentum p and helicity $\lambda/2$ in the initial state, a factor $u_\lambda(p)$ on the right; in the final state, a factor $\bar{u}_\lambda(p)$ on the left.

(*iv*) For a spin-1/2 antifermion with momentum p and helicity $\lambda/2$ in the initial state, a factor $\bar{v}_\lambda(p)$ on the left; in the final state, a factor $v_\lambda(p)$ on the right.

Internal Lines:

A propagator is associated with each internal line representing a particle with four-momentum q and mass m.

(*i*) For each internal spin-zero boson, a factor

$$\frac{i}{q^2 - m^2 + i\varepsilon}.$$

(*ii*) For each internal photon, a factor (in Feynman gauge)

$$\frac{-ig_{\mu\nu}}{q^2 + i\varepsilon}.$$

(*iii*) For each internal massive vector boson, a factor (in unitary gauge)

$$\frac{-i(g_{\mu\nu} - q_\mu q_\nu/M_V^2)}{q^2 - M_V^2 + i\varepsilon},$$

where M_V is the boson mass.

(*iv*) for each internal fermion line, a factor

$$\frac{i(\not{q} + m)}{q^2 - m^2 + i\varepsilon}.$$

An internal antifermion is regarded as a fermion of negative four-momentum $(-q)$.

Loops and Combinatorics:

(*i*) For each momentum k not fully determined by the requirement of momentum conservation at the vertices, a factor

$$\int \frac{d^4 k}{(2\pi)^4},$$

where the integral runs over all values of the loop momentum. Techniques for dealing with the resulting integrals are reviewed below.

(*ii*) For each closed loop of fermions, a factor -1.

(*iii*) A factor $1/n!$ for each closed loop containing n identical boson lines.

B.3   Feynman Integrals

In general, the integral over an undetermined loop momentum k in a Feynman diagram takes the form

$$\mathcal{I} = (2\pi)^{-4} \int \frac{d^4 k\, F(k; p_i; m_j)}{a_1 a_2 \cdots a_n}, \tag{B.3.1}$$

where

$$a_k = (k - s_k)^2 - m_k^2 + i\varepsilon, \tag{B.3.2}$$

s_k is a linear combination of external momenta p_i, m_j are the (internal and external) masses in the problem, and the function F is a polynomial in the components of k. Convenient general methods for evaluating such integrals were devised by R. P. Feynman.[1] Careful elaborations appear in many textbooks (see in particular Jauch and Rohrlich[2]). The form of the k-space integral may be simplified by the introduction of auxiliary parameters. This may be done in two ways, either of which may be advantageous in particular applications.

The first method relies upon the identity

$$\xi_n \equiv [a_1 a_2 \cdots a_n]^{-1} = (n-1)! \int_0^1 dz_1 \int_0^1 dz_2 \cdots \int_0^1 dz_n \delta\left(\sum_{i=1}^n z_i - 1\right) \Big/ [\textstyle\sum a_i z_i]^n, \tag{B.3.3}$$

or the related forms

$$\xi_n = (n-1)! \int_0^1 dz_1 \int_0^{z_1} dz_2 \cdots \int_0^{z_{n-2}} dz_{n-1} [a_1 + z_1(a_2 - a_1)$$
$$+ \cdots + z_{n-1}(a_n - a_{n-1})]^{-n}, \tag{B.3.4}$$

and

$$\xi_n = (n-1)! \int_0^1 dz_1 z_1^{n-2} \int_0^1 dz_2 z_2^{n-3} \cdots \int_0^1 dz_{n-1}$$
$$\times [z_1 z_2 \cdots z_{n-1}(a_1 - a_2) + z_1 z_2 \cdots z_{n-2}(a_2 - a_3) + \cdots + a_n]^{-n} \tag{B.3.5}$$

Some other useful relationships of the same kind are

$$\frac{1}{nA^n B} = \int_0^1 \frac{dx\, x^{n-1}}{[Ax + B(1-x)]^{n+1}}; \tag{B.3.6}$$

$$\frac{1}{A^n} - \frac{1}{B^n} = -\int_0^1 \frac{dx\, n(A-B)}{[(A-B)x + B]^{n+1}}. \tag{B.3.7}$$

By such devices an integral characterized by (B.3.1) can always be brought to the schematic form

$$\mathscr{I} = \int \cdots \int_{\substack{\text{(auxiliary} \\ \text{parameters)}}} (2\pi)^{-4} \int \frac{d^4 k F(k; p_i; m_j)}{[(k-R)^2 - a^2]^n}, \qquad \text{(B.3.8)}$$

the precise form depending on the identities employed. If the k-space integral is at worst logarithmically divergent, a change of variable

$$k' = k - R \qquad \text{(B.3.9)}$$

will not affect the value of the integral (nor add any finite number to a logarithmic divergence). Hence the k-space integral can be rewritten as

$$\int \frac{d^4 k F(k+R; p_i; m_j)}{(2\pi)^4 (k^2 - a^2)^n}. \qquad \text{(B.3.10)}$$

Because of the symmetry of the range of integration, the odd powers of k_μ in F do not contribute. Symmetric integration requires an average over the directions of k_μ, which amounts to the substitutions

$$k_\mu k_\nu = \tfrac{1}{4} g_{\mu\nu} k^2, \qquad \text{(B.3.11a)}$$

$$k_\mu k_\nu k_\rho k_\sigma = \frac{(k^2)^2}{24} (g_{\mu\nu} g_{\rho\sigma} + g_{\mu\rho} g_{\nu\sigma} + g_{\mu\sigma} g_{\nu\rho}), \qquad \text{(B.3.11b)}$$

etc. Therefore the general form to be evaluated is

$$\mathscr{I}_{mn} = \int \frac{d^4 k (k^2)^{m-2}}{(k^2 - a^2)^n} = \frac{B(m, n-m)}{16\pi^2 i (a^2)^{n-m}}, \qquad \text{(B.3.12)}$$

where $B(x, y) = \Gamma(x)\Gamma(y)/\Gamma(x+y)$, which is finite provided $n > m > 0$. It remains to perform the integrals over auxiliary parameters. In many calculations this last step accounts for the bulk of the labor, and an extensive calculus has been developed.

An alternative procedure for combining denominators is based upon the representation

$$\frac{i}{q^2 - m^2 + i\varepsilon} = \int_0^\infty d\alpha \exp[i\alpha(q^2 - m^2 + i\varepsilon)], \qquad \text{(B.3.13)}$$

which leads to

$$\xi_n = \frac{1}{a_1 a_2 \cdots a_n} = \prod_{j=1}^n \left[-i \int_0^\infty d\alpha_j \exp(i\alpha_j a_j) \right]$$

$$= (-i)^n \int_0^\infty d\alpha_1 \cdots \int_0^\infty d\alpha_n \exp\left(i \sum_{j=1}^n \alpha_j a_j \right). \qquad \text{(B.3.14)}$$

By completing the square in the exponential, one may always bring the integral over loop-momentum to the form

$$(2\pi)^{-4}\int d^4k F(k + R; p_i; m_j)\exp(i\lambda k^2), \tag{B.3.15}$$

which may be evaluated using standard Gaussian integral techniques. (Simple illustrations of these manipulations are given in Section 8.2 and Problem 8-6.) Note in particular that

$$(2\pi)^{-4}\int d^4k \exp(i\lambda k^2) = \frac{1}{16\pi^2 i\lambda^2}. \tag{B.3.16}$$

Differentiation with respect to λ then yields the general result

$$(2\pi)^{-4}\int d^4k (k^2)^m \exp(i\lambda k^2) = \frac{(m+1)!}{16\pi^2 i\lambda^2(-i\lambda)^m}. \tag{B.3.17}$$

B.4   Regularization Procedures

It is frequently the case that the Feynman graphs to be calculated suffer from severe ultraviolet divergences. Several procedures have been developed to regularize the integrals without losing the gauge invariance of the theory. The simplest such method is the Pauli–Villars procedure[3] described in Section 8.2 in the calculation of the vacuum polarization in electrodynamics. One adds unphysical fields of adjustable mass to the Lagrangian in a gauge-invariant manner. After gauge-invariant renormalization, the unphysical masses are taken to be infinite, and the renormalized quantities are shown to be finite.

This device is inadequate to deal with non-Abelian theories. An alternative scheme, known as dimensional regularization, was applied to this problem by 't Hooft and Veltman.[4] Stated simply, the 't Hooft–Veltman technique consists in defining the loop integral in exponential form in $(n - 1) > 3$ space dimensions and one time dimension, while restricting the external momenta and polarization vectors to the physical $(3 + 1)$-dimensional space. After performing the $(n - 4)$-dimensional integration in the space orthogonal to the physical space, one analytically continues the result in n. For sufficiently small values of n, or for complex values of n, the subsequent 4-dimensional integrals are convergent. Divergences may then be isolated systematically as singularities of the form $(n - 4)^{-1}$, etc. The following results are sufficient to deal with integrals encountered in this book:

$$(2\pi)^{-n}\int d^n k \exp(i\lambda k^2) = \frac{-\exp(-i\pi n/4)}{(4\pi\lambda)^{n/2} i}, \tag{B.4.1}$$

$$(2\pi)^{-n}\int d^n k (k^2)^m \exp(i\lambda k^2) = \frac{-\exp(-i\pi n/4)\Gamma(m + n/2)}{(4\pi\lambda)^{n/2} i(-i\lambda)^m \Gamma(n/2)}, \tag{B.4.2}$$

where symmetric integration yields

$$k_\mu k_\nu \rightarrow (1/n)g_{\mu\nu}k^2,$$ (B.4.3)

etc.

B.5   Feynman Rules: Electrodynamics

The Feynman rules for spinor and scalar electrodynamics are needed for several illustrative examples and problems. Complete rules and their derivations are given in many places; only the elementary vertices are required here.

For spinor electrodynamics of a fermion with charge Q the only interaction is

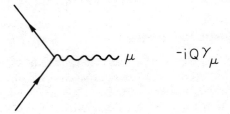

$$-iQ\gamma_\mu$$

In the case of a spinless boson with charge Q the vertices are

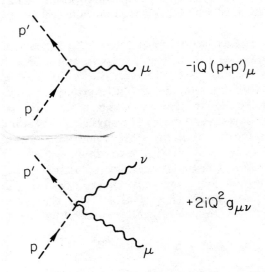

$$-iQ(p+p')_\mu$$

$$+2iQ^2 g_{\mu\nu}$$

The needed Feynman rules for other theories are given in the text where the theories are introduced.

For Further Reading

Most of the material in this Appendix is standard fare for textbooks in relativistic quantum mechanics and quantum field theory. One exception is the method of dimensional regularization, here introduced only briefly. A further discussion is to be found in

> G. 't Hooft and M. Veltman, "Diagrammar," CERN Yellow Report 73-9; reprinted in *Particle Interactions at Very High Energies*, 1973 Summer Institute on Elementary Particle Physics, Louvain, Belgium, edited by D. Speiser, F. Halzen, and J. Weyers, Plenum, New York, 1974, part B, p. 177.

The problems of regularization and renormalization of non-Abelian gauge theories are treated at some length in the recent books by

> T.-P. Cheng and L.-F. Li, *Gauge Theory of Elementary Particle Physics*, Oxford University Press, Oxford, 1983.
>
> L. D. Faddeev and A. A. Slavnov, *Gauge Fields*, Benjamin, Reading, Massachusetts, 1980.
>
> C. Itzykson and J.-B. Zuber, *Quantum Field Theory*, McGraw-Hill, New York, 1980.
>
> P. Ramond, *Field Theory: A Modern Primer*, Benjamin, Reading, Massachusetts, 1981.

References

[1] R. P. Feynman, *Phys. Rev.* **76**, 769 (1949).

[2] J. M. Jauch and F. Rohrlich, *The Theory of Photons and Electrons*, second edition, Springer-Verlag, New York, 1976.

[3] W. Pauli and F. Villars, *Rev. Mod. Phys.* **21**, 434 (1949); see also S. N. Gupta, *Proc. Phys. Soc. (London)* **66A**, 129 (1953).

[4] G. 't Hooft and M. Veltman, *Nucl. Phys.* **B44**, 189 (1972).

APPENDIX C: PHYSICAL CONSTANTS[1] AND DEFINITIONS

Speed of light *in vacuo*	$c = 2.99792458(1.2) \times 10^{10}$ cm/sec
Planck's constant/2π	$\hbar = 6.582173(17) \times 10^{-22}$ MeV·sec
	$\hbar c = 1.9732858(51) \times 10^{-11}$ MeV·cm
	$= 197.32858(51)$ MeV·fm
	$(\hbar c)^2 = 0.3893857(20)$ GeV^2·mb
Fine structure constant	$\alpha = e^2/\hbar c = 1/137.03604(11)$
Fermi constant	$G_F = 1.16632(4) \times 10^{-5}$ GeV^{-2}
	$G_F^2 \simeq 5.29 \times 10^{-38}$ cm^2/GeV^2
Gravitational constant[2]	$G_N = 6.6726(5) \times 10^{-8}$ cm^3/g·sec^2
	$\simeq 1.32 \times 10^{-55}$ MeV·cm/(MeV/c^2)^2
1 year	3.1558×10^7 sec
1 fm	10^{-13} cm
1 mb	10^{-27} cm$^2 = 0.1$ fm^2

References

[1] Particle Data Group, *Phys. Lett.* **111B**, 1 (1982).
[2] G. G. Luther and W. R. Towler, *Phys. Rev. Lett.* **48**, 121 (1982).

AUTHOR INDEX

Numbers in parentheses indicate the numbers of the references when these are cited in the text without the name of the author.

Numbers set in *italics* designate those page numbers on which the complete literature citations are given.

SUBJECT INDEX

Abelian Higgs model, 74–76, 80, 136
Aharonov-Bohm effect, 43–45, 51, 52
α_s,
  determined in e^+e^- annihilations, 228
  Q^2-evolution, 223–224
Altarelli-Parisi method, 233–242, 261
Alternatives to gauge theories, 191
Angular momentum, conservation of, 35
Anomalies, 137–139
  cancellation of, 138, 145
    and quark color, 139
    in $SU(5)$, 275
Antiscreening of color charge, 223–224, 263
Antisymmetric structure constants of $SU(3)$, 197
Asymptotic freedom 193–194, 224–225, 263, 265
  in $SO(3)$ gauge theory, 261
Axial charge, of nucleon, 139
Axion, 266

Baryon number
  conservation, 51–52
  long-range interaction coupled to, 51–52
  of the Universe, 290–293, 297
  sum rule, 177

Baryon-to-photon ratio, of Universe, 290
BCS theory of superconductivity, 78
Big bang cosmology, 290, 297
Bjorken scaling, 171
  approximate, in QCD, 242
  power-law violations, in fixed-point theories, 242
  violations of, See scaling violations
b-quark, 152

Cabibbo angle, 6, 150
Cabibbo current, 149
Cabibbo mixing, generalized, 152, 189
Cabibbo universality hypothesis, 17
Charge
  conservation of, 37
    local, 42
  quantization of, 271, 275
  renormalized, 210
Charge asymmetry
  in $\bar{p}p \rightarrow W^\pm$ + anything, 187
  of hadrons, in e^+e^- annihilations, 160
Charged-current weak interactions
  in parton model, 174–178, 190
    differential cross sections, 174–175, 178
  of leptons, in Weinberg-Salam model, 111, 113

RETURN TO: PHYSICS LIBRARY

351 LeConte Hall   510-642-3122

LOAN PERIOD 1	2	3
1-MONTH		
4	5	6

ALL BOOKS MAY BE RECALLED AFTER 7 DAYS.
Renewable by telephone.

DUE AS STAMPED BELOW.

DEC 1 5 2005		
OCT 0 1 2010		